Contents

1.1	What do living things do?	7.8	Che
1.2	Organs and organ systems	7.9	Rat
1.3	Human reproduction		
1.4	Blood and circulation	8.1	The weather and weather forecasting
1.5	Respiration	8.2	Weather science
1.6	Breathing	8.3	Climate, farming and catastrophes
1.7	Good health	8.4	Air masses and Charles' law
1.8	When things go wrong	8.5	Weathering and landscaping
1.9	Micro-organisms	8.6	The water cycle
1.10	Soil decay and compost	8.7	The rock cycle
1.11	Reproduction in flowering plants	8.8	Earthquakes and volcanoes
1.12	Leaves and photosynthesis		
2.1	Living organisms	9.1	How things work
2.2	Grouping animals and plants	9.2	Go with the flow
2.3	Variation	9.3	Electricity in your home
2.4	Why do we look like our parents?	9.4	The life of Michael Faraday
2.5	Extinction – the end of a species	9.5	Electromagnetism
		9.6	Numbers and words
3.1	People – a threat to the Earth	9.7	Microelectronics
3.2	The problem of pollution	9.8	Logic gates
3.3	Are we doing enough?		
3.4	Living things and their environment	10.1	Fuels
3.5	Why do animals and plants live in different places?	10.2	Making use of energy
		10.3	On the move
3.6	Populations within ecosystems	10.4	Energy conservation
		10.5	Human life and activity – the energy we need
4.1	Recycle it		
4.2	Passing on energy	11.1	What is a force?
4.3	The carbon and nitrogen cycles	11.2	Going down and slowing down
		11.3	Starting and stopping
5.1	Types of materials	11.4	Sinking and floating
5.2	Properties, uses and development of material	11.5	Some forces in action
		11.6	Turning forces
5.3	Separating and purifying	11.7	Work and power
5.4	Acids, alkalis and indicators	11.8	Under pressure
5.5	The Periodic Table		
5.6	Reactivity of metals	12.1	All done with mirrors
		12.2	Light and shadow
6.1	Solids, liquids and gases	12.3	Light
6.2	Atoms and temperature scales	12.4	Recording pictures
6.3	Particles and changes of state	12.5	Seeing and hearing
6.4	Atoms, ions and molecules	12.6	How sound is made
6.5	Radioactivity	12.7	Frequency, pitch and volume
6.6	Solvents, glues and hazards	12.8	The speed of sound
		12.9	Noise in the environment
7.1	Physical and chemical changes	12.10	A brief history of sound recording
7.2	Changes for the better from land and air		
7.3	Changes for the better from water and living things	13.1	Voyage to the Moon
		13.2	The planets
7.4	Chemicals from oil	13.3	The Sun as a star
7.5	Changing our food	13.4	Our views of the universe
7.6	Oxygen – the gas of life		
7.7	Oxidation	**Answers**	

Introduction

This resource of photocopiable material has been designed to provide a comprehensive 'homework bank' with up to 3 years' material for Years 7–9.

The activities are free-standing and they cover a wide interpretation of the statutory Programme of Study for Key Stage 3. We have followed the organisation and content of our textbook, Key Stage 3: Science Companion which has similar aims with respect to the National Curriculum.

We expect teachers will find many other uses for the materials besides an 'off the shelf' homework resource, for example cover and consolidation of Key Stage 3 topics and revision.

There are two styles of material and 90 sets of each. The first offers two differentiated versions of the same (or a similar) activity. The strategy we have used is to differentiate (but not to discriminate) by task and reading age. In both cases, we have tried to provide on average about half an hour of work but we hope you will adjust the amount of work for your pupils.

The differentiated homework is provided at A5 size and doubled up to save photocopying budget. It is ideal for sticking in exercise books if you want the pupils to keep their copy. Alternatively, it can be enlarged to A4 size.

The second style is less flexible but hopefully just as useful. The sheets are intended for pupils to write on. The questions are less formal and follow a proven and popular style with a wide variety of question types.

There are suggested answers at the end of the pack. The structure of the 'write on' sheets has lent itself to our providing a filled-in copy which we hope will save time.

ANDREW PORTER
MARIA WOOD
TREVOR WOOD

1.1.1 — What do living things do?

1 Match the characteristic to the activity.

Feeding	A plant turns its leaves towards the light.
Moving	Breathing out carbon dioxide and water.
Growing	A woman having a baby.
Reproducing	Pulling your hand away from a hot surface.
Sensitivity	Getting breathless after playing football.
Excretion	Eating a packet of crisps.
Respiration	A young child having a new pair of shoes. (7)

2 Make two lists from these statements, one for animals and one for plants.

unable to make food
have no chlorophyll
do not move around
have nerves and muscles
have roots
move around
have leaves
have chlorophyll
have no roots
have no nerves or muscles
have no leaves
make their own food (6)

3 Explain the differences between how a snake and plant move. (4)

4 a) What is a stimulus? (2)
  b) To which stimuli do humans respond? (2)
  c) To which stimuli do plants respond (2)

5 What is meant by the term 'excretion'? (2)
  a) What are the products of excretion in humans. (2)
  b) Which process is responsible for these products? (2)

6 Name the five senses present in humans. (5)

7 Draw a typical plant and animal cell. Label both cells. (4)

8 Explain how a cell from a leaf would be different from a cell from leg muscle. (2)

1.1.2 — What do living things do?

1 Match the characteristic to the activity.

Feeding	A plant turns its leaves to the light.
Moving	Breathing out carbon dioxide and water.
Growing	A woman having a baby.
Reproducing	Pulling your hand away from a hot surface.
Sensitivity	Getting breathless after playing football.
Excretion	Eating a packet of crisps.
Respiration	A young child having a new pair of shoes. (7)

2 Explain how a sheep and a rose feed. (4)

3 A snake and a geranium do not have legs but both move. Draw the diagrams and add notes to them to explain how they move. (14)

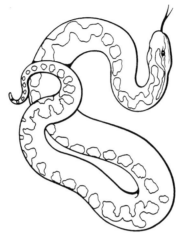

4 Why do plants not need to move from place to place? (2)

5 Copy and label the diagram of a plant cell. Highlight those features which are present in plant cells only. (12)

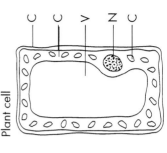

Plant cell

1.1.3 What do living things do?

1 Look at the diagrams below and answer the questions. Then take the letters from the shaded boxes and arrange them into a characteristic of living things.

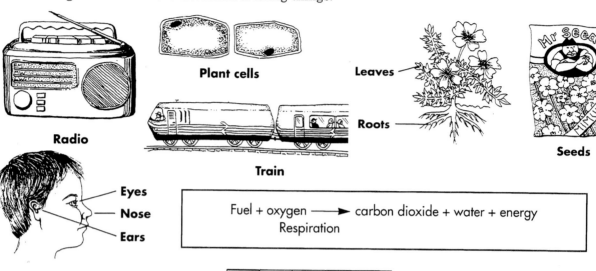

Radio, Plant cells, Train, Leaves, Roots, Seeds, Eyes, Nose, Ears

Fuel + oxygen → carbon dioxide + water + energy
Respiration

a) Some plants reproduce by producing [S][ ][ ][ ]

b) A [ ][ ][ ][I][ ] makes noise but does not move and is not living.

c) The [ ][ ][ ][S][ ] are the organs of hearing.

d) The [ ][ ][V][ ] of a plant make food.

e) A [T][ ][ ][ ][ ] has wheels but cannot move by itself.

f) Making energy using fuel plus oxygen is [ ][ ][ ][ ][ ][ ][I][ ][ ][ ][I][ ]

g) [ ][ ][N][ ] cells have walls made of cellulose.

h) The [ ][ ][ ][E][ ] is the organ of smell.

i) [ ][ ][T][ ] grow downwards.

j) The [ ][Y][ ][ ] are the organs of sight.   (10)

The characteristic is [ ][ ][ ][ ][ ][ ][ ][ ][ ][ ][ ]   (3)

2 Find 27 words to do with living things in the wordsearch.

N	B	M	F	E	E	D	I	N	G	G	H	E	A	R	I	N	G	T	A
O	A	O	U	A	E	I	D	F	R	E	S	P	I	R	A	T	I	O	N
I	C	V	E	T	R	E	P	R	O	D	U	C	T	I	O	N	X	U	I
S	M	E	L	L	O	P	R	Q	W	R	N	N	T	A	S	T	E	C	M
E	X	C	R	E	T	I	O	N	T	P	L	A	N	T	S	L	M	H	A
C	L	E	A	V	E	S	O	W	H	M	I	N	E	R	A	L	S	K	L
E	S	T	S	I	G	H	T	V	L	I	G	H	T	A	D	U	L	T	S
L	U	P	H	O	T	O	S	Y	N	T	H	E	S	I	S	J	I	H	G
L	M	E	M	B	R	A	N	E	X	Y	T	Z	A	B	W	A	T	E	R	(27)

Name: Class: Date:

Science Companion © A Porter, M Wood, T Wood and Stanley Thornes (Publishers) Ltd 1995

1.2.1 — Organs and organ systems

1. Define the terms, cell, organ and tissue. (3)
2. Give a name to the organ or organs described below.
   a) Two bean-shaped organs near the spine. They filter the blood. (1)
   b) Two spongy bags in the chest which enable us to breathe. (1)
3. Write a description for these organs: **heart liver** (2)
4. What is the function of an organ system? (2)
5. Look at the structures found in cells. Match them to the sentences below:

   **chromosomes   chloroplasts   starch grains
   mitochondria   vacuoles**

   a) Small sausage-shaped cells which release energy.
   b) Thread-like structures in the nucleus of a cell.
   c) These contain chlorophyll for photosynthesis.
   d) Plant cells store starch in these.
   e) Fluid-filled cavities which help cells keep shape. (5)
6. Name two organs in plants (2)
7. Look at the diagram of the digestive system below:

   a) Identify organs E to J. (5)
   b) Explain what happens to food as it passes along the digestive system? (2)
   c) Where is water absorbed? (1)
   d) Where are the products of digestion absorbed? (1)
   e) Which part stores glucose? (1)
   f) Which part produces insulin? (1)

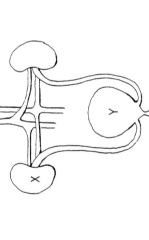

8. a) Identify organs X and Y on the diagram here. (1)
   b) Explain their functions. (1)
   c) What do tubes Z do? (1)

Science Companion © A Porter, M Wood, T Wood and Stanley Thornes (Publishers) Ltd 1995

1.2.2 — Organs and organ systems

1. What is the smallest unit of a living thing? (2)
2. What is an organ? (2)
3. Name three major organs in the human body. (3)
4. Which major organs are protected by bones? (3)
5. Give a name to the organ or organs described below.
   a) Two bean-shaped organs near the spine. They filter the blood.
   b) Two spongy bags in the chest which enable us to breathe.
   c) A hollow organ which pumps blood around the body.
   d) The largest organ inside the body. It controls blood sugar.
   e) The centre of the nervous system. (5)
6. Look at the diagram of the digestive system below:

   a) Identify parts E to J. (5)
   b) How long is tube H? (2)
   c) Where is water absorbed? (2)
   d) Where are digested substances absorbed? (2)
   e) Which part stores glucose? (2)

7. a) Name organs X and Y on the diagram here. (1)
   b) What do they do? (1)

Science Companion © A Porter, M Wood, T Wood and Stanley Thornes (Publishers) Ltd 1995

1.2.3 Organs and organ systems

1 Below are three organs cut up into pieces. Choose diagrams from the boxes to draw the complete organs.

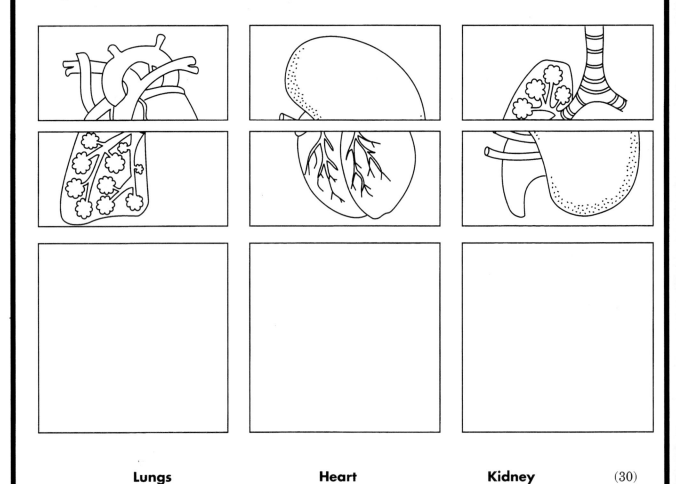

Lungs   **Heart**   **Kidney**   (30)

2 Add labels and arrows to the diagrams below to show how the digestive system is made up.

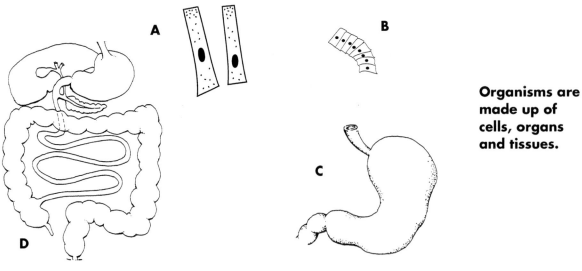

Organisms are made up of cells, organs and tissues.

Labels to use:

| Organs make up organ systems, e.g. digestive system | Individual cells | Tissues make up organs, e.g. stomach | Layers of cells make tissues, e.g. stomach lining |

(20)

Name: **Class:**   **Date:**

Science Companion © A Porter, M Wood, T Wood and Stanley Thornes (Publishers) Ltd 1995

1.3.1 — Human reproduction

1. Name the male and female gametes and where they are produced. (4)
2. Copy and label the diagrams of egg and sperm, describe their structure. (8)

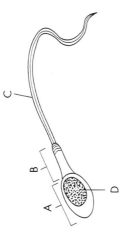
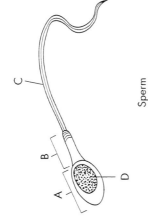

Egg   Sperm

3. Which part of sex cells contain genetic information? (1)
4. Why is an egg cell much larger than a sperm cell? (2)
5. What do we call the stage at which sex organs become mature? (1)
6. At what age do these changes occur in boys and girls. (2)
7. Describe the secondary sexual characteristics in boys and girls. (4)
8. How often are eggs released by a woman's ovaries? (1)
9. What is meant by the term "fertilisation"? (2)
10. On the diagram, where do the following occur? (4)

    a) sperms are released during intercourse
    b) eggs are released
    c) fertilisation takes place
    d) fertilised eggs become implanted

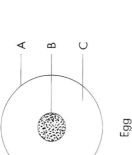

11. How many sperms are needed to fertilise an egg? (1)

Science Companion © A Porter, M Wood, T Wood and Stanley Thornes (Publishers) Ltd 1995

1.3.2 — Human reproduction

1. Name the female sex cells. (1)
2. Name the male sex cells. (1)
3. Copy and label the diagrams of egg and sperm. (7)

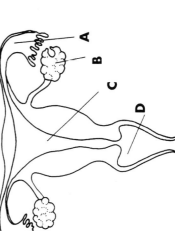

Sperm   Egg

4. Which of the cells above:
   a) looks like a tadpole (1)
   b) is produced by the ovaries (1)
   c) contains a large store of food (1)
5. Name some of the changes which happen to boys and girls when they change into adults. (3)
6. Write out the sentences giving the correct answer chosen from the word in brackets. (14)

   a) Eggs are released by a woman's ovaries
      (once every 28 days) (once a week) (all the time)
   b) The joining of eggs and sperms is called
      (fertilisation) (intercourse) (sterilisation)
   c) The place where fertilisation takes place is the
      (uterus) **(fallopian tubes)** (vagina)
   d) A fertilised egg is implanted in the
      (ovary) **(uterus)** (fallopian tubes)
   e) Sperms are produced by the
      (penis) (bladder) **(testes)**
   f) How many sperms are needed to fertilise an egg?
      (one) (two) (four)
   g) A fertilised egg is called an
      (ovary) **(embryo)** (ovum)

7. How long does a baby take to develop? (1)

Science Companion © A Porter, M Wood, T Wood and Stanley Thornes (Publishers) Ltd 1995

1.3.3 Human reproduction

1 Label the male and female reproductive organs. (5)

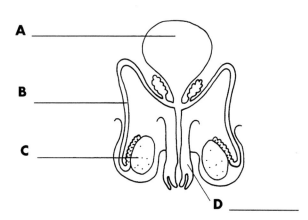

E Male organs

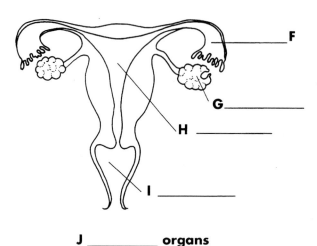

J _____ organs

2 Fill in the missing words in the passage below.

Fertilisation occurs when an e_____ meets a s_____. An egg is released from the ovaries once every _____ days. The sperms are released into the v_____ during sexual i_____. The sperms swim in the fluid of the u_____ to the fallopian tubes. If an egg is present, only o_____ sperm is needed to fertilise it. The rest die. The egg travels down to the lining of the u_____ where it is i_____. The egg is now called an e_____ and takes nine months to develop. (10)

3 Use the clues to fill in the words on the grid.

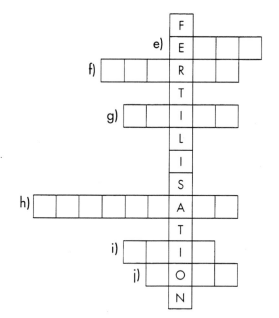

(10)

CLUES

a) This makes the male sperm move.
b) Sperms are produced by these in the male.
c) The outside layer of the female egg cell.
d) The centre of the female egg cell.
e) These are produced by the ovaries in the female.
f) A baby is called this in its early development.
g) This deepens as boys develop into adults.
h) This contains the supply of food in the female egg.
i) As boys develop into adults this grows on their bodies.
j) A baby develops in this part of the female's body.

Name: Class: Date:

Science Companion © A Porter, M Wood, T Wood and Stanley Thornes (Publishers) Ltd 1995

1.4.1 Blood and circulation

1 What are the constituents of blood? (1)

2 Name three functions of blood. (3)

3 How is blood carried around the body? (1)

4 a) Draw and name the blood cells below:

A   B   C

b) Which cell is i) biconcave in shape ii) is the largest iii) is made in the bone marrow iv) clots blood v) has a nucleus vi) is a leucocyte? (6)

5 Define the term "immunity". (2)

6 What is a vaccination? (2)

7 a) Which organ pumps blood around the body? (1)

b) Which type of muscle makes up the heart? (1)

8 Complete the table to show three types of blood vessel. It has been started for you. (6)

VESSEL	STRUCTURE	FUNCTION
Arteries	Thick muscular walls	Carry oxygen and carbon dioxide.

9 The diagram shows the human circulatory system.

a) Draw it (1)

b) Colour the path of oxygenated blood red. (1)

c) Colour the path of de-oxygenated blood blue. (1)

d) Put in more arrows to show the direction of the blood flow. (1)

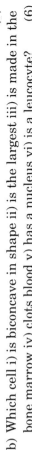

Head
Lungs
Heart
Liver
Intestines
Kidneys
Limbs

Veins   Arteries

1.4.2 Blood and circulation

1 Copy and finish these sentences:

a) Blood is made up of a liquid called plasma and c _____ which float in it.

b) Plasma contains d _____ substances.

c) Red cells carry o _____ to every cell of the body.

d) White cells protect the body from d _____ . (4)

2 a) Draw and name the blood cells below:

      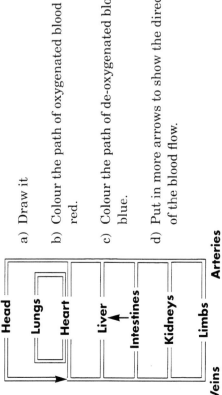

A   B   C

b) Which cell i) is biconcave in shape ii) is the largest iii) is made in the bone marrow iv) clots blood v) has a nucleus vi) is a leucocyte? (6)

3 Match the words to the descriptions below

antibodies   antigens   heart   immunity   vaccination

a) Being protected from a disease

b) Substances which are produced by lymphocytes to kill microbes

c) Substances produced by disease causing microbes

d) An injection of a mild form of a disease to produce antibodies for that disease

e) An organ which pumps blood around the body (10)

4 The diagram shows a cross-section of an artery.

a) Which type of blood do arteries carry? (1)

b) Why are artery walls thick and muscular? (2)

c) Do arteries carry blood towards or away from the heart? (1)

d) Name two ways in which veins are different to arteries. (2)

e) Name the smallest type of blood vessel? (1)

Thick muscle wall

1.4.3 Blood and circulation

1 The diagram shows a cross section of the heart.

a) Put in **arrows** to show the direction in which the blood flows through the heart (4)
b) Label chambers **A–D**. (4)
c) Label one of the valves with the letter **E**. (2)
d) Label the following places with these letters:
   F Blood leaves the heart to pick up oxygen
   G Oxygenated blood comes from the lungs
   H Blood is pumped to the rest of the body. (3)
e) Colour the oxygenated blood **red** and de-oxygenated blood **blue**. (2)

2 Fill in the spaces in the following passage:

The heart is an o_____ which pumps b_____. It is made of a special type of muscle called c_____ muscle. The heart beats _____ times a minute. It has f_____ chambers, two a_____ and two v_____. The r_____ side of the heart pumps blood to the l_____ to pick up o_____. The left side of the heart pumps blood to the r_____ of the b_____. (12)

Group	% British population	Can receive blood group
A	41%	A & O
B	9%	B & O
AB	3%	A, B, AB, O
O	47%	O

3 The table shows the four blood groups:

a) How many blood groups are there? _____
b) Which blood group is the most common? _____
c) Which blood group is the least common? _____
d) What percentage of the population is blood group B? _____ (4)
e) A person with group AB is called a universal recipient. What does this mean? _____ (2)
f) A person with group O is a universal donor. What does this mean? _____

Name: Class: Date:

Science Companion © A Porter, M Wood, T Wood and Stanley Thornes (Publishers) Ltd 1995

1.5.1 Respiration

1. Explain the difference between internal and external respiration. (2)
2. Name the structures inside cells which release energy. (1)
3. Name the two types of internal respiration. (2)
4. Which type of respiration releases energy without the use of oxygen? (1)
5. The diagram below represents what occurs during aerobic respiration. Study it and answer the questions.

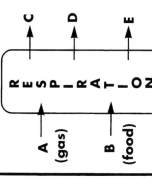

   a) Name   i) the gas at A
               ii) the food at B
               iii) products C, D and E (5)
   b) Which are the waste products? (2)
   c) Give a use for the energy produced during respiration. (1)
   d) Write out the word equation for the reaction in the diagram. (2)

6. a) What do the letters A.T.P. stand for? (2)
   b) What does A.T.P. do? (2)

7. Study the reactions below, then answer the questions which follow.

   A Fuel + Oxygen $\xrightarrow[\text{(Fire)}]{\text{High temperature}}$ Energy

   B Glucose + Oxygen $\xrightarrow[\text{(Enzymes)}]{\text{Low temperature}}$ Energy

   a) Which reaction shows respiration and which shows combustion? (2)
   b) Give one example of each type of reaction (2)

8. The diagram below shows two foods made using yeast.
   a) What type of organism is yeast? (2)
   b) How does yeast respire and what is the name of the process? (2)
   c) How do the products of respiration help to make these foods? (2)

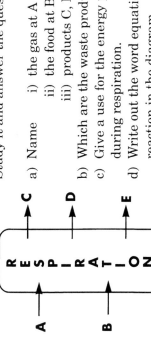

1.5.2 Respiration

1. Name the two types of internal respiration. (2)
2. What are mitochondria? (2)
3. Study the diagram below which represents aerobic respiration.

   a) Name   i) the gas at **A**.
               ii) the food at **B**.
               iii) products **C, D** and **E**. (5)
   b) Which organs in the body take in oxygen? (2)
   c) Where does the body get glucose? (2)
   d) What does the body use energy for? (2)

4. Complete the word equation for aerobic respiration below:

   Glucose + O____ $\longrightarrow$ C____ D____ + Energy + W____ (3)

5. Look at the following processes:
   internal respiration   photosynthesis   combustion   breathing alcoholic fermentation

   Which process   a) produces a rapid increase in heat
                           b) occurs in all living tissues
                           c) is another word for external respiration
                           d) does not require oxygen (4)

6. The diagram below shows two foods made using yeast.
   a) Is yeast a fungus or a bacterium? (2)
   b) Yeast respires by a process called? (2)
   c) i) Yeast breaks down sugar into which two products? (2)
      ii) Which of these makes the bread rise and beer fizz? (1)
      iii) Which gives beer its taste? (1)

1.5.3 Respiration

1 Complete the passage below about baking bread. Use the following words to help you.

**carbon dioxide   rest   cooked   dough
kneaded   trapped   evaporates
alcohol   ferment   salt   yeast
rise   water**

Bread is made by mixing flour, s_____,

w_____ and y_____. It is

k_____ into a d_____ and

then left in a w_____ place to r_____. The

dough rises because the yeast cells

f_____ producing a_____

and c_____ d_____. The

dough is elastic and carbon dioxide bubbles are

t_____ making it rise. The dough is then

c_____ in an oven, the alcohol

e_____ during baking. (13)

2 The experiment below shows the effect of sugar on yeast. Study it and answer the questions.

a) Why was the water in tube **A** boiled?____

_____ (1)

b) What colour would the limewater be at the start of the experiment?

_____ (1)

c) Name and describe the reaction between yeast and sugar.

_____
_____
_____
_____
_____
_____ (4)

d) What would the colour of the limewater be at the end of the experiment?

_____ (1)

e) Account for this colour

_____
_____ (1)

f) What might you smell in tube **A** at the end of the experiment?

_____ (1)

g) Account for this smell

_____
_____ (1)

Name: **Class:**   **Date:**

Science Companion © A Porter, M Wood, T Wood and Stanley Thornes (Publishers) Ltd 1995

1.6.2 Breathing

1 Name the organs of breathing (1)

2 Match the organs below to their common names.
  i) Larynx   A. chest
  ii) Trachea   B. voice box
  iii) Thorax   C. throat
  iv) Pharynx   D. windpipe (4)

3 Place the words which follow in the correct order for the path of air into the lungs. (5)

  alveoli   blood   bronchioles   bronchi   mouth

4 The tiny air sacs in the lungs are called alveoli. Which of these statements about them is true? (6)
  a) One air sac is called an alveolus
  b) They are found at the end of bronchi.
  c) They have thick walls.
  d) They have moist walls
  e) They cover a large surface area.
  f) They are surrounded by veins.

5 The table shows two samples of air. Read it and answer the questions.

GAS	Air breathed in	Air breathed out
OXYGEN	21%	16%
CARBON DIOXIDE	0.04%	4%
NITROGEN	78%	78%
OTHER GASES	1%	1%
WATER VAPOUR	small amount	saturated

  a) How much oxygen is breathed in? (1)
  b) How much oxygen is breathed out? (1)
  c) Why is more carbon dioxide breathed out than breathed in? (2)
  d) Why is a lot of water vapour breathed out? (1)
  e) What does the word 'saturated' mean? (2)
  f) Give a use of oxygen in the body. (2)

1.6.1 Breathing

1 Identify the organs of breathing and the system to which they belong. (2)

2 Give the common name for the following parts: (4)
  a) Thorax   b) Pharynx   c) Larynx   d) Trachea

3 Name:
  a) the organs which protect the lungs
  b) the sheet of muscle under the lungs
  c) the muscles in the chest which aid breathing (3)

4 a) Identify the parts of the lungs where gaseous exchange takes place (1)
  b) How are these parts adapted for gaseous exchange? (2)
  c) Explain how gases are exchanged. Use these words in your explanation: (4)

  **red   cells   blood plasma   alveoli   capillary
  oxygen   carbon dioxide   diffuse**

5 Arrange the following words into the correct order for the path of air into the lungs. (1)

  alveoli   blood   bronchioles   bronchi   mouth

6 The table shows two samples of air, one from inspiration and one from expiration. Read it and answer the questions which follow.

GAS	SAMPLE A	SAMPLE B
OXYGEN	21%	16%
CARBON DIOXIDE	0.04%	4%
NITROGEN	78%	78%
OTHER GASES	1%	1%
WATER VAPOUR	small amount	saturated

  a) Which sample is inspiration and which is expiration? (1)
  b) Give two reasons for your answer: (2)
  c) Why is less oxygen expired than inspired? (1)
  d) Why is more carbon dioxide expired than inspired? (1)
  e) Why is expired air saturated? (1)
  f) Why are the lungs moist? (1)
  g) Where in the body is oxygen taken and what is it used for? (1)

1.6.3 Breathing

1 The diagram below shows the organs of breathing.

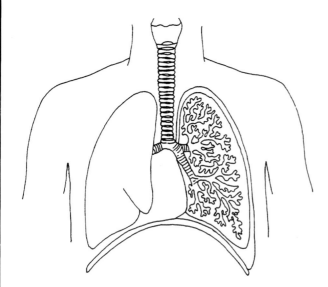

a) Label the parts of the diagram with the following letters:
- **A** Larynx
- **B** Trachea
- **C** Bronchi
- **D** Bronchioles
- **E** Lung
- **F** Alveoli
- **G** Diaphragm (7)

2 Look at the part of the lungs shown below then answer the questions.

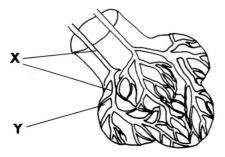

a) Name this part of the lungs _____ (1)

b) What are parts **X**? _____ _____ (1)

c) What passes in and out of the blood between **X** and **Y**? _____ _____ (2)

d) Which cells transport oxygen? _____ (1)

e) Name the process by which gas is exchanged. _____ (1)

3 The diagram below represents model lungs:

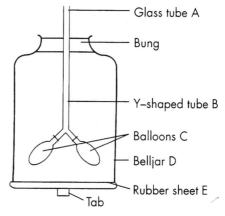

a) Name the parts represented by the letters A–E.

A _____ B _____
C _____ D _____
E _____ (5)

b) Explain what happens when the tab is pulled down.

_____
_____
_____
_____
_____ (2)

4 Transfer the figures below about lung capacity, on to the measuring cylinder.

a) Lungs always hold 1.5 l of air. Colour this part yellow. (1)

b) Lung capacity for normal breathing 3.0 l. Draw a blue line. (1)

c) Lung capacity for faster breathing 5.0 l. (e.g. exercising) Draw a red line. (1)

d) Why is there always air left in the lungs? _____ _____ _____ _____ (1)

Name: **Class:** **Date:**

Science Companion © A Porter, M Wood, T Wood and Stanley Thornes (Publishers) Ltd 1995

1.7.1 Good health

1 What do we mean by the term 'balanced diet'? (1)

2 Which of the following three food groups should be eaten in moderation and why? (4)

carbohydrates   sugar   fat   salt   protein

3 Foods can be divided into the five broad groups below.

A starchy foods   B fruit and vegetables   C meat and fish or meat alternative   D dairy foods   E fatty foods

From the groups say which you recommend should:
a) make up most of our diet
b) be eaten only occasionally
c) be eaten at least once a day
d) make up five portions of the meals eaten each day
e) be eaten in moderate amounts (5)

4 The chart below shows the change in the consumption of meat products over the last ten years.

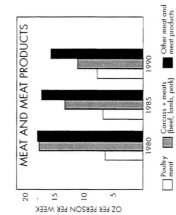

a) Comment on the figures for poultry and carcass meats. (2)
b) Give two reasons for the changes in meat consumption. (2)
c) What type of foods would be included in the 'other meat' products section? (2)

5 Define the term 'germ'. (2)

6 Give one example of a disease caused by each of the following organisms: **bacteria   viruses   fungi** (3)

7 Explain how doing regular exercise can help:
a) the heart and circulation   b) general health (2)

8 Name two alcohol related diseases. (2)

9 Write out ten tips for a healthy lifestyle **or** design a poster to show people how to live a healthy lifestyle. (5)

Science Companion © A Porter, M Wood, T Wood and Stanley Thornes (Publishers) Ltd 1995

1.7.2 Good health

1 Give reasons why doing each of the following keeps you healthy.
a) Keeping clean
b) Watching your weight
c) Taking regular exercise
d) Not smoking (4)

2 On the list of foods below indicate with the letters given whether a person should **A = eat plenty, B = eat in moderation or C = cut down on**

i) starchy foods   vi) vegetables
ii) dairy foods   vii) burgers and pies
iii) fruits   viii) sweets and cakes
iv) milk and yogurt   ix) salads
v) crisps and chips   x) fish and poultry (10)

3 The chart below shows the change in the consumption of meat products over the last ten years.

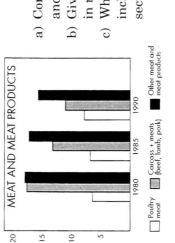

a) Give a reason why people are now eating more poultry? (1)
b) Are people eating more or less carcass meats now? (1)
c) Which of the groups of meats will contain the most fat? (1)
d) What does the chart mean by the term 'meat products'? Give examples. (3)

4 What are the four types of microbe? (1)

5 Which of the following sentences best describes a germ?
a) A microbe   b) A harmful microbe   c) A bacterium (1)

6 Name two diseases caused by germs. (2)

7 Design a poster which shows people how to keep healthy. (6)

Science Companion © A Porter, M Wood, T Wood and Stanley Thornes (Publishers) Ltd 1995

1.7.3 Good health

1 Examine the table below which shows the changes in nutrient intake over a twenty year period, then answer the questions.

Energy per person per day	1970	1975	1980	1985	1992
Total energy (kcal)	2560	2290	2230	2020	1960
Total fat (g)	119	107	106	96	87
Saturated fat (g)	52.0	51.7	46.8	40.6	34.3

a) What was the total calorie intake per day in 1970 and 1992? _____ (2)

b) Suggest two reasons for the reduced intake.

   i) _____

   ii) _____ (2)

c) What is saturated fat? _____
_____ (2)

d) What can happen to your body if too much saturated fat is eaten? _____
_____
_____ (2)

e) Suggest ways in which people have reduced their intake of saturated fat. _____
_____
_____
_____ (2)

f) Identify the organ below:

_____ (1)

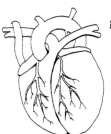

g) How does eating less fat help to keep this organ healthy?
_____
_____

_____
_____
_____
_____
_____
_____
_____
_____ (3)

2 a) From these foods, underline those which keep you healthy in blue, those bad for you in red.

burgers, sausages, lean meat, pies, fish, chicken, crisps, milk, vegetables, sweets, fruit, pasta, chocolate, rice, bread, chips

(4)

b) Look at the foods below and say why they should be eaten only occasionally.

_____
_____
_____
_____
_____
_____
_____
_____ (2)

Name: Class: Date:

Science Companion © A Porter, M Wood, T Wood and Stanley Thornes (Publishers) Ltd 1995

1.8.1 — When things go wrong

1 Look at the information in the table and answer the questions.

Causes of death	%
Heart disease	31
Cancer	28
Accidents	17
Strokes	9
Respiratory illness	8
Other diseases	7

a) Which disease kills most people in Britain? (1)
b) What is the major cause of heart disease? (1)
c) What is the major cause of cancer? (1)
d) How can people reduce the risk of getting heart disease and cancer? (1)
e) Draw a graph to show the causes of death. (5)

2 The diagram below shows what harmful substances are produced when a cigarette is smoked.

Tar	Dust	Carbon Monoxide	Nicotine	4000 Other Chemicals
1	2	3	4	5

a) For each of the chemicals 1–4 say how they damage the body. (8)
b) Design a symbol for cigarette packets and posters to stop people smoking. (2)

3 The diagram shows a damaged tooth.

a) Explain what is happening to the tooth. (2)
b) What causes this damage? (1)
c) How might this tooth affect the health of this person? (1)
d) What is a major cause of this disease in children? (1)
e) How can people keep their teeth healthy? (2)
f) Write a slogan for an advertising campaign to improve the health of our teeth. (1)

4 Comment on each of these statements:
a) Coronary heart disease can be avoided.
b) You can only get heart disease if you smoke more than ten cigarettes a day.
c) Plaque helps to cause tooth decay. (3)

1.8.2 — When things go wrong

1 Look at the graph and answer the questions.

Deaths %: Other diseases 7%, Respiratory illness 8%, Strokes 9%, Accidents 17%, Cancer 28%, Heart disease 31%

a) Which disease kills most people in Britain? (1)
b) Why do a lot of people in Britain get heart disease? (2)
c) Why do a lot of people get lung cancer? (2)
d) Make a list of things that people can do to reduce the risk of getting heart disease and lung cancer. (2)

2 Read the information below about cigarettes and answer the questions.

TAR	CARBON MONOXIDE	NICOTINE	DUST	4000 OTHER CHEMICALS
Irritates damages lungs. Causes cancer and bronchitis.	is the same gas as in car exhausts. Damages red cells.	Addictive drug. Affects brain. Raises blood pressure. Narrows arteries.	Causes coughs and irritates lungs. Causes bronchitis.	Cause damage to every organ in the body.
1	2	3	4	5

a) Name the four main chemicals in cigarettes. (2)
b) Which chemical in cigarettes
 i) damages red cells   ii) raises blood pressure   iii) irritates lungs
 iv) is a poisonous gas   v) is addictive? (5)
c) Design a symbol for cigarette packets to stop people smoking. (2)

3 The diagram shows a decayed tooth.

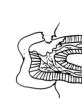

a) Which organisms causes tooth decay? (2)
b) How do they cause decay? (2)
c) How might this tooth affect the health of a person? (2)
d) Which foods cause the most tooth decay in children? (2)
e) Which foods and drinks should we all avoid to reduce tooth decay. (2)
f) Explain how to keep teeth healthy. (2)
g) Write a slogan for an advertising campaign to improve the health of our teeth. (2)

1.8.3 — When things go wrong

1 Read the newspaper article and answer the questions which follow.

> **Have a Heart**
> Heart disease is still responsible for most deaths in Britain today. One cause of heart disease is eating too much saturated fat. This is the fat from animal tissues which includes beef, pork, lard, dripping butter, cheese and cream. Saturated fat is also found as 'hidden fat' in foods like pies, sausages, cakes and biscuits. This type of fat contains the substance cholesterol. This blocks arteries which causes heart attacks. The Minister for Health is today launching a huge campaign to make the public aware of the importance of eating less fat.
>
> Dr. P. Vein 24/11/94

a) What is saturated fat? _____ (1)

b) Give two examples of foods high in saturated fat. _____ (2)

c) What is 'hidden fat'? _____ (1)

d) What is cholesterol? _____ (1)

e) How does cholesterol cause heart disease? _____ (2)

f) Suggest ways of cutting down on the amount of saturated fat we eat. _____ (2)

g) Which other factors increase the risk of getting heart disease? _____ (2)

2 Study the table showing damage caused to the body by smoking.

Organ	Disease caused
Throat, mouth, nose, bladder and kidneys	Cancer
Lungs and trachea	Cancer. Bronchitis. Emphysema.
Oesophagus and stomach	Ulcers. Cancer.
Female reproductive organs	Cancer of the cervix. Premature babies. Low birth weights in babies.
Heart	Blocked arteries. Narrowing of the blood vessels. Build up of fat. Heart attack.
Legs	Buerger's disease. (blocked blood vessels, leads to amputations.)

Now label the diagram to show how smoking damages the body.

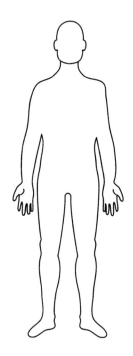

(12)

Name: **Class:** **Date:**

Science Companion © A Porter, M Wood, T Wood and Stanley Thornes (Publishers) Ltd 1995

1.9.1 Micro-organisms

1. Define the term 'microbe'. (1)
2. Name four types of microbe. (1)
3. Examine the diagrams below.

   A   B

   mushrooms   sore throat (*Streptococcus*)

   a) Name the two types of microbe. (2)
   b) Which is the pathogen and why? (2)
   c) Where do bacteria and fungi live? (1)
   d) What are the three types of bacteria? (1)
   e) Name two types of fungi. (1)
   f) How do bacteria reproduce? (1)
   g) Name three conditions which prevent the growth of bacteria (2)

4. Examine the food labels below and answer the questions (1)

   Woodlands Farms
   Fresh Cream
   Pasteurised
   Keep in fridge at 2° to 4°C
   USE BY 20 Oct

   a) What is the product? (1)
   b) How is the food treated to preserve it? (1)
   c) What type of microbe is present in this product? (1)
   d) How do you know when the product is not safe to eat? (1)
   e) How should the product be stored? (1)
   f) How does this method of storage keep it fresh longer? (1)

5. For each disease below explain how their spread could be avoided.

Disease	How spread
a) *Salmonella* poisoning	Raw foods and incorrect cooking
b) Athlete's foot	Fungal infection
c) Cholera	Dirty water and sewage
d) Malaria	Protist carried by mosquito
   (4)

6. Design symbols which could be put on food labels to show that they had been treated in the following ways: a) Pasteurised b) Frozen c) Sterilised (3)

1.9.2 Micro-organisms

1. What is a microbe? (2)
2. Which of these illnesses are caused by microbes?

   heart disease   **athlete's foot**   **flu**   **diabetes** (2)

3. Look at the diagram of a type of microbe and answer the questions.

   a) Are they bacteria or fungi? (1)
   b) Which conditions do they live in? (2)
   c) How do they reproduce? (1)
   d) Name an illness caused by this type of microbe. (1)
   e) Name a medicine made with this type of microbe. (1)

4. Look at the food label and answer the questions.

   Woodlands Farms
   Fresh Cream
   Pasteurised
   Keep in fridge at 2° to 4°C
   USE BY 20 Oct

   a) What is the food? (1)
   b) Which word on the label means that it has been treated to kill some of the microbes? (1)
   c) At what temperature should it be kept? (1)
   d) Which microbes are present in this food? (1)
   e) How should you keep this product? (1)
   f) How do you know when it is no longer safe to eat? (1)

5. Match the word to the definitions which follow in brackets.

   a) Means can only be seen under a microscope. (**micro**) (**mini**) (**macro**) (1)
   b) Increase in number. (**more**) (**multiply**) (**divide**) (1)
   c) Organisms which feed on the dead remains of plants and animals. (**aphids**) (**saprophytes**) (**sulphytes**) (1)
   d) A fertile part of the soil made by bacteria and fungi. (**humus**) (**minerals**) (**mulch**) (1)
   e) Break down or rot. (**decay**) (**swollen**) (**broken**) (1)

6. Draw symbols which could be put on to food labels to show that they had been treated in the following ways: a) pasteurised b) kept in the fridge (4)

1.9.3 Micro-organisms

1 Complete the passage on microbes.

There are four types of microbe, b_____, v_____, f_____ and p_____. Bacteria are microscopic organisms which live e_____. There are three types: r_____, rod-s_____ and s_____. They multiply by d_____ into two. Not all bacteria are harmful; those which cause diseases are called p_____. Fungi are made of tiny m_____ threads called h_____. Examples include y_____ and m_____. Fungi contain no chlorophyll and feed on the d_____ remains of plants and a_____. Viruses are smaller than bacteria but are not l_____. They can only live on living tissues and cause diseases. Protists are s_____-celled organisms which live in m_____ places. (19)

2 Label and complete the diagram below to show how minerals are recycled in nature.

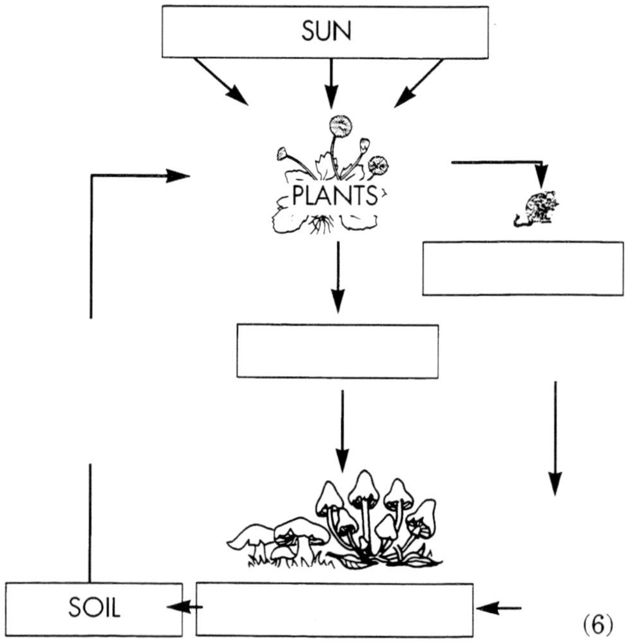

SUN → PLANTS → ... → SOIL (6)

3 Read the article below and answer the questions

Robert Koch 1843–1910

Robert Koch was a German bacteriologist. He became interested in the disease anthrax which was killing thousands of cows. He grew the anthrax bacteria in his laboratory on flat dishes which he named after his assistant J.R. Petri. The dishes were filled with a gel called agar which fed the bacteria. Koch found a way of staining the bacteria so that they could be seen easily under the microscope. He also found that steam rather than dry heat was better at killing bacteria, this was a great help to the medical profession. Robert was also responsible for identifying the T.B. bacillus and worked on many other diseases including cholera, malaria and sleeping sickness.

a) What is anthrax?_____ (1)

b) How did Koch grow bacteria?_____ (1)

c) How did he make bacteria easier to see?_____ (1)

d) How did he help the medical profession?_____ (1)

e) What is a T.B. bacillus?_____ (1)

f) Which other diseases did Koch work on?_____ (1)

Name: Class: Date:

Science Companion © A Porter, M Wood, T Wood and Stanley Thornes (Publishers) Ltd 1995

1.10.1 Soil, decay and compost

1 Define the term 'soil'. (1)

2 Read the information about soil and answer the questions which follow:

> **The Formation of Soil**
>
> Weathering breaks down rocks into very small pieces which become part of the soil. Grains differ in size. A sandy soil drains well because it has large grains with large spaces between them. A clay soil drains badly because the grains are a flat shape with small spaces between them. A soil profile shows the layers of different materials in soil down to the parent rock. The layers in a soil profile are called horizons. A fertile soil contains a lot of rotten vegetation called humus and is also called loam. This contains grains of different sizes and is ideal for gardening.

a) What is weathering? (1)
b) Which soil contains large grains and drains well? (1)
c) Why does a clay soil drain badly? (1)
d) What does a soil profile show? (1)
e) What is humus? How is it formed? (2)
f) Name the type of soil which is ideal for gardening. (1)

3 A sample of soil was shaken up with water and left to separate into layers.

[diagram showing layers: Organic matter, Water, Clay, Silt, Sand, Gravel]

a) Is the soil a clay or a sandy soil? (1)
b) Explain your answer to question 1. (2)
c) Which particles are the biggest, clay or gravel? How do you know? (1)
d) What is organic matter? (1)

4 The two labels below are from plants. Which type of soil would you plant each into? Give reasons for your answers. (6)

A	*Flora bundicum*	B	*Purple piecus*
	Hardy evergreen. Roots must not sit in water. Prefers dry soil with large grains. Do not overfeed with organic matter.		Fast-growing shrub. Keep moist at all times. Water in dry weather. Feed well in summer. Plant in well-rotted manure.

1.10.2 Soil, decay and compost

1 Read the passage about soil and answer the questions.

> **The Formation of Soil**
>
> Weathering breaks down rocks into very small pieces which become part of the soil. Grains differ in size. A sandy soil drains well because it has large grains with large spaces between them. A clay soil drains badly because the grains are a flat shape with small spaces between them. A soil profile shows the layers of different materials in soil down to the parent rock. The layers in a soil profile are called horizons. A fertile soil contains a lot of rotten vegetation called humus and is also called loam. This contains grains of different sizes and is ideal for gardening.

a) Name the process which breaks soil into smaller pieces? (1)
b) Which type of soil contains large grains? (2)
c) Why does a clay soil drain badly? (1)
d) What do we call soil layers in a soil profile? (1)
e) What is humus? (2)
f) Which type of soil is best for gardening and why? (1)

2 Look at the two plant labels, then answer the questions.

A	*Flora bundicum*	B	*Purple piecus*
	Hardy evergreen. Roots must not sit in water. Prefers dry soil with large grains. Do not overfeed with organic matter.		Fast-growing shrub. Keep moist at all times. Water in dry weather. Feed well in summer. Plant in well-rotted manure.

a) Which plant should be planted in i) sandy soil ii) loamy soil?
b) Give reasons for your answers. (7)

3 Which of the following words is described by the sentences below?

fertile   humus   sandy   leaching   soil profile

a) A cross-section of soil showing layers.
b) Nutrients being washed out of the soil.
c) Organic matter from dead plants and animals.
d) A soil rich in organic matter.
e) A type of soil with large air spaces in between the grains. (5)

1.10.3 Soil, decay and compost

1 Use the following words to complete the passage below:

**recycles   mix   oxygen
heat   warm   invertebrates
fertile   decays   bacteria
nutrients   respire
humus   moist**

When plant material from the garden is put into a compost bin it _____ to make _____. This is rich in _____ and makes soil _____. Plants decay best in a _____ and _____ environment. A compost bin provides such a place. Plants in a compost bin have _____ and fungi on them. A lid on the bin keeps the _____ in. The bacteria and fungi break down plants while feeding on them. A compost bin has no bottom so that _____ and _____ can get inside.

Microbes need oxygen to _____ and the invertebrates _____ up the compost.

Putting compost back on to the garden _____ nutrients. (12)

2 A pupil carried out an experiment to find out which soil drains the fastest, clay or sand.

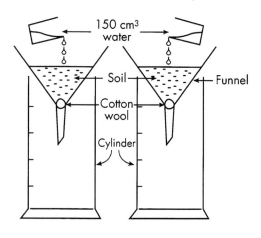

Results Table	A	B
Time for water to drain into cylinder (mins)	1.30	5.00
Volume of water collected after 30 minutes	145 cm³	20 cm³

a) Which sample drained the fastest? _____ (1)

b) How long did it take sample B to drain? _____ (2)

c) Suggest which sample was clay and why. _____ _____ (2)

d) How was the experiment kept fair? Give two ways. _____ _____ _____ _____ (2)

e) Explain how you could improve the drainage in a heavy wet soil. _____ _____ _____ (2)

f) Draw the particles in a clay and sandy soil.

sandy soil	clay soil

(4)

Name: Class: Date:

Science Companion © A Porter, M Wood, T Wood and Stanley Thornes (Publishers) Ltd 1995

1.11.1 — Reproduction in flowering plants

1 Explain the function of the following plant organs:
a) stem b) leaves c) roots d) flower (4)

2 Name two ways in which plants can reproduce. (1)

3 Look at the diagram below, then answer the questions.

a) Draw the diagram and label parts **A–G**. (7)

b) Identify the following organs by name:
 i) male part (1)
 ii) female part (1)
 iii) parts which attract insects (1)
 iv) parts which protect the flower while in bud (1)

4 Examine the plant organs below and answer the questions.

a) Name organs **A** and **B**. (2)
b) Name parts **E–H** on the diagrams. (4)
c) Which part
 i) is a sticky landing pad for pollen grains (1)
 ii) contains the male sex cells (1)
 iii) contains the female sex cells (1)
 iv) supports the stigma (1)
 v) grows into a seed (1)
 vi) grows into a fruit? (1)

5 Explain what happens during pollination. (2)

6 Some plants are 'wind pollinated'. What does this mean? (1)

7 Use the diagrams and words below to explain how pollinated flower becomes a fruit. (4)

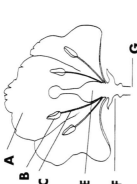
Blackberry flower

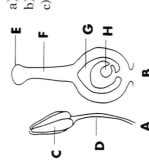
Blackberry fruit

**stamens   carpels   ovaries   flower
seeds   pollen   male   female   sex
cell   ripen   fertilisation   fruit
petals   join**

1.11.2 — Reproduction in flowering plants

1 Match the plant to their functions below.

stem   flower   roots   leaves

a) To make food
b) Contain the reproductive organs
c) Hold the plant in the soil
d) Support the plant and contain the transport system. (4)

2 Look at the diagram of a flower below and answer the questions.

a) Draw the diagram. (1)
b) Label parts **A–G** using the following names.

**stem   petal   anther   stamen
carpel   filament   sepal** (7)

c) Which part
 i) is the male part (1)
 ii) is the female part (1)
 iii) attracts insects (1)
 iv) protects the flower while in bud? (1)

3 The diagrams below show the stamen and the carpel of a plant.

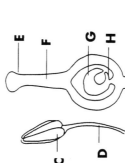

stamen   carpel

Which part
a) makes pollen (2)
b) holds up the anther (2)
c) is where pollen lands (2)
d) contains unripe seeds (2)
e) grows into a fruit? (2)

4 What do we call these processes?
a) taking pollen from one plant to another (2)
b) a male and female cell join to make a seed (2)

5 After fertilisation an ovary forms a fruit. This can be a seed pod or a fleshy fruit. Which of the plants below produces a seed pod?

**runner bean   strawberry   lupin   peas   apple
tomato   blackberry   poppy   pansy   plum** (5)

1.11.3 Reproduction in flowering plants

1 a) Label the diagram of a broad bean seed. Use the words given.

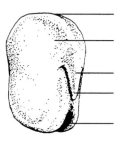

Testa
Cotyledon
Scar
Plumule
Radicle

(5)

b) Give the functions of each part next to the letters below.

A_____ (1)
B_____ (1)
C_____ (1)
D_____ (1)
E_____
_____ (1)

b) Complete the table to show how you expect the seeds to look, giving reasons for your answers.

TUBE	RESULTS	REASON
A		
B		
C		
D		

2 The diagram below shows an experiment to find the conditions needed for germination of seeds.

a) Draw what the seeds might look like after a few days in the test tubes.

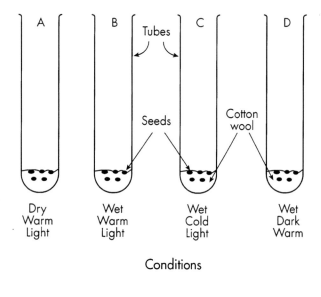

Conditions

3 Seeds are dispersed in different ways. Look at the diagrams below and explain how each is dispersed.

a)
Blackberry fruit

b) Parachutes / Seeds / Dandelion

c) Dry seed pods
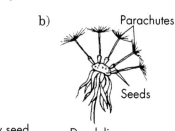
Lupin

(6)

Name: **Class:** **Date:**

1.12.1 — Leaves and photosynthesis

The diagram shows a vertical section through a leaf. Examine it and answer the questions which follow.

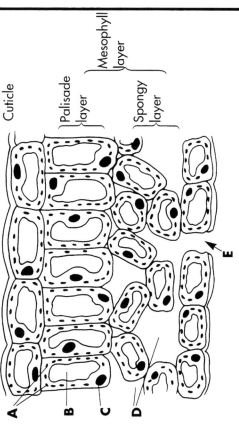

1. What is the function of a leaf? (1)
2. Give two ways in which leaves are adapted for this function? (2)
3. Identify parts **A–D** on the diagram. (4)
4. Which layer a) is a protective covering (1)
   b) is the top layer of cells (1)
   c) contains the most chloroplasts (1)
   d) contains the upper palisade layer (1)
   e) has large spaces between the cells? (1)
5. a) Write out the word equation for photosynthesis. (2)
   b) Which molecules in a plant are able to absorb sunlight energy? (1)
   c) In which structures are these molecules found? (1)
   d) Explain in detail what happens to the products of photosynthesis. (4)
6. a) Identify the structure at **E** on the diagram. (1)
   b) What is its function? (1)
   c) Which gas is needed for photosynthesis? (1)
   d) Which gas from photosynthesis is removed by these cells? (1)
   e) Why are these structures closed at night? (1)

1.12.2 — Leaves and photosynthesis

The diagram shows a vertical section through a leaf. Examine it and answer the questions which follow.

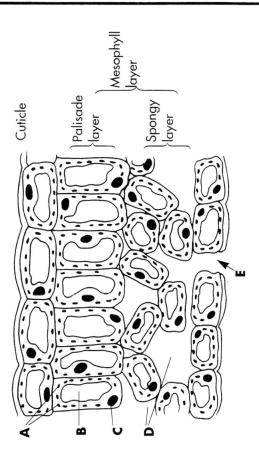

1. What job does the leaf do for the plant? (2)
2. Why are leaves a) broad and flat b) thin? (2)
3. a) Name parts **A–D** on the diagram. Use the words given below (4)

   vacuole   air spaces   nucleus   chloroplasts

   b) Which parts contain the green pigment which absorbs sunlight energy? (1)
   c) Name the green pigment (1)
4. a) Complete the word equation for photosynthesis.

   $$\text{carbon} \_\_\_\_\_ + \_\_\_\_\_ \xrightarrow[\text{C}]{\text{sunlight}} \text{glucose} + \_\_\_\_\_$$ (2)

   b) What are the raw materials for photosynthesis? (2)
   c) What are the products of photosynthesis (2)
   d) Plants are unable to store the glucose they make, it is changed into another substance. Name this other substance. (2)
   e) Name one use of glucose in plants. (2)
5. a) Name the hole at **E** which lets gases in and out of the leaf. (2)
   b) Which gases enter and leave through this hole? (2)
   c) Why do these holes close at night? (2)

1.12.3 Leaves and photosynthesis

1 Complete the passage about photosynthesis.

Photosynthesis is the process by which plants make f_____. Most photosynthesis occurs in the l_____ of a plant. The raw materials c_____ _____ and w_____ diffuse into the c_____ where light e_____ is absorbed. They react to form the products g_____ and o_____. Glucose m_____ cannot be stored in plants because they are small and s_____ in water, so they are changed into s_____ molecules which are l_____ and insoluble in water. O_____ gas is a w_____ product of photosynthesis and is removed from the plant. (14)

2 Label the boxes on the diagram to show the raw materials and products of photosynthesis.

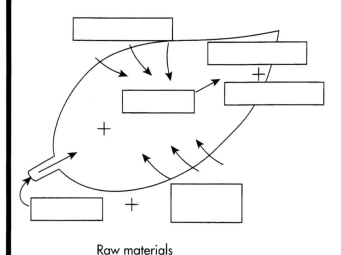

Raw materials

3 a) The diagrams below show how to test a leaf for starch. Complete the labels.

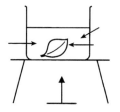

Place leaf in b_____ water for two minutes to k_____ it.

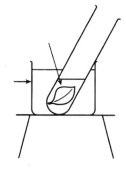

Place leaf in a tube of a_____ and place in a w_____ bath. This d_____ the leaf.

Wash the leaf to soften it. Alcohol makes leaves b_____.

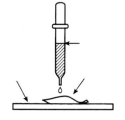

Test with i_____ solution. A b_____ b_____ result indicates s_____ is present. (10)

b) Which parts of these leaves would turn blue black after a starch test? Colour those areas blue. (2)

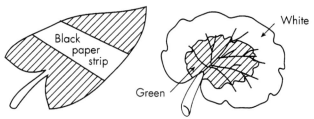

C = kept in light after destarching

D = variegated leaf in light

Name: Class: Date:

Science Companion © A Porter, M Wood, T Wood and Stanley Thornes (Publishers) Ltd 1995

2.1.2 Living organisms

1 What are the two biggest groups of living things on earth? (2)

2 Place the living things below into the two groups. (2)

shark pine tree wasp frog apple tree jellyfish
algae dandelion

3 Look at the groups of living things below. Name the odd one out in each group and give your reason.
a) frog newt buttercup toad
b) grass seaweed oak tree mouse
c) rose daisy mushroom dandelion
d) snail worm newt jellyfish (4)

4 Below are four animals from different groups.

Robin

Wasp

Snake

Frog

a) Which
i) can fly   ii) is a reptile   iii) is an insect   iv) are vertebrates? (2)
b) Write a sentence about each animal. (2)

5 Look at the pictures of animals below:

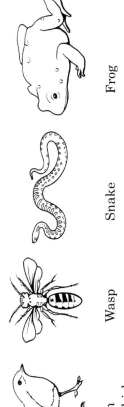

A   B   C   D

a) Which features put **A** and **B** into the same group? (2)
b) Which features put **C** and **D** into the same group? (2)
c) Give a difference between **A** and **B** (2)
d) Give a difference between **C** and **D** (2)

Science Companion © A Porter, M Wood, T Wood and Stanley Thornes (Publishers) Ltd 1995

2.1.1 Living organisms

1 What is meant by the term 'classification'? (1)

2 a) Name the two largest groups of organisms. (1)
b) Give two differences between the organisms in these groups. (2)

3 What seven features do living things have in common? (1)

4 Look at the plants below and answer the questions which follow:

   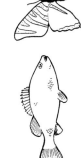

a) Name two ways in which they are alike. (1)
b) Name two ways in which they are different? (1)

5 Look at the animals below and answer the questions which follow:

A

B

a) Which animal is an invertebrate? (1)
b) Which animal is a vertebrate? (1)
c) Describe their features (2)

6 Arrange the plants below into two groups, those with flowers and those without flowers. (2)

seaweed grass rose moss oak tree
horse chestnut tree daffodil fern liverwort

7 Look at the organisms below.

honeysuckle pine tree oak tree
chlamydomonas wasp snake

Which organism is:
a) A climbing plant   b) A deciduous tree
c) An insect   d) A microscopic plant
e) A vertebrate   f) A conifer? (3)

8 Describe the skin coverings in the following animals. (4)

Science Companion © A Porter, M Wood, T Wood and Stanley Thornes (Publishers) Ltd 1995

2.1.3 Living organisms

1 Place the letter of the animals below next to their descriptions.

A bat   B robin   C octopus   D seal
E spider   F trout   G jellyfish

a) A nocturnal mammal. _____

b) A bird _____

c) An arachnid with eight legs _____

d) A mollusc with eight legs _____

e) A freshwater fish _____

f) A sea mammal _____

g) A sea animal with a transparent body. _____ (7)

2 Examine the plant in the following diagram.

a) What type of plant is it? _____ (1)

b) This plant is evergreen. What does this mean?
_____
_____
_____ (1)

c) Describe the leaves of this plant. _____
_____ (2)

d) Where are the seeds produced in this plant?
_____ (1)

3 The diagram below shows an alligator. Answer the questions about it.

a) To which group of animals does the alligator belong? _____ (1)

b) What is its habitat? _____ (1)

c) Describe its skin. _____
_____ (2)

d) Explain how the following features of the alligator help it to survive.

i) Powerful tail _____
_____ (1)

ii) Webbed feet _____ (1)

iii) Sense organs remain above the water when it is swimming along and its body is submerged. _____
_____
_____ (2)

4 Locusts belong to the largest group of invertebrates, the arthropods. Their features include:
**compound eyes   segmented body parts
antennae   body in three parts: head,
thorax and abdomen   jointed limbs
exoskeleton**

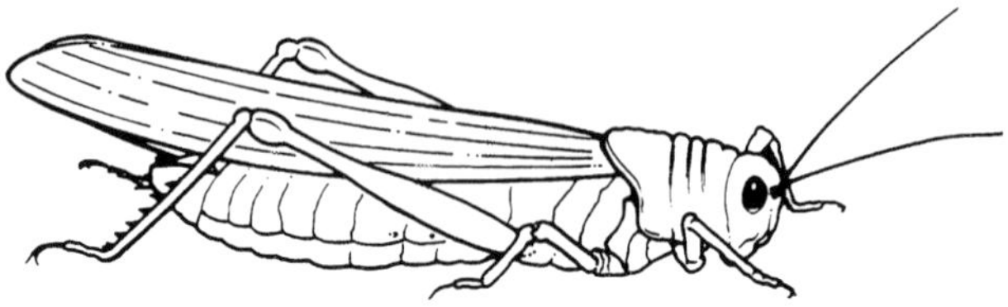

(6)

a) Label those parts on the locust.

b) The following insects are also from the same group.

butterfly   beetle   wasp

Match them to their descriptions below:

i) Narrow waist between abdomen and thorax. _____ (1)

ii) Body covered with velvety fur and wings with scales. _____ (1)

iii) Wings protected by wing cases.
_____ (1)

Name: Class: Date:

2.2.2 Grouping animals and plants

1 Name **five** animals with backbones. (1)

2 Name **five** animals without backbones. (1)

3 Arrange the following organisms into these groups: (5)

Birds   Fish   Amphibians   Reptiles   Mammals

parrot   **toad**   **salamander**   **newt**
halibut   **chameleon**   **lizard**   **eagle**
dormouse   **bat**   **eel**   **elephant**
penguin   **salmon**   **python**

4 Which of the following sentences describes the features of which vertebrate group? (5)

a) Tough waterproof skins made of horny scales.
b) Smooth moist skin, lay eggs in water.
c) Breathe through gills, have fins and scales.
d) A body covered in hair, feed young on milk.
e) Body covered in feathers, wings and a beak.

5 Place the plants below into the groups, flowering plants and non-flowering plants. (2)

liverwort   **buttercup**   **oak tree**   **moss**   **blue bell**
fern   **pine tree**   **rose**

6 A key can be used to identify organisms. Look at the plant and animal below and write down the features which could be used to separate them in a key. One example has been done for you. (8)

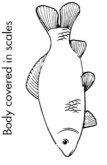

Body covered in scales

7 a) Give each flower a name   b) Draw a key to the flowers. (8)

Science Companion © A Porter, M Wood, T Wood and Stanley Thornes (Publishers) Ltd 1995

2.2.1 Grouping animals and plants

1 Name the **five** vertebrate groups. (5)

2 Arrange the organisms below into those five vertebrate groups. (5)

parrot   **toad**   **salamander**   **newt**
halibut   **chameleon**   **lizard**   **eagle**
dormouse   **bat**   **eel**   **elephant**
penguin   **salmon**   **python**

3 Which of the **five** vertebrate groups are warm blooded? (1)

4 Which of the following sentences describes the features of which vertebrate group? (5)

a) Tough waterproof skins made of horny scales.
b) Smooth moist skin, lay eggs in water.
c) Breathe through gills, have fins and scales.
d) A body covered in hair, feed young on milk.
e) Body covered in feathers, wings and a beak.

5 Name **two** invertebrate groups. (2)

6 Name the **two** largest plant groups. (2)

7 A key can be used to identify organisms. Look at the plant and animal below and write down the features which could be used to separate them in a key. (2)

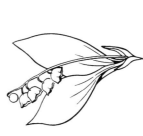

8 Give each of the flowers below a name, then draw a key to identify them. (8)

Science Companion © A Porter, M Wood, T Wood and Stanley Thornes (Publishers) Ltd 1995

2.2.3 Grouping animals and plants

1 Follow the key to identify the reef fish. Write the names under the diagrams.

Key

1 Has it a striped body... go to question 2.
   Has it no stripes on its body? .... go to question 4.
2 Black and white markings over eye and upper body. .... **Raccoon Butterfly**.
   Dorsal fin is elongated to a point. .... go to question 3.
3 Body has two broad dark stripes diagonally across it. .... **Pennant Fish**.
   Narrow stripes on body and spot on dorsal fin. .... **Threadfin Butterfly**.
4 Tail fins divided into two points. .... go to question 5.
   Single tail fin. .... go to question 6.
5 Large flat body. Tail fin tapers into two long points. .... **Orange-spined Tang**.
   Streamlined body. Two dorsal fins. Two barbels on mouth. .... **Yellowstriped Goatfish**.
6 Elongated snout. Serrated dorsal fins. .... **Longnose Butterfly**.
   Spotted body. First dorsal fin curved. Tail fanshaped. .... **Fantail Filefish**. (14)

2 The following are all fungi. Draw the correct fungus in the box which has its description.

  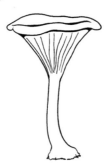 

(16)

| FALSE CHANTERELLE Funnel-shaped cap curled at edges. | ANTABUSE INK CAP Ribbed bell shaped cap. Dark gills. | FALSE DEATH CAP Convex cap. White spots on cap. Thick stem with bulb at base. | PARASOL MUSHROOM Umbrella shaped cap. Scales on cap. |

Name: **Class:** **Date:**

Science Companion © A Porter, M Wood, T Wood and Stanley Thornes (Publishers) Ltd 1995

2.3.1 Variation

1 What is the name given to the differences between species? (1)
2 Name the two types of variation? (2)
3 Which type of variation is caused by genes only? (1)
4 The diagram shows the two types of ear shapes that people have.

   Some people have ear lobes, others do not. There is no 'in between'.

   a) What type of variation is shown by this characteristic? (1)
   b) Name two other characteristics shown by this type of variation. (2)
   c) Explain how statistics showing this type of variation would not show a normal distribution curve. (2)

5 Explain how the environment can influence the following.
   a) Birth weight in babies   b) Fruit yields in crops. (2)(2)
6 What is meant by the term "natural selection"? (2)
7 Examine the diagrams of the apes' hands below.

   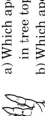

   Gorilla   Orang-utan

   a) Which ape's hands are adapted for living in tree tops? (1)
   b) Which ape is more suited to living on the ground? Why? (2)
   c) Which ape would be more likely to survive if the only source of food was in the tree tops? Why? (2)

8 Read the information below about the dodo and answer the questions which follow.

   The dodo

   The dodo was a large flightless bird with short legs. It lived on the island of Mauritius over 300 years ago. It became extinct soon after people and their pets came to inhabit the island.

   a) Define the term 'extinct'. (1)
   b) Describe the dodo bird. (2)
   c) What does the word 'flightless' mean? (1)
   d) Give three reasons why the bird became extinct. (3)

2.3.2 Variation

1 People of the same age have differences in hair colour and skin type. What do we call these differences? (1)
2 a) There are two types of variation, continuous and discontinuous. Weight and height are examples of which type? (1)
   b) Which of the following are examples of discontinuous variation?

   **hair length   blood type   handspan   eye colour
   footsize**
   (1)
   c) Which type of variation cannot be influenced by the environment? (2)
3 In nature animals with the features best suited to their environment are more likely to survive and those not suited may die. This process is called:

   (**natural breeding**)   (**natural selection**)   (**artificial selection**) (2)

4 Read the information about the dodo below, then answer the questions which follow.

   The dodo

   The dodo was a large flightless bird with short legs. It lived on the island of Mauritius over 300 years ago. It became extinct very soon after people and their pets like cats and dogs came to live on the island.

   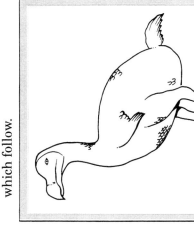

   a) What does the word extinct mean? (2)
   b) Describe the dodo bird. (2)
   c) What does the word flightless mean? (2)
   d) How long ago did the dodo live on earth? (2)
   e) Where did the dodo live? (2)
   f) Give three reasons why the dodo became extinct. (3)

2.3.3 Variation

1 The heights of pupils in a class was measured. The results are shown in the table.

Height group (cm)	134–40	141–2	143–4	145–6	147–8	149–50	151–2	153–4	155–6
Number of students	1	3	3	5	7	5	3	1	1

a) Draw a block graph to show the results.

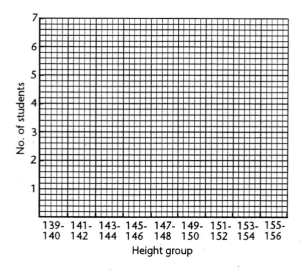

(8)

b) How many students were in the class? _____ (1)

c) Which groups holds the most students? _____ (1)

d) How many students are 142 cm in height or less? _____ (1)

e) Do the results show continuous or discontinuous variation? (1)
_____ (1)

f) What do we call a graph with this shape?
_____ (1)

g) Which two of the following characteristics show the same type of variation as those in the graph?

**blood group   eye colour   foot size
handspan   sex** (2)

2 Study the passage below and then answer the questions which follow:

Industrial melanism

Some insect species in Britain show **industrial melanism**. Melanism is the word used to describe a darker form of a naturally light-coloured species. Melanism occurs in natural conditions where organisms adapt to changing environments. Industrial melanism can happen where species adapt to live in sooty environments caused by industrialisation. The most common example of this is the peppered moth, *Biston betularia*. It has two forms, one black, the other speckled grey.

Before the Industrial Revolution most peppered moths in Britain were the grey form. This is because they rested on lichen-covered trees which were grey, and provided camouflage from predators. During the Industrial Revolution, soot killed the lichens on trees and the light moths could clearly be seen on the dark tree trunks. In 1848 the percentage of grey moths was 99%. In contrast, in 1894 the percentage of dark moths was 99%. The percentage of grey moths remained high in unpolluted areas of Britain. Other species showing industrial melanism are the oak eggar moth, the two-spot ladybird and the zebra spider. In each case the melanic form became more prevalent during the Industrial Revolution.

Dark peppered moth (the melanic form), and light peppered moth

a) What is melanism? _____
_____
_____ (1)

b) Explain why the number of dark peppered moths increased during the industrial revolution. _____
_____
_____
_____
_____ (1)

c) What kind of predators might catch the moths? _____ (1)

Name: **Class:**   **Date:**

2.4.2 Why do we look like our parents?

1. When a baby is born it inherits features from it's parents. Who do scientists call this?
   heredity   heraldry   height   passing on   (2)

2. Look at the diagram below:

   a) Name three facial features which can be inherited from parents.   (3)
   b) Which of the following are not inherited from parents?
   blood type   weight   eye colour   hairstyle   (2)

3. The structures below are called chromosomes.

   a) In which part of a cell are they found?
   vacuole   cell wall   nucleus   cytoplasm   (2)
   b) How many chromosomes are there in most human cells?
   23   45   46   48   (2)
   c) How many chromosomes are found in sex cells?
   23   45   46   48   (2)

4. The diagram below shows a baby and his parents. Which features did Sam inherit from which parent?   (2)

    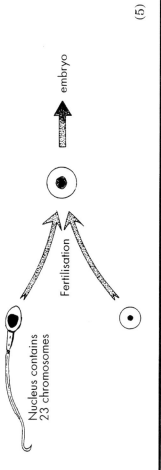

   Mrs Spot — Blonde curly hair, pointed chin
   Mr Spot — Brown curly hair, pointed ears, round chin
   Sam Spot — Curly blonde hair, straight nose, pointed chin, pointed ears

Science Companion © A Porter, M Wood, T Wood and Stanley Thornes (Publishers) Ltd 1995

2.4.1 Why do we look like our parents?

1. Define the term 'heredity'.   (1)
2. Look at the picture below.

   a) Name two facial features which are inherited from our parents.   (2)
   b) A person has parents both with blue eyes. Why could that person not have brown eyes?   (2)

3. Examine the structures below and answer the questions.

   a) What are the structures in the diagram called?   (1)
   b) Where are these structures found?   (1)
   c) Name the parts of the structures which contain genetic information.   (1)
   d) How many of these structures do most human cells contain?   (1)
   e) How many of these structures are found in sex cells.   (1)

4. Complete the diagram below to explain how many chromosomes are found in embryo cells and how they contain genetic information from both parents.

   Nucleus contains 23 chromosomes
   Fertilisation
   embryo   (5)

Science Companion © A Porter, M Wood, T Wood and Stanley Thornes (Publishers) Ltd 1995

2.4.3 Why do we look like our parents?

1 Read the following information about how sex is determined in a baby, then answer the questions which follow.

SEX OF A BABY
Sex chromosomes determine the sex of a baby. There are two types of sex chromosome called the Y chromosome which is short and the X chromosome which is long. Males contain an X and a Y chromosome and females contain two X chromosomes. Male sex cells, sperms have either an X or a Y chromosome, (half are X and half are Y) female sex cells all have X chromosomes. When fertilisation takes place an egg joins with a sperm which contains either an X or a Y chromosome. A combination of two X chromosomes will produce a girl and a combination of a Y and X chromosome will produce a boy.

a) How many types of sex chromosomes are there?_____ (1)

b) Which sex chromosome is a long one?
_____ (2)

c) Which sex chromosome do males have?
_____ (2)

d) What percentage of male chromosomes are X?
_____ (2)

e) If a sperm has an X chromosome and an egg has an X chromosome, what sex will the baby be?
_____ (2)

2 Solve the clues to find the words in the grid. Then find the word in column 6 which links them.

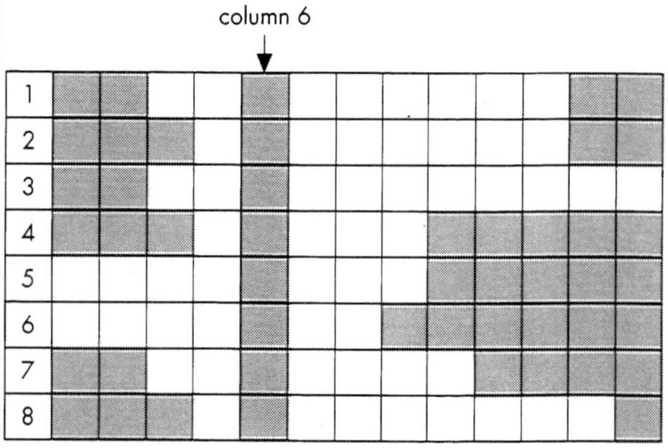

(8)

CLUES
1. Features passed on to offspring are this.
2. The study of heredity and variation.
3. Thread-like structures in the nucleus.
4. Parts of clue 3 which carry genetic information.
5. Creating new individuals.
6. Chromosomes are found like this.
7. The opposite of artificial selection.
8. Jelly-like substance found in a cell.

The word in column 6 is_____ (1)

3 Quiz

1 Differences between individuals are called
_____ (1)

2 An example of discontinuous variation is
_____ (1)

3 The process by which species which most suited to an environment will survive is called _____ (1)

4 Passing on features from parents to offspring is called _____ (1)

5 Features like hair colour and ear shape are called _____ (1)

6 Label parts **A–C** on the diagram below.

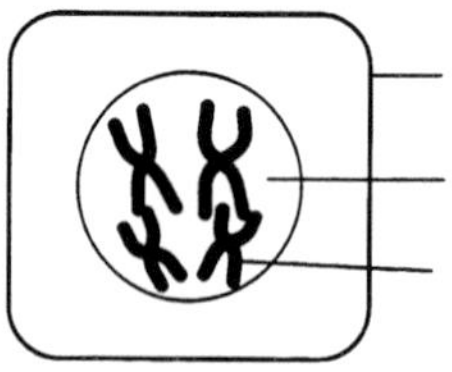

A _____ (1)
B _____ (1)
C _____ (1)

7 Parts of chromosomes which contain genetic information are called
_____ (1)

8 Most human cells contain _____ chromosomes (1)

9 How many chromosomes do these cells contain?
____(1)  ____(1) 

Name: **Class:** **Date:**

Science Companion © A Porter, M Wood, T Wood and Stanley Thornes (Publishers) Ltd 1995

2.5.2 Extinction – the end of a species

1 Look at the picture of a dinosaur and answer the questions.

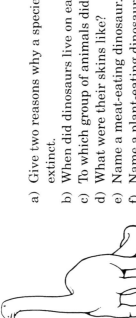

a) What does the word extinct mean? (1)
b) When did they live on earth? (1)
c) Were they reptiles or amphibians? (1)
d) Name a meat-eating dinosaur. (1)
e) The nearest living relation to the dinosaur is the crocodile. How are they similar to the dinosaurs? (2)

2 The diagram below shows a fossil.

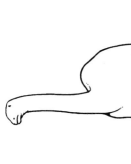

a) What is a fossil?
  i) the hardened remains of an animal or plant
  ii) the bones of an animal
  iii) imprints of animal remains. (2)
b) Name two things that scientists can learn by looking at fossils. (2)
c) What do we call scientists that study fossils? (2)

3 There are different ideas about why dinosaurs became extinct. One idea is below. Look at the diagram and answer questions about that idea.

Dust cloud

The Earth's sky carried the dust for years. The Sun's light could not get through. The temperature of the Earth fell and stayed cold for years. The cold killed the dinosaurs and many other species

A meteorite hit the Earth
Earth
The meteorite threw up dust and rocks, and caused earthquakes, landslides, tidal waves and global fires.

a) What was supposed to have hit the Earth? (2)
b) What changes in the weather were caused? (2)
c) What was carried in the Earth's sky for years? (2)
d) What was thought to have killed the dinosaurs? (2)

Science Companion © A Porter, M Wood, T Wood and Stanley Thornes (Publishers) Ltd 1995

2.5.1 Extinction – the end of a species

1 Look at the diagram of an extinct species below and answer the questions which follow.

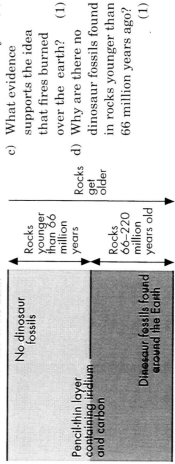

a) Give two reasons why a species might become extinct. (2)
b) When did dinosaurs live on earth? (1)
c) To which group of animals did dinosaurs belong? (1)
d) What were their skins like? (1)
e) Name a meat-eating dinosaur. (1)
f) Name a plant-eating dinosaur. (1)

2 The only remains of dinosaurs are fossils.

a) What is a fossil? (1)
b) What do we call a scientist who studies fossils? (1)
c) What can scientists learn by studying fossils? (2)

3 One theory about why dinosaurs became extinct is that the earth was hit by a giant meteorite.

a) Briefly explain how this could have killed the dinosaurs. (2)
b) What visual evidence is there that the earth was hit by a meteorite about the time that marine species became extinct? (2)

4 Look at the diagram below and answer the questions.

a) How old are the rocks in which the dinosaur fossils have been found? (1)
b) How does the presence of the metal iridium on earth support the meteorite theory? (2)
c) What evidence supports the idea that fires burned over the earth? (1)
d) Why are there no dinosaur fossils found in rocks younger than 66 million years ago? (1)

Surface of the Earth

No dinosaur fossils	Dinosaur fossils found around the Earth
Rocks younger than 66 million years	Rocks 66–220 million years old

Pencil-thin layer containing iridium and carbon

Rocks get older →

Science Companion © A Porter, M Wood, T Wood and Stanley Thornes (Publishers) Ltd 1995

2.5.3 Extinction – the end of a species

1 From the list below, choose eight features which belong to *Tyrannosaurus rex* and write them on the diagram.

(8)

Meat-eater
Plant-eater
Walks on two legs
Walks on four legs
Huge head
Tiny head
Serrated teeth
Small teeth
Huge curved teeth
Tiny clawed hands
Long neck
Short powerful neck
Long whip-like tail
Short strong tail

2 Explain how *T. rex* was adapted to meat eating. _____
_____
_____
_____
_____ (2)

3 Underline the correct answers below:
a) Dinosaurs belonged to the group of animals called:
(birds) (amphibians) (reptiles) (mammals)
b) Which of the following hitting the earth could have caused the extinction of dinosaurs?
(comet) (meteorite) (star) (lightning)
c) What does a paleontologist study?
(planets) (comets) (fossils) (plates)
d) A rare metal present in meteorites is called:
(carbon) (copper) (iridium) (iranium)
e) A species which is found in only one place on earth is called:
(epidemic) (endemic) (igneous) (indigenous) (7)

4 Read the passage below and answer the questions which follow.

Extinction in the Hawaiian islands

Hawaiian Goose

It took over 70 million years for plants and animals to colonise the Hawaiian islands. Large animals like reptiles and mammals could not reach the islands because they were unable to cross the Pacific ocean. The living things which arrived first were those which could travel by sea currents, insects, seeds and snails. Birds arrived by air currents. Life in the islands evolved slowly but flourished because there were no predators or competition for food. For this reason many species evolved without protective mechanisms like stingless nettles and flightless birds. The arrival of people to the island brought extinction to many species and many more are now in danger like the Hawaiian Goose and the Happyface Spider. People destroyed the forests to plant food crops and brought predators like rats, cats, and pigs. Because 90% of Hawaii's flora and fauna is endemic, that means that it is not found anywhere else on Earth, species which die are therefore made extinct in the world.

Happyface Spider

a) Why were large animals unable to reach the Hawaiian islands when they were being colonised? _____
_____ (2)

b) How did life get to the islands?
_____ (2)

c) Why did some species evolve without protective mechanisms? _____
_____
_____ (2)

d) Name one of those species.
_____ (1)

e) Name two predators brought by man to the islands. _____ (2)

f) Why did man destroy forests?
_____ (2)

g) What does the word 'endemic' mean?
_____
_____ (2)

Name: **Class:**   **Date:**

3.1.1 People – a threat to the Earth

1 In the diagrams below are the resources which people need to live. For each give two reasons why we need them. (8)

Air Water Food Shelter

2 Give two reasons why the Earth's resources are being exhausted. (2)

3 Explain how the activities below change the appearance of and damage the Earth.
 a) Electricity production
 b) Mining
 c) Deforestation (6)

4 Read the article below and answer the questions which follow.

Gardeners destroying rare plant species

Peat is formed in wetland bogs. It is cut out of the ground, dried and sold in bags. Peat was used as a fuel in Britain in medieval times but is now used mainly for gardening. Peat is used to plant seeds and cutting and for improving garden soil.

Peat bogs are covered with vegetation which is rich in wildlife and rare plants which only grow in peat bogs. When it was cut for fuel by hand the wildlife could re-establish itself but modern methods destroy the vegetation completely because they use machines to cut the peat. Peat bogs took 10 000 years to form but could be exhausted by the end of the next century if used at the present rate. Alternatives to peat for gardening are coconut fibre or animal slurry mixed with straw which can be used in the same way as peat.   **Pete Plant. Gardening Daily.**

 a) Where is peat formed? (1)
 b) How was peat used in medieval times? (1)
 c) How is peat used today? (1)
 d) Why do modern methods of cutting peat destroy vegetation and wildlife?
 e) Why could supplies of peat be exhausted by the end of the next century? (1) (1) (1)
 f) Explain how gardeners could help to preserve peat.
 g) Imagine this plant is a rare species living in peat bogs. Use it to draw a poster informing gardeners why it is in danger and how they could help to protect it's habitat. (3)

Spiny Arrowtop

3.1.2 People – a threat to the Earth

1 Complete the paragraph below by filling in the missing words.

food   water   shelter   earth

People need resources to live. They come from the _____. Some resources we need are air, _____, _____ and water. (4)

2 Give an example of the following:
 a) A gas we need for respiration (1)
 b) A food we grow (1)
 c) A liquid we need (1)
 d) A fuel we use to heat homes (1)
 e) A type of shelter (1)

3 Draw and complete the table which shows which resources are extracted by which method?   **limestone   wood   coal**

METHOD	RESOURCE
Mining	
Quarrying	
Deforestation	

(3)

4 Look at the power station and answer the questions.

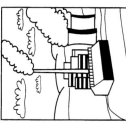

 a) Which resource is used at many power stations in Britain? (1)
 b) How does the power station affect the appearance of the environment? (1)
 c) How does the power station damage the environment? (2)
 d) What type of energy do power stations produce? (1)
 e) Name some other ways in which this type of energy can be produced. (2)

5 Look at the diagram and answer the questions about peat.

 a) How was peat used in the past? (1)
 b) What is peat used for today? (1)
 c) How does peat cutting destroy wildlife and plants? (2)
 d) What can gardeners use instead of peat? (2)

3.1.3 People – a threat to the Earth

1 Place the resources below next to their uses. Use each one only once.

**carbon dioxide   coal   flour
hardwood   heating homes   oil
limestone   oxygen   water   wood**

a) Respiration _____

b) Fire extinguishers _____

c) Petrol and diesel fuels _____

d) Window frames _____

e) Paper products _____

f) Bread and baking _____

g) Cement _____

h) Drinking and washing _____

i) Producing electricity _____

j) Heating homes _____

(10)

2 Look at the illustration below and answer the questions which follow.

Heavy industry

a) How is this industry affecting the environment?
Give at least two ways

_____
_____
_____ (1)

b) How might the health of people be damaged?

_____
_____ (1)

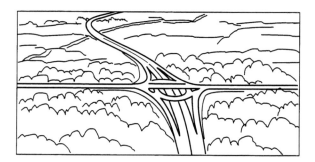

c) Give three ways in which a motorway can damage the environment.

1. _____ (1)

2. _____ (1)

3. _____ (1)

3 Solve the clues to complete the word puzzle. Then find the resources in the boxes.

	A	B	C	D	E	F	G	H	I
1									
2								░	░
3								░	░
4									
5									
6									
7								░	
8						░	░	░	
9									
10						░	░	░	

(10)

CLUES
1. Things we need to live
2. A house is a form of this.
3. Taking ores and minerals from the ground.
4. When resources run out the supply is this.
5. Damage to the earth.
6. Products are made using raw _____ .
7. We need water for this.
8. Our planet.
9. Peat is used mainly for this.
10. A gas we need for respiration.

| 3F | 4D | 2A | | 5B | 5G | 5C | | 6B | 7E | 8C |

| 7A | 8B | 5F | 2C | 2G | | 7A | 9B | 9C | 6A | 5F | 2B |

(5)

3.2.2 The problem of pollution

1 Match the words with their meanings below.

**environment   biodegradable   pollutant   resource
toxic   non-biodegradable**

a) A supply of raw materials in an area.
b) A poisonous substance is this.
c) The conditions around a living thing which affect the way which it lives.
d) A substance which can be broken down by nature.
e) Substances which harm the Earth and living things.
f) A substance which cannot be broken down by nature. (6)

2 Sort the items below into the two groups, biodegradable and non-biodegradable.

acrylic jumper	lead	sewage
cardboard	leaves	teabags
crisp bag	newspaper	tights
compost	plastic food tray	tin can
glass bottle	polyester shirt	woollen jumper
grass cuttings	plastic bottle	wood

3 Look at the diagram below and answer the questions which follow.

Acid rain — Oxides of sulphur and nitrogen produced — Fossil fuels burnt — Forests killed

a) Which fuels are burnt in power stations? (2)
b) What substances are produced when these fuels are burnt? (2)
c) What do these substances make when mixed with water in the air? (2)
d) How does this substance affect plants? (2)

4 Which of the following pollutes air, land or water? Write out the substance and **A**, **L** or **W** after it.

**Bleach   chlorofluorocarbons   metals   paper   petrol
plastic   phosphates   raw sewage   sulphur dioxide** (2)

3.2.1 The problem of pollution

1 Define the following terms:
a) Pollution b) Pollutant c) Biodegradable d) Non-biodegradable (4)

2 Place the items below into the two groups, biodegradable and non-biodegradable.

acrylic jumper	lead	sewage
cardboard	leaves	teabags
crisp bag	newspaper	tights
compost	plastic food tray	tin can
glass bottle	polyester shirt	woollen jumper
grass cuttings	plastic bottle	wood

(4)

3 a) Use the diagram alongside to explain how the burning of fossil fuels causes acid rain. (2)

b) How does acid rain
  i) affect vegetation
  ii) create 'dead' lakes? (4)

Acid rain — Oxides of sulphur and nitrogen produced — Fossil fuels burnt — Forests killed

4 a) What is ozone? (1)
b) How does ozone protect the Earth? (2)
c) How has the ozone around the Earth been damaged? (2)
d) How can skin be damaged by overexposure to the sun's rays. (1)

5 Explain how the following pollutants get into the water cycle and what kind of damage they cause.
a) Raw sewage
b) Chemicals from industry
c) Chemicals from landfill sites. (3)

6 Match the pollutants to their causes. Write them in the correct order.

POLLUTANT	CAUSE
Acid rain	Use of phosphates in detergents
Global warming	Use of CFC's
Lead poisoning	Burning coal
Depletion of the ozone layer	Burning petrol
Dead lakes and rivers	Carbon dioxide in the atmosphere

(2)

3.2.3 The problem of pollution

1 The Bloggs family have had a picnic and have rubbish to take home with them. Use a highlighter pen or coloured pencil to shade each item of rubbish to show how they should dispose of it.

Blue [B] = recycle **Green [G] = compost heap** **Red [R] = dustbin**

aluminium foil   cola bottle   plastic cup
apple core   crisp packet   plastic carrier bag
banana peel   egg shell   teabags
burger tray   fish and chips paper   wine bottle
beer can   yogurt pot

2 Look at the diagram about global warming and complete the passage which follows.

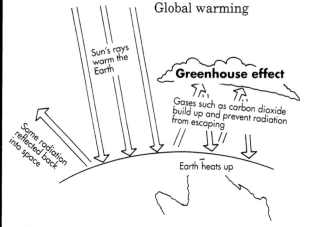

The S_____ heats up the Earth by r_____. Normally some radiation is r_____ back into space. We produce c_____ _____ gas which builds up around the Earth and prevents r_____ from escaping. This heats up the earth and is called the g_____ _____. This has increased the temperature of the Earth's a_____. (7)

3 Look at the label from the product below and answer the questions.

WHIZ WASHING POWDER
Active ingredients
Anionic surfactant Phosphates Enzymes

a) Which ingredient causes pollution in rivers and lakes?

_____ (2)

b) These ingredients cause algae to grow in lakes and rivers. How does this kill fish and plants?

_____
_____
_____
_____
_____
_____ (2)

c) How can we help to stop this substance getting into the water cycle?

_____
_____ (2)

4 Use arrows to match the words to their meanings below.

Any poisonous substance	Nitric
A gas which protects the Earth	Mercury
A type of acid	Toxic
A poisonous metal	Fossil
Fuel formed over millions of years	Ozone

(5)

Name: Class: Date:

3.3.1 Are we doing enough?

1. Give three ways in which governments can reduce pollution. (1)
2. What percentage of litter could be recycled instead of dumped? (1)
   50%   70%   90%
3. Name two substances which pollute the sea. (1)
4. a) Define the word 'toxic'. (1)
   b) Name two pollutants which are toxic. (1)
5. Which substance can be added to acid lakes to make them less acid again? (1)
6. Which type of products can be used by householders to reduce pollution in rivers? (1)
7. a) What is smog? (1)
   b) Why do we no longer have smog in Britain? (1)
8. Give three ways in which industry can reduce the emissions of sulphur dioxide gas. (1)
9. Drax power station in Yorkshire burns coal but has an F.G.D. plant.
   a) What does F.G.D. stand for? (1)
   b) Which pollutant is removed from by the F.G.D. plant? (1)
   c) How is the pollutant removed from the waste gases? (1)
   d) What useful by-product is produced by the plant? (1)
   e) Give a use for this by-product. (1)
10. Examine the activities below and explain how doing them helps to protect the environment.

Use CFC-free aerosol

Look for the ozone friendly symbol.

Take paper and bottles to recycling bins.

Change to organic gardening

Use phosphate-free detergents and bleaches

(4)

3.3.2 Are we doing enough?

1. Give three ways in which governments can reduce pollution. (2)
2. What percentage of litter could be recycled instead of dumped? (2)
   50%   70%   90%
3. Name two substances which pollute the sea. (2)
4. What is the meaning of the word 'toxic'. (2)
5. Which substance can be added to acid lakes to make them less acid? (2)
6. Which type of products can be used by householders to reduce pollution in rivers? (2)
7. a) What is smog? (1)
   b) Why do we no longer have smog in Britain? (1)
8. Drax power station in Yorkshire has an F.G.D. plant. Read the information about it and answer the questions.

Sulphur dioxide produced by burning coal in power station

F.G.D. plant removes sulphur dioxide by 'washing' gases with limestone

The gases are cleaned and a solid produced

Gases with 90% of sulphur dioxide removed

Cleaner gases released

The solid produced is called gypsum

It is used to make plasterboard

F.G.D. plant at Drax power station

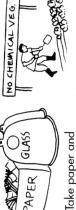

a) Which fuel is burnt at this power station? (1)
b) Which gas is removed by the plant? (1)
c) Which pollutant is caused by this gas (1)
d) Which substance is used to remove this gas? (1)
e) Which solid is produced when gases are cleaned in this plant? (1)
f) Name a product which can be made by the solid produced? (1)

3.3.3 Are we doing enough?

1 Read the information about animals caught in oil spills and answer the questions which follow.

Animals swallow oil and lungs, kidneys and liver are damaged.

Mammals are blinded by oil.

Birds preen to clean feathers and swallow oil. This kills them.

Fur of animals gets matted. Insulating properties of fur are lost. Animals die of cold.

The structure of birds' feathers is destroyed. Feathers are no longer waterproof and no longer provide insulation against cold.

Oil is sticky and does not dissolve in water.

Oil destroys red blood cells in birds. They become anaemic.

a) What is crude oil? _____ (1)

b) Which properties of oil make it difficult to clean up? _____ (2)

c) Name three animal organs damaged by oil. _____ (3)

d) How do the birds swallow the oil? _____ (1)

e) How does oil damage fur and feathers? _____ (1)

f) Which cells in birds are damaged by oil? _____ (1)

g) Suggest ways in which oil spills at sea could be reduced. _____ (2)

2 Complete the table about poisonous gases.

GAS	SYMBOL	FORMED	POLLUTION CAUSED
Sulphur dioxide	SO_2	Burning compounds containing sulphur	
Carbon dioxide			Greenhouse effect
	CO		Toxic exhaust fumes

3 Underline the correct definitions of the words below.

a) TOXIC **(edible) (poisonous) (pleasant)**

b) COMBUSTION **(burning) (mixing up) (pollution)**

c) RAW **(cooked) (untreated) (unnatural)**

d) LEACHED **(washed away) (bleached) (blocked)**

e) CONVERTED **(separated) (changed) (purified)**

f) ALTERNATIVE **(different) (useful) (changing)**

g) LEVELS **(heights) (amounts) (stages)**

h) REMOVE **(take out) (replace) (travel)**

i) EMITTED **(given out) (shouting) (burnt)**

j) FILTERED **(smoothed) (cleaned) (untreated)**

(10)

Name: Class: Date:

Science Companion © A Porter, M Wood, T Wood and Stanley Thornes (Publishers) Ltd 1995

3.4.1 Living things and their environment

1. Why is it important that living things are able to detect changes in their environment? (1)
2. Name two daily and two seasonal changes in the environment. (2)
3. The organisms in the pictures below are nocturnal. Answer the questions about them.

Moonflower Owl

a) What does the word 'nocturnal' mean? (1)
b) Give reasons why they are nocturnal. (2)
c) Give an advantage of being nocturnal. (2)

4. For each of the organisms below, explain how they survive the winter.

Squirrel   Wolf   Poppy   Arctic Tern

(4)

5. a) Define the term 'hibernate' (2)
   b) What is the main reason that animals hibernate in winter? (1)
   c) Give three ways in which the animals' bodies change during hibernation. (3)
   d) The lungfish spends the summer sleeping to escape the drought. What is this called? (1)

6. a) Give two reasons why animals migrate. (2)
   b) Name four animals which migrate. (2)
   c) What is man's theory about how animals are able to navigate for long distances? (2)

Science Companion © A Porter, M Wood, T Wood and Stanley Thornes (Publishers) Ltd 1995

3.4.2 Living things and their environment

1. Give three ways in which animals spend their time. (3)
2. Give two reasons why plants do not grow well in winter. (2)
3. Look at the organisms below and answer the questions about them.

Part 1
a) This tree is deciduous. What does that mean? (1)
b) Name four deciduous trees. (2)
c) How do other trees like conifers survive the winter? (2)

Part 2
d) How do plants like the poppy survive the winter? (1)
e) Name two other plants which spend winter in this way. (2)
f) Some plants like pansies live only in the summer, then die. How do they keep their species going? (2)

4. Look at the animals below.

Squirrel   Wolf   Arctic Tern   Ladybird

a) Which animal
   i) Migrates to a warmer place in winter (2)
   ii) Stores food so that it will survive the winter (2)
   iii) Grows thick fur to keep warm in winter (2)
   iv) Spends winter in a pupal stage? (2)

5. Place the animals below into two lists, those which migrate and those which hibernate.

Arctic Tern   bear   hamster   hedgehog   Magellanic Penguin   mouse   porpoise   salmon   turtle   vole   whale (2)

Science Companion © A Porter, M Wood, T Wood and Stanley Thornes (Publishers) Ltd 1995

3.4.3 Living things and their environment

1. Fill in the missing words to complete the passage below.

    Living things can d_____ changes in their environments. They need to do this in order to f_____ and protect themselves from p_____. Organisms react to changes in the environment like the seasons and the w_____. The seasons affect the temperature and the length of d____ and n_____. In winter animals and plants find it difficult to find f_____. Plants cannot g_____ well because the ground is f_____ and there is less light. Many plants spend the winter in a d_____ state. Animals h_____, m_____ or store up food for the winter. (12)

2. The list below shows how these animals spend the winter. Write the correct letter in the box.

    Arctic Fox   Ladybird
    Penguin   Walrus
    Tortoise   Arctic Tern
    (6)

    A Grows a dense woolly coat.

    B Huddle together in a sleeplike torpor.

    C Migrate to a warmer place.

    D The cold makes their bodies slow down so they sleep most of the winter.

    E Has an extra thick layer of blubber.

    F Migrates by swimming to warmer places.

    You may colour the diagrams if you wish.

3. Read the information about the hedgehog, then answer the questions which follow.

    The Hedgehog
    Hedgehogs are animals which hibernate during the winter. Food is scarce during the winter and they would probably starve if they did not hibernate. They feed themselves to make extra fat before the winter. Their bodies change in a number of ways during hibernation. Their body temperature and heart rate slow down so that they conserve energy. Curling up into a tight ball reduces the surface area of the hedgehog and also conserves energy. When the warm weather returns the animal's body returns to normal and it wakes up. If you have a pile of twigs or leaves in your garden which is left undisturbed during the winter you might get a hedgehog hibernating in your garden.

    a) Why does the hedgehog hibernate in winter?
    _____ (1)

    b) How does the hedgehog prepare for winter?
    _____ (1)

    c) How does the heart rate change during hibernation?
    _____ (1)

    d) How does curling up into a tight ball conserve energy?
    _____
    _____ (1)

    e) Write down the names of four other hibernating animals.
    _____
    _____ (1)

Name: Class: Date:

Science Companion © A Porter, M Wood, T Wood and Stanley Thornes (Publishers) Ltd 1995

3.5.1 Why do animals and plants live in different places?

1 Explain why animals and plants live in different places around the earth. (3)

2 Name four physical factors which affect where organisms live. (2)

3 Look at the map below, then answer the questions which follow.

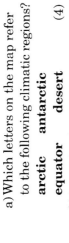

a) Which letters on the map refer to the following climatic regions? (4)

arctic   antarctic
equator   desert

b) Describe the weather conditions in each place. (4)

c) Which organisms below live in the areas **A–D** on the map? (2)

cactus   camel
flying frog   head-standing
beetle   lichens   penguin
polar bear   orchid
toucan

4 Read the adaptations of the organisms below and answer the questions which follow.

- no leaves
- sharp spines
- shallow roots
- small surface area

- large feet
- fur
- does not sweat
- stores fat in hump
- bushy eyebrows and eyelids

a) Name the organisms and say where they live. (2)

b) Explain how each adaptation above protects the organism from its environment. (2)

5 a) In which region of the world are the rainforests? (2)

b) Give two reasons why people cut down rainforests. (1)

c) How might the Earth's climate change if rainforests were cut down. (2)

d) Draw a poster about the endangered rainforests. (1)

Science Companion © A Porter, M Wood, T Wood and Stanley Thornes (Publishers) Ltd 1995

3.5.2 Why do animals and plants live in different places?

1 Different factors, like temperature, affect where an animal lives. Write down three more of these factors. (3)

2 Look at the map below and answer the questions which follow.

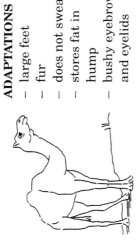

a) Which letter on the map refers to the places with these weather conditions?

i) Very hot and very wet. No seasons.

ii) Hot during the day and cold at night. Very little rain.

iii) Freezing all year round. Long days or nights. (6)

b) Where do these animals live? Choose from **A–D** on the map. (4)

camel   penguin
polar bear   orang-utan

3 Read the facts about the cactus and camel below and answer the questions.

ADAPTATIONS
- no leaves
- sharp spines
- shallow roots
- small surface area

ADAPTATIONS
- large feet
- fur
- does not sweat
- stores fat in hump
- bushy eyebrows and eyelids

a) In which type of climate do these organisms live? (2)

b) Which of the adaptations above

i) Prevents the camel from sinking into the sand (2)

ii) Reduces water loss from the cactus (2)

iii) Helps the cactus to capture water from the surface of the soil (2)

iv) Protects the camel's eyes from sandstorms? (2)

c) Name four other animals which live in this type of place. (1)

Science Companion © A Porter, M Wood, T Wood and Stanley Thornes (Publishers) Ltd 1995

3.5.3 Why do animals and plants live in different places

1 Complete the passage about penguins below.

> WORDS: Antarctica   birds   blubber   feathers   fly   freezing   warm   predators
>
> ### Penguins
>
> Penguins are b_____ but they do not f_____. Penguins like the emperor penguin live only in A_____. The temperature is nearly always below f_____. It can fall to minus forty degrees celsius. Penguins' skin is protected by dense f_____ and a thick layer of bl_____ under their skin. They stand in groups to keep w_____, taking it in turns to stand on the inside of the group. Keeping together also protects them from p_____. (8)

2 The table below shows the different layers in a tropical rainforest. Answer the questions about the information given.

LAYER	LIFE
Emergent layer	Giant trees grow above the canopy. Large predators like Harpo eagle live here and spy on their prey below.
Canopy	Trees form a dense roof of leaves. Most animals are found here because there is protection and food. Animals found include toucans, spider monkey, Lar gibbon and flying frog.
Understorey layer	This layer has shade-loving plants only. Creepers grow here and reach to the forest floor.
Shrub layer	No light here so sparse vegetation like ferns. Many insects live here.

a) Which layer contains the least vegetation?

_____ (1)

Why?_____

_____ (1)

b) Why do most animals live in the canopy layer?

_____

_____ (1)

c) How are animals adapted to living in this layer?

_____
_____
_____
_____
_____
_____
_____
_____ (1)

d) What type of plants live in the understorey layer?

_____
_____ (1)

e) Write a sentence about what it might be like on the forest floor.

_____
_____
_____
_____ (1)

3 Unscramble the names below of some animals which inhabit rain forests.

a) **nyfilg orfg** _____
b) **irpta** _____
c) **alutnaart** _____
d) **tecolo** _____
e) **auoten** _____
f) **anaigu** _____
g) **rbgmmuhinid** _____
h) **yloolw ekomny** _____ (8)

Name: Class: Date:

Science Companion © A Porter, M Wood, T Wood and Stanley Thornes (Publishers) Ltd 1995

3.6.2 Populations within ecosystems

1 Match the words below to their definitions.

ecology   ecosystem   food web   habitat   population

i) A group of one species living together.
ii) A diagram showing how energy is passed from one organism to another.
iii) The place where an organism lives.
iv) The place where organisms react with their environment.
v) The study of where organisms live. (5)

2 Match the organisms below to their habitat (where they live)

ORGANISM	HABITAT
Frog	Forest floor
Blackbird	Cornfield
Mouse	Old tree trunk
Woodlouse	Forest canopy
Blackberry	Pond

(5)

3 The chart shows the plants which live in different layers in the forest.

LAYER	VEGETATION
Canopy	Oak, Beech,
Shrub	Rhododendron, Hawthorne, Holly, Elder.
Herb	Bluebells, Foxgloves.
Ground	Mosses, Ferns, Fungi Blackberry.

a) Name the layers in a forest. (1)
b) Name two plants living in the shrub layer. (1)
c) Why do plants below the canopy layer grow shorter than those in the canopy layer? (1)
d) Which plants in the chart flower in spring? (1)
e) Which organism on the forest floor rots the leaves which fall from the trees? (1)

4 Look at the food chain below and answer the questions.

Fox ← Rabbit ← Plants

a) Name the predator in the food chain. (1)
b) What do rabbits eat? (1)
c) If foxes die, will the number of rabbits get bigger or smaller? (1)
d) If the number of rabbits gets bigger, how will their food supply be affected? (1)

3.6.1 Populations within ecosystems

1 Define the following terms: i) ecosystem   ii) ecology   iii) habitat   iv) population (4)

2 Name one living and one non-living factor which determines the size of a population. (2)

3 Name places which might be a habitat for the following animals:
i) mice   ii) woodlice   iii) hawk   iv) crab (2)

4 The chart shows the type of vegetation found at different layers in a forest. Read the information and answer the questions.

LAYER	VEGETATION
Canopy	Oak, Beech,
Shrub	Rhododendron, Hawthorne, Holly, Elder.
Herb	Bluebells, Foxgloves.
Ground	Mosses, Ferns, Fungi, Blackberry.

a) How do trees in the canopy layer affect light levels in the forest? (1)
b) What do we call the layer beneath the canopy? (1)
c) Why is it an advantage for plants like bluebells to grow and flower in spring? (1)
d) Give two reasons why fungi flourish on the forest floor? (1)
e) Name two plants found growing in the shrub layer. (1)

5 Look at the food chain alongside and answer the questions.

Fox ← Rabbit ← Plants

a) Name the predator in the food chain. (1)
b) How do predators prevent the overpopulation of rabbits? (1)
c) How would killing foxes affect the rabbit population? (1)
d) What happens to a species when they become overpopulated? (1)
e) How could a shortage of plants affect the fox? (1)

6 Explain how lichens can be used as indicators of population. (2)

3.6.3 Populations within ecosystems

1 Look at the pictures below and answer the questions which follow.

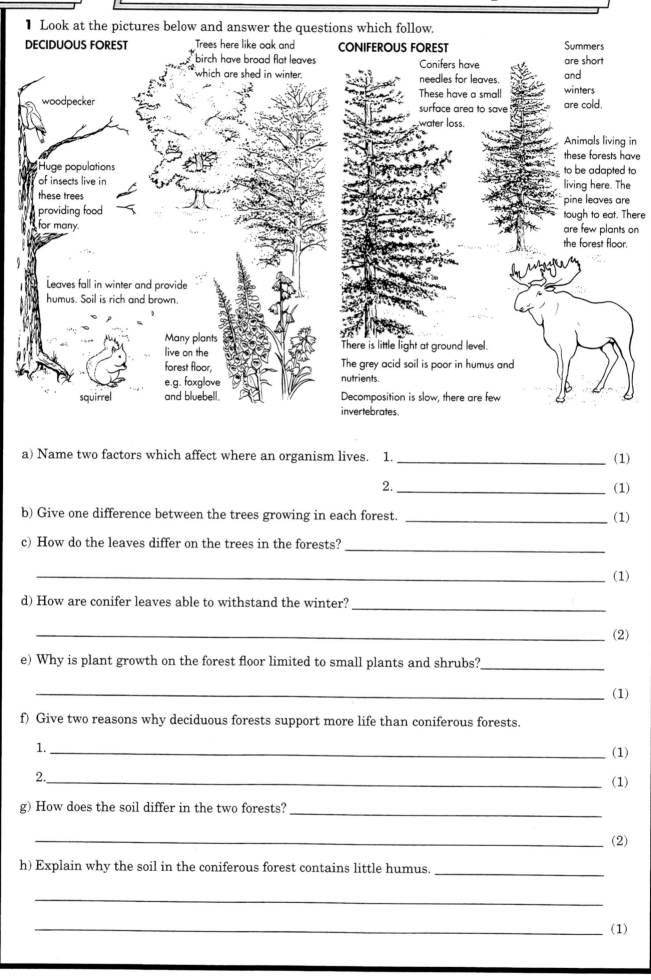

DECIDUOUS FOREST — Trees here like oak and birch have broad flat leaves which are shed in winter. woodpecker. Huge populations of insects live in these trees providing food for many. Leaves fall in winter and provide humus. Soil is rich and brown. squirrel. Many plants live on the forest floor, e.g. foxglove and bluebell.

CONIFEROUS FOREST — Conifers have needles for leaves. These have a small surface area to save water loss. Summers are short and winters are cold. Animals living in these forests have to be adapted to living here. The pine leaves are tough to eat. There are few plants on the forest floor. There is little light at ground level. The grey acid soil is poor in humus and nutrients. Decomposition is slow, there are few invertebrates.

a) Name two factors which affect where an organism lives.   1. _____ (1)

   2. _____ (1)

b) Give one difference between the trees growing in each forest. _____ (1)

c) How do the leaves differ on the trees in the forests? _____

   _____ (1)

d) How are conifer leaves able to withstand the winter? _____

   _____ (2)

e) Why is plant growth on the forest floor limited to small plants and shrubs? _____

   _____ (1)

f) Give two reasons why deciduous forests support more life than coniferous forests.

   1. _____ (1)

   2. _____ (1)

g) How does the soil differ in the two forests? _____

   _____ (2)

h) Explain why the soil in the coniferous forest contains little humus. _____

   _____

   _____ (1)

Name: **Class:**   **Date:**

4.1.1 Recycle it

1 a) What do people mean by the word recycle? (1)
   b) Name two types of products which can be made from recycled materials (1)

2 a) What is a non-renewable resource? (1)
   b) Name a non-renewable resource? (1)

3 a) Most household rubbish is taken to landfill sites. What is a landfill site? (1)
   b) Name three problems caused by landfill sites. (3)

4 Explain how recycling materials helps to protect the environment. (2)

5 Suggest how these items of rubbish should be disposed of instead of thrown into the dustbin. (2)

   textiles   glass   plastic   paper

6 Explain how the following activities can pollute the environment.
   a) Pouring used car oil down the drain. (1)
   b) Burning car tyres. (1)
   c) Dumping chemical containers on landfill sites. (1)

7 Look at the packaging of two products below. For each say how they
   a) save the customer money (1)
   b) save resources (1)

8 The symbols below are taken from products with recyclable packaging.

   a) What could be the contents of packages **A – D**? (2)
   b) Name two ways in which symbols like these help people to protect the environment. (2)
   c) Design a symbol which shows that a product does not have excess packaging. (2)

9 Make a list of things that we can do regularly to protect resources and save the environment. (2)

Science Companion © A Porter, M Wood, T Wood and Stanley Thornes (Publishers) Ltd 1995

4.1.2 Recycle it

1 Name two things which can be recycled. (2)

2 Most household rubbish is taken to a rubbish tip.
   a) Name two ways that a rubbish tip causes damage to the environment. (2)
   b) Which harmful gases can be produced by rubbish on a tip? (2)
   c) Name two animals which can spread diseases from rubbish tips. (2)

3 Look at the packaging of the products below.

   A   **B**

   a) How can the packaging of **A** harm the environment? Name two ways. (2)
   b) Give two ways in which package **B** helps the environment. (2)

4 Look at the symbols below which come from recyclable packaging.
   a) What are packages **A** and **B** made from? (2)

   A   **B**
   PLEASE RECYCLE

   b) How would you dispose of each package? (2)
   c) Name two sorts of products packaged in **A**. (2)
   d) Name two sorts of products packaged in **B**. (2)
   e) Which package is biodegradable? (2)

5 Make a list of things that we can do regularly to protect resources and save the environment. (3)

Science Companion © A Porter, M Wood, T Wood and Stanley Thornes (Publishers) Ltd 1995

4.1.3 Recycle it

1 Complete the passage below about recycling aluminium. Use each word once.

**household   environment   ore   litter
stick   recycle   magnetic
aluminium   energy   resources**

Millions of drinks cans are sold in Britain each year. Half of these are made of _____. You can find out if a can is aluminium by using a magnet. Aluminium is a non-ferrous metal so it is not _____ and a magnet will not _____ to its side. It is cheaper to _____ aluminium metal because it takes 95% less _____ than producing the metal from its _____. Recycling aluminium saves _____ and helps the _____ because it reduces the amount of _____ buried in the ground. Drinks cans make up about 10% of _____ waste. (10)

2 Read the passage about plastic and answer the questions which follow.

> **Many bottles are now made of plastic instead of glass. Plastic bottles are lighter in weight than glass or metal which makes them easier to carry. Plastic is made from oil and is non-biodegradable. It is not easily recycled because there are many different types. Some manufacturers are now labelling their plastic containers with the symbols opposite which identify the type of plastic used. They can then be sorted into the different types and recycled.**

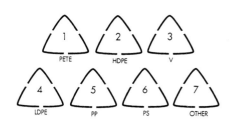

a) Name one advantage of using plastic instead of glass to make bottles.

_____

_____ (1)

b) From which material is plastic made?

_____ (2)

c) Why does plastic cause pollution?

_____

_____ (2)

d) What do the symbols below the passage mean?

_____

_____

_____ (2)

e) Find four products with these symbols on them.

_____

_____

_____

_____ (2)

Name: Class: Date:

Science Companion © A Porter, M Wood, T Wood and Stanley Thornes (Publishers) Ltd 1995

4.2.1 Passing on energy

1 Write a definition for the following terms:

producer   consumer   decomposer (3)

2 Why is the first organism in a food chain always a plant? (1)

3 Construct three separate food chains using the following organisms:
  a) lettuce
  b) mouse
  c) owl (3)

4 What happens to energy as it passes along a food chain? (1)

5 The diagram shows a pyramid of numbers:

number of organisms decreases →

owl	4th trophic level
shrews	3rd trophic level
woodlice	2nd trophic level
green plants	1st trophic level

  a) What is a pyramid of numbers (2)
  b) Why does the number of organisms decrease at each level? (1)
  c) How is energy lost from the animals in a food chain? (3)

6 How is a pyramid of biomass different to a pyramid of numbers? (2)

7 From the list below, match the predators with their prey.

**lioness   lizard   mongoose   mouse   owl   snake
wildebeest   aardvark   ant   chicken   dingo
fox   ground beetle   hedgehog** (7)

8 Read the information below about wood ants and answer the questions.

> **Wood ants** (*Formica rufa*) build nests in forests by making huge mounds of soil which they tunnel into. They eat large numbers of caterpillars and other insects which would otherwise do great damage to trees. These predators are encouraged and are a protected species in some forests.

  a) Where do wood ants live? (1)
  b) How do they protect trees? (1)
  c) What is a predator? (1)
  d) What is a protected species? (1)
  e) Make a food chain from the organisms in the passage. (1)
  f) What would happen if a predator of wood ants was introduced into the forest? (2)

4.2.2 Passing on energy

1 Look at these types of organism:

producer   consumer   decomposer

Which
  a) eats food (2)
  b) makes food (2)
  c) breaks down dead material? (2)

2 Draw two separate food chains using these organisms:

**dandelion   blackberries   slug   owl   mouse
hedgehog** (6)

3 Energy is passed along a food chain as animals feed. Which of the following statements is true?
  a) Producers are always plants
  b) Energy increases as it passes along a food chain.
  c) Energy made by plants comes from the sun.
  d) Omnivores eat animal flesh only.
  e) Energy is lost as it passes along a food chain (3)

4 Read about wood ants in the box then do the questions.

> **Wood ants** (*Formica rufa*) build nests in forests by making huge mounds of soil which they tunnel into. They eat large numbers of caterpillars and other insects which would otherwise do great damage to trees. These predators are encouraged and are a protected species in some forests.

  a) Where do wood ants live? (2)
  b) How do they protect trees? (2)
  c) Why are they a protected species? (2)
  d) What is a predator? (2)
  e) What does the word 'encouraged' mean? (2)

5 Look at the two pond water animals below: guess which clings to plants and has better protection against predators. Give reasons for your answer.

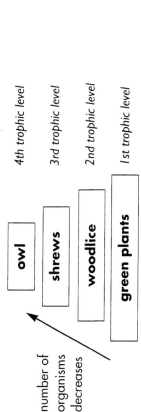

A   B

(5)

4.2.3 Passing on energy

1 Look at the food web and answer the questions.

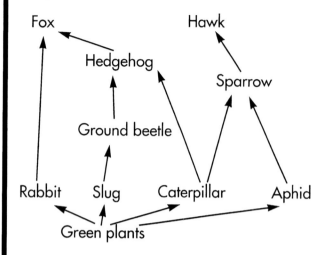

a) Draw three food chains from the food web.

(3)

b) Name an organism eaten by foxes _____
c) What eats caterpillars? _____
d) Name a prey of sparrows _____
e) Name a predator of hedgehogs _____
f) Name a carnivore _____
g) Name the producers _____ (6)

2 Fill in the gaps to complete the definitions:

A type of energy stored in green plants

c_ _mi_ _l e_ _ _g_

Organisms which make food

p_ _ _u_ _r_

Animals which eat only animal flesh

_a_ _ _v_ _ _ _

Animals which eat only plants

h_ _b_ _ _r_s

Many food chains drawn together

f_ _d w_ _

How plants make food

_ _ot_s_ _th_ _ _s (6)

3 A new predatory mite has been found feeding on cabbages. Read the information and draw it in the box.

> **New mite! Description**
> Round body
> Four eyes
> Two long antennae
> Pointed tail
> Ten jointed hairy legs
> Zig-zag markings on its back
>
> **Give this new species a name.**

Name:

(10)

Name: Class: Date:

Science Companion © A Porter, M Wood, T Wood and Stanley Thornes (Publishers) Ltd 1995

4.3.1 — The carbon and nitrogen cycles

1
a) What kind of element is carbon? (1)
b) Give an example of a pure form of carbon. (1)

2 Name three ways in which carbon dioxide is released into the atmosphere. (3)

3 Draw the diagram of the carbon cycle below.

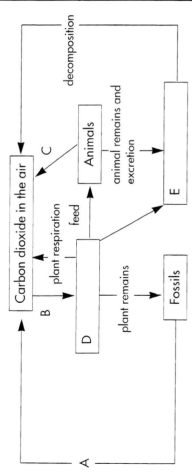

a) Label the processes at A, B and C. (3)
b) Name two fossil fuels. (1)
c) Label the organisms at D. (1)
d) Explain why the organisms at D are vital to the carbon cycle. (1)
e) Label the organisms at E. (1)
f) Why are they important to the carbon cycle? (1)
g) Animal tissues contain carbon. How do they take it in. (1)
h) How does carbon dioxide contribute to the greenhouse effect? (1)

4
a) How is carbon dioxide gas prepared in a laboratory? (1)
b) Which test below is the test for carbon dioxide gas? (1)

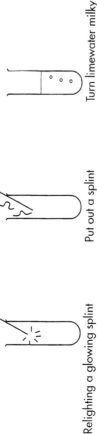

Relighting a glowing splint   Put out a splint   Turn limewater milky

c) Is carbon dioxide an acid or an alkaline gas? (1)
d) Give two uses of carbon dioxide. (2)

Science Companion © A Porter, M Wood, T Wood and Stanley Thornes (Publishers) Ltd 1995

4.3.2 — The carbon and nitrogen cycles

1 Write out the correct answers from the words in brackets below.
a) Carbon is which type of element?
   (**metallic**)   (**non-metallic**)   (**liquid**)   (**gas**) (2)
b) The symbol for carbon dioxide is:
   (**CO**)   (**CO₂**)   (**cO₂**)   (**Co**) (2)
c) A pure form of carbon is:
   (**carbon dioxide**)   (**diamond**)   (**chalk**)   (**sand**) (2)
d) Carbon dioxide is carbon combined with which gas?
   (**nitrogen**)   (**oxygen**)   (**helium**)   (**oxide**) (2)

2 Copy the diagram of the carbon cycle below.

Use these words to finish the sentences below:
combustion   **photosynthesis**   **respiration**   **plants**
decomposers

a) The process at A is called _____. (2)
b) Animals give out carbon dioxide by the process at C called _____. (2)
c) Plants make food by the process at B called _____. (2)
d) Organisms which break down dead material at E are called _____. (2)
e) The organisms at D are _____. (2)

3 Which test alongside is the correct test for carbon dioxide gas?

Relight a glowing splint   Turn limewater milky

Science Companion © A Porter, M Wood, T Wood and Stanley Thornes (Publishers) Ltd 1995

4.3.3 The carbon and nitrogen cycles

1 Complete the passage below by filling in the missing words.

**atmosphere   combine   elements
gaseous   fixing   plants   proteins
reactive**

Nitrogen is a _____ element. It makes up 78% of the Earth's _____. Living things need nitrogen to make _____ in their cells. Most _____ are unable to take nitrogen directly from the air because it is an unreactive gas. It will not _____ easily with other _____. Nitrogen is changed into a more _____ form by processes called nitrogen _____. (8)

2 The diagram shows the nitrogen cycle.

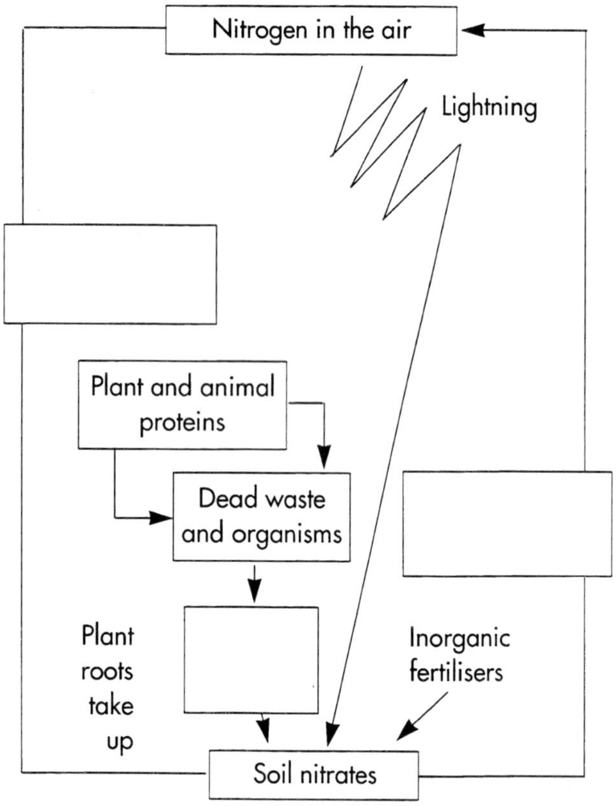

[Diagram: Nitrogen cycle showing Nitrogen in the air, Lightning, Plant and animal proteins, Dead waste and organisms, Plant roots take up, Inorganic fertilisers, Soil nitrates]

a) Write the following on the diagram:
   Nitrogen fixing bacteria
   Denitrifying bacteria
   Nitrifying bacteria (3)

b) Finish the following sentences:
   i) Nitrogen fixing bacteria are found in root n_____ and turn nitrogen into n_____. (2)

   ii) Nitrifying bacteria act on a_____ from decaying organisms and animal w_____ and turn n_____ into nitrates. (3)

   iii) Denitrifying bacteria turn nitrates from the soil back into nitrogen g_____ and release it into the air. (1)

3 Look at the diagram below.

a) Why are fertilisers added to soil?
   _____
   _____ (2)

b) Which substance in fertilisers is converted into nitrates in soil?
   _____ (2)

c) Name an advantage of farmers using inorganic fertilisers.
   _____ (1)

d) What are organic fertilisers?
   _____
   _____ (2)

e) Excess fertilisers leach into rivers. What does the word 'leaching' mean?
   _____
   _____ (2)

f) Explain how leached minerals affect the plants and animals in rivers.
   _____
   _____
   _____
   _____
   _____
   _____ (4)

Name: **Class:**   **Date:**

Science Companion © A Porter, M Wood, T Wood and Stanley Thornes (Publishers) Ltd 1995

5.1.1 — Types of materials

1 Your school wants to put together a children's dictionary for use in class. With the help of a textbook or dictionary, write out your own definitions for:
a) natural material   b) manufactured material
c) raw material   d) ceramic   e) glass   (10)

2 The 'plastic revolution' took place in the 1960s. Until then plastics were rare. See if you can find out what materials were used before plastics. For example, dustbins and buckets used to be made from metal (galvanised steel which is steel coated with zinc). Make a list for display in the classroom. (5)

3 Knitting yarn comes in balls. Below are the labels from two different balls of yarn. Study them carefully and then answer the questions which follow.

A

B

a) Which manufactured materials are found in these yarns? (2)
b) What percentage of natural material is found in each yarn? (2)
c) Why are yarns a mixture of fibres? (1)
d) Give two reasons why the manufacturers give the fibre contents of the yarns. (2)
e) Why are pure wool jumpers warmer than ones made from acrylics? (1)
f) Each label has on it a series of symbols. What does the first symbol mean on the top left-hand corner of each label? (2)

Science Companion © A Porter, M Wood, T Wood and Stanley Thornes (Publishers) Ltd 1995

5.1.2 — Types of materials

1 Here are five types of materials:
   natural material   manufactured material   raw material
   ceramic   glass

Below are five meanings **A–E**. Match each meaning to a material.
A The starting material for making items in factories. (1)
B Made by heating clay in an oven. This removes water. (1)
C A solid, often clear and brittle. (1)
D Made in factories, by people, from raw materials. (1)
E Found in the world and unchanged by people. (1)

2 Plastic is a new material. Look at the questions (a)–(e) below. Say why you think plastic is better than the old material, in each case.

ITEM	OLD MATERIAL USED BEFORE PLASTIC	
a) bucket	metal	(1)
b) knife and fork	metal	(2)
c) television bodies	wood	(2)
d) cups	ceramic	(2)
e) car bumper	metal	(1)

3 Knitting yarn comes in balls. Below are the labels from two different balls of yarn. Study them carefully and then answer the questions which follow.

A   **B**

a) Who makes yarn **A**, and where? (2)
b) Who makes yarn **B**, and where? (2)
c) How much cotton is in yarn **A**? (1)
d) How much cotton is in yarn **B**? (1)
e) Which yarn contains the most acrylic fibre? (1)
f) Which fibre is present only in yarn **B**? (1)
g) What do you think is the English meaning of BAUMWOLLE? (1)
h) What mass of yarn was originally in each ball? (1)
i) What is the meaning of the symbol 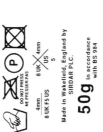 for yarn **A**? (2)

Science Companion © A Porter, M Wood, T Wood and Stanley Thornes (Publishers) Ltd 1995

5.1.3 Types of materials

1 Use the pictures below to help you complete the name of the product, and the raw material from which it is made. Then place an N in the box if the raw material is natural or an M if it is manufactured.

a)
PRODUCT   RAW MATERIAL □

P _ _ _ L   CR _ _ _ _ _ L   (3)

b)
PRODUCT   RAW MATERIAL □

P _ _ _ _ R   _ O O _   (3)

c)
PRODUCT   RAW MATERIAL □

T _ _ _ _ _ _ _ T   C _ T _ _ _   (3)

2 Sort out the following mixed up words. They are all to do with materials.

a) RWA _____ (1)

b) CRIBAF _____ (1)

c) ROE _____ (1)

d) YORTACF _____ (1)

e) RREEOUSC _____ (1)

f) CLEERCY _____ (1)

g) LEALB _____ (1)

3 Aluminium is found in the ground as an ore.

a) The ore of aluminium is _____ (1)

b) Aluminium is often used to make drink _____ (1)

c) Recycling aluminium can save W_____ and the _____ (2)

4 Use the clues below to complete the maze. A new word starts in each numbered square.

CLUES

1 across
This is the stuff which things are made up of.

2 down
Found in the world and unchanged by people.

3 down
The raw material for pottery.

4 across
This word comes from the Greek word for pottery.

5 down
These are hair-like strands.

6 across
Aluminium is an example of this.

7 down
A modern material used to make buckets.

8 across
Most gold is found in the south of this continent.

9 down
Used with limestone and sodium carbonate to make 10 down.

10 down
You see through this material out of the classroom.

(10)

Name: Class: Date:

5.2.2 Properties, uses and development of materials

1 The Mohs scale of hardness is shown alongside. The harder substances will scratch softer ones. Diamond is the hardest natural rock (mineral) known to us. Using the Mohs scale, place the following minerals in order from hardest to softest.
**apatite   calcite   corundum
diamond   gypsum**   (5)

Grade	Name of mineral
10	Diamond
9	Corundum
8	Topaz
7	Quartz
6	Orthoclase feldspar
5	Apatite
4	Fluorite
3	Calcite
2	Gypsum
1	Talc

2 Why is talc used rather than calcite in talcum powder for babies bottoms? (1)

3 Still using the Mohs scale, say if mineral **A** will scratch mineral **B** in the questions below. Answer Yes or No.

	Mineral A	Mineral B
a)	Topaz	Fluorite	(1)
b)	Calcite	Quartz	(1)
c)	Gypsum	Corundum	(1)
d)	Fluorite	Calcite	(1)

4 Tensile strength is strength against a pulling force. Compressive strength is strength against a squeezing force. Look at the diagrams (**X**), (**Y**) and (**Z**) alongside, and answer the questions below. In which diagram is

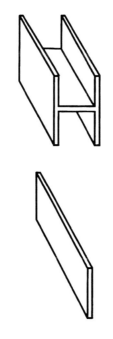

(X) Unbent
(Y) Bent one way
(Z) Bent the other way

a) side AB being stretched? (2)
b) side CD being stretched? (2)
c) side CD being compressed? (2)
d) side AB under the same force as side CD? (2)
e) side AB being compressed? (2)

5.2.1 Properties, uses and development of materials

1 In 1959 Pilkington Brothers invented the float glass process for making flat sheets of glass. In it, the glass leaves the furnaces at 1500°C and floats on the surface of a bath of molten tin. Why is the glass made in this way flat? (2)

2 What is the difference between tensile strength and compressive strength? (2)

3 Thinking of tensile and compressive strengths, what makes a material flexible? (2)

4 Look at the shapes shown below. They are all beams used to support floors in buildings.

A   B   C

Beam **A** in narrow but deep.
Beam **B** is wide but shallow.
Beam **C** is known as an I-shaped beam and is widely used in buildings. Explain why beam C is used rather than beams A or B. (3)

5 Suggest why metals are ideally suited for car bodies. (3)

6 Use your library, either in school or town, to research **one** of the following in preparation for giving either a short talk in class or writing a 200 word essay.
**Bakelite   Polythene   Nylon
Leo Hendrich Baekeland (1863–1944)
Wallace Carothers (1896–1937)**   (8)

5.2.3 Properties, uses and development

Below are 19 clues to help you find 19 words. Most of the words are linked with 'properties, uses and developments of materials'. The number of letters in each word is shown in brackets after the space for you to write your answer.

_____(8) Sand, cement, pebbles and water when mixed, dry into this material.

_____(4) Opposite of closed.

_____(4) The last name of the man who invented the scale of hardness.

_____(10) These 'brothers' invented the float glass process.

_____(3) First colour of the rainbow.

_____(5) Our planet.

_____(7) Hard minerals will do this to soft minerals.

_____(3) The star of our solar system.

_____(9) You make this when you invent something.

_____(5) There are five of these in our alphabet.

_____(10) Hopefully, you do lots of these in science.

_____(4) This fibre is made by silkworms.

_____(7) Strength which resists pulling.

_____(5) An early manufactured fibre for use in ladies' stockings.

_____(6) This is saved by recycling things.

_____(5) Fibre invented by Wallace Carothers.

_____(5) Window panes are made of this.

_____(4) The softest mineral.

_____(8) The last word in the third clue up above   (19)

Use the first letter of each answer to name one property of materials.

_____ (1)

Name: **Class:**   **Date:**

Science Companion © A Porter, M Wood, T Wood and Stanley Thornes (Publishers) Ltd 1995

5.3.1 Separating and purifying

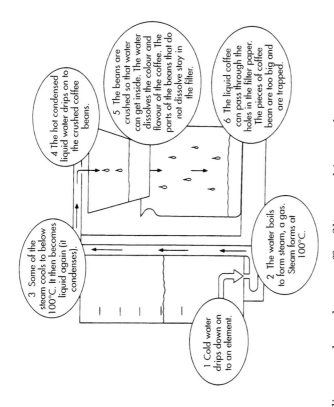

The diagram shows how a coffee filter machine works.

1. Describe how the particles in the cold water move. (2)
2. What is the difference between particles in cold water and the particles in hot water? (1)
3. What is the difference between particles in hot water at 100°C and particles in steam at 100°C? (1)
4. a) At what point on the diagram (1 to 6) do particles which are far apart begin to join together? (1)
   b) Explain your answer. (1)
5. What happens to any solid substance dissolved in the cold water? (2)
6. The water at point 4 is distilled water. What does distilled mean? (2)
7. The coffee powder is a mixture of soluble and insoluble substances. What is the difference between soluble and insoluble substances? (2)
8. What happens to the soluble part of the coffee powder as the water drips on to it? (1)
9. Explain why the soluble part of the coffee ends up in the jug but the insoluble part stays in the filter paper. (2)

Science Companion © A Porter, M Wood, T Wood and Stanley Thornes (Publishers) Ltd 1995

5.3.2 Separating and purifying

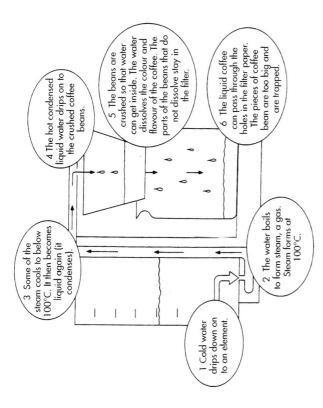

The diagram shows how a coffee filter machine works.

1. What happens to the cold water when it drips on to the element? (1)
2. What value of the boiling point of water is given in the diagram? (1)
3. We say the water at point 4 is distilled. What two things have happened to the water? (2)
4. a) The cold water may be impure. What does impure mean? (1)
   b) Distilled water is pure. What has happened to the impurities? (1)
5. The water which drips on to the coffee is also described as condensed. What does this mean? (2)
6. a) How would you describe the colour of the condensed water before it touches the coffee? (1)
   b) What colour is the water which drips through the filter paper? (1)
   c) What has happened to change the colour of the water? (1)
7. How can the filter paper tap the solids in the crushed coffee bean but let the water pass through? (2)
8. Why would high quality paper not work in the filter? (1)
9. What would happen to the coffee if all the water in the coffee jug evaporated? (1)

Science Companion © A Porter, M Wood, T Wood and Stanley Thornes (Publishers) Ltd 1995

5.3.3 Separating and purifying

Chromatography

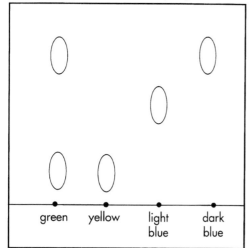

The diagram above shows how the ink from a green pen compares with three other pens (yellow, light blue and dark blue). All four pens were tested using the technique called chromatography.

1. Use coloured pens or crayons to show where the colours are on the diagram. (5)
2. Complete the following paragraph which describes how the chromatogram above was made. (8)

A square of paper was cut out. The paper must be able to _____ water. A line was drawn two centimetres from the bottom of the paper using a _____. Four marks were made along this line. A dot of ink from the _____ pen was put on the first mark. The yellow on the second mark and so on. The paper was stood in a tank with _____ used as the liquid in the tank. It was important that the liquid level was _____ the line. This was to stop the colours from _____ in the liquid. The liquid travels up the paper and takes the colours with it. The colours travel at different _____ which means they separate out. The green pen is made up of _____ different colours.

Separating a sand and salt mixture

This diagram shows the steps needed in separating a mixture of sand and salt. There are four steps. The mixture is added to water. Then it is stirred and then passed through a filter paper. Finally it is heated to evaporate the water.

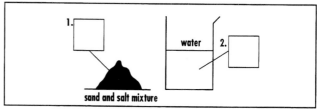

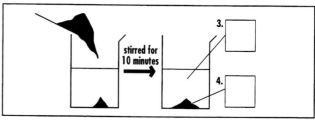

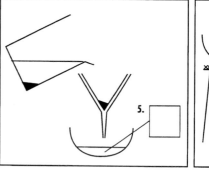 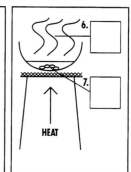

There are empty squares on the diagram numbered 1 to 7. You must fill in the squares using drawings to represent the particles at that point.
Use the diagrams below which represent particles of water ◯ sand ◇ and salt ●.

You will need to use one of the diagrams more than once. (7)

Name: **Class:** **Date:**

Science Companion © A Porter, M Wood, T Wood and Stanley Thornes (Publishers) Ltd 1995

5.4.2 Acids, alkalis and indicators

1 Study the following newspaper article, and then answer the questions which follow.

> **WASHING POWDER BURNS**
> World News 23-10-94
>
> Detergent powders are used every day by lots of people to wash their clothes. 'Wonderwhite' is new, and has been taken out of shops by police in Coventry after it caused severe burns to eight people. Police said that these people had been treated in West Coventry Hospital for burns caused by an acidic substance. The burns were caused by sodium hydroxide which was in the boxes along with soap powder.

a) What was the name of the new detergent? (1)
b) What are detergents normally used for? (1)
c) How many people were burned? (1)
d) Which hospital treated the people for their burns? (1)
e) Which chemical caused the burns? (1)
f) From your science lessons, is this chemical which caused the burns an acid or alkali? (1)
g) According to the newspaper, what mistake did the police make? (1)
h) If you dissolved some 'Wonderwhite' carefully in water, how could you test it to see whether it was acidic or alkaline? (3)

2 Oven cleaners contain a strong alkali. Packets of oven cleaners carry a corrosive hazard sign suggesting protection for eyes, hands and arms.
a) Name the strong alkali usually found in oven cleaners. (1)
b) What can this alkali do to the skin of the user? (2)
c) What protection do you suggest for the hands and arms of the person cleaning the oven? (2)

3 Two bottles from your school science laboratory have lost their labels. One contains sulphuric acid, the other calcium hydroxide solution. Using an indicator, say how you could tell what was in each bottle, and so be able to add new labels. (5)

5.4.1 Acids, alkalis and indicators

1 Study the following newspaper article, and then answer the questions which follow.

> **WASHING POWDER BURNS**
> World News 23-10-94
>
> Boxes of 'Wonderwhite', a new detergent powder for washing clothes, have been removed from a warehouse in Coventry after some bought from a local market stall caused severe burns to several users. Police report that over the last four days, eight people have been admitted to West Coventry Hospital with severe burns caused by the acid-like substance caustic soda in the powder. The boxes carried no hazard sign and appear to have been filled with sodium hydroxide and soap powder.

a) Give the common name of the substance which seems to have caused the burns. (1)
b) What is the chemical name for this substance? (1)
c) Is this substance acidic or alkaline in water? (1)
d) What mistake seems to have been made by the police? (1)
e) Suggest a simple test, with results, that the police could have performed to correct their apparent error. (3)

2 How can you make a neutral solution using only an acid and an alkali? (3)

3 Our teeth are made in part from calcium carbonate.
a) What type of substances attack teeth? (1)
b) Suggest why toothpastes usually have a pH of about 8. (1)

4 Some laboratory bottles have lost their labels. They are known to contain sulphuric acid, pure water, ethanoic acid and calcium hydroxide solution. Suggest a method by which you could identify the contents of each bottle and so be able to label it correctly. (5)

5 Many items at home such as drinks, cleaning fluids and stomach preparations contain acids and alkalis. Make a list from products you have at home of the name of the product and the acid or alkali it contains. Bring some products into school for an exhibition. (3)

5.4.3 Acids, alkalis and indicators

1 Complete the nine horizontal words using clues (a) to (i). Then answer part (j).

	1	2	3	4	5	6	7	8	9	10
a)										
b)										
c)										
d)										
e)										
f)										
g)										
h)										
i)										

CLUES

a) Sulphuric is a type of this substance. (1)

b) This liquid is often added to fish and chips. (1)

c) A liquid which is pumped around our bodies. (1)

d) A white liquid often delivered to our doors. (1)

e) This acid is found in lemons. (1)

f) The opposite of an acid. (1)

g) These liquids all have a pH of seven. (1)

h) Cooking equipment which needs a strong alkali to clean it. (1)

i) When pure, this common liquid has a pH of 7. (1)

j) Now find the word in column 5. (1)

2 In the letters below are hidden the names of seven acids and hydroxides. The names are printed left to right or read vertically from the top downwards. Circle the names of the seven chemicals. (7)

T	F	S	O	D	I	U	M	Y	A
U	H	U	R	S	G	K	X	J	M
C	A	L	C	I	U	M	O	E	M
R	I	P	C	S	W	Z	A	G	O
E	T	H	A	N	O	I	C	D	N
R	H	U	P	F	C	A	V	E	I
E	H	R	F	J	I	U	M	M	U
A	C	I	T	R	I	C	A	B	M
G	Y	C	A	R	B	O	N	I	C

3 Use the words in the list below to complete the passage. You may use the words once, more than once or not at all.

**alkalis   caustic   citric
hydrochloric   more   neutral
seven   sodium   rubber**

All acids have a pH of less than _____. All alkalis have a pH of _____ than seven. Acids can react with exactly equal amounts of _____ to make a _____ solution of pH 7. Strong acids include _____ acid. Weak acids include _____ acid. The strongest common alkali in the home is _____ hydroxide which is used as oven cleaner. The common name for this alkali in oven cleaner is _____ soda. The word caustic means burning. When cleaning an oven it is wise to use _____ gloves and eye protection. (9)

4 Complete the pH number maze using the following clues. All the answers go left to right or downwards. In the maze, the pH number is written as a word. So, if the answer is 6, in the maze this is written as SIX.

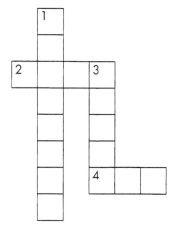

(4)

CLUES
1 *down* pH of soap flakes
2 *across* pH of soda water
3 *down* pH of toothpaste
4 *across* pH of lemon juice

Name: Class: Date:

5.5.1 The Periodic Table

1 The boiling points of the first 19 elements in the Periodic Table are shown on the graph below:

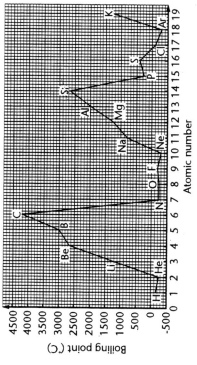

a) Name the element with the highest boiling point. (1)
b) Name the element at the top of the second peak on the graph. (1)
c) In which group are the elements in answers (a) and (b) above? (1)
d) Predict the element which you would expect to find at the top of the next major peak, and explain your choice. (2)
e) In which group are 3 out of the 4 elements with the lowest boiling points? (2)
f) Calcium is element number 20. Would you expect it to have a higher or lower boiling point than potassium? Give a reason for your answer. (2)

2 Give the name and symbol of:
a) the element which has atoms with the electron structure 2,8,1. (2)
b) ONE other metal which is in the same period as magnesium. (2)
c) the element which has atoms containing only one proton. (2)
d) the most reactive element in group VII. (2)
e) the only metallic element to be a liquid under normal lab conditions. (2)

3 The element francium (atomic number 87), is in group I.
a) Is francium a metal or non-metal? (2)
b) Is it harder or softer than sodium? (1)
c) Is it more or less reactive than sodium? (1)
d) How many electrons are there in its outermost electron shell? (2)

5.5.2 The Periodic Table

1

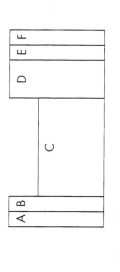

On this rough outline of the Periodic Table, which area represents
a) Group VII (1) b) Group I (1)
c) Group 0 (1) d) Group II? (1)

2 One of the elements in the Periodic Table may be represented by the letter **E**. In the diagram alongside the crosses represent electrons.
a) What is the name of this element? (1)
b) What is the symbol for this element? (1)
c) How many electrons does it have? (1)
d) How many of these electrons are in the outside shell? (1)
e) Which group is it in? (1)

3 Alongside is a table showing three elements found on another planet, together with their electron arrangements.

Batlium	2.8.2
Woodium	2.8.8.2
Portium	2.8.18.8.2

a) In which group of the Periodic Table would we put these elements? (1)
b) Which element has the largest atoms? (1)
c) Which element would be the most reactive? (1)
d) What would be the electron arrangement for the element above Batlium in the group? (1)

4 a) Name the only 2 elements which are liquids at normal conditions. (2)
b) Name any 6 of the 11 elements which are gases at normal conditions. (6)
c) If 109 elements are known, how many of these are normally solids in the laboratory? (1)
d) Element number 95 is americium, number 98 is californium and number 99 is einsteinium. Suggest after whom or where they were named. (3)

5.5.3 The Periodic Table

A copy of The Periodic Table will be needed to answer these questions. It will need names, symbols and the atomic numbers of the elements.

1 Below is a word square. The letter m is in the square already at square (a1).

	1	2	3	4	5	6	7	8	9
a)	M								
b)									
c)									
d)									
e)									
f)									
g)									
h)									
i)									

PART 1
Find the nine words to fill the rows (a) to (i). The clues are all atomic numbers. The answers are the names of the elements.

CLUES
a) 25, b) 28, c)12, d) 80, e) 53, f) 20, g) 18, h) 47, i) 55. (9)

PART 2
From the word square, find the letters which fit in these spaces.

a4	i3	h9	c1	g1	b1	e5	f7	d3

(1)

PART 3
a) What is the atomic number of the element you have discovered in part 2? _____ (1)
b) What country is it named after?
_____ (1)

2 Below are the melting points of elements 10 to 18 of the Periodic Table.

Atomic number	Element	Melting point at atmospheric pressure (°C)
10	Neon	−249
11	Sodium	98
12	Magnesium	650
13	Aluminium	660
14	Silicon	1410
15	Phosphorus	44
16	Sulphur	113
17	Chlorine	−101
18	Argon	−190

a) Plot a graph, on the graph paper below, of melting point against atomic number.
(3 for points)
(1 for line)

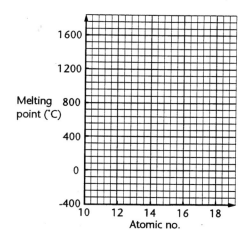

b) In which group of the Periodic Table is the element with the highest melting point? ____ (1)

3 Use the following words, once, more than once, or not at all, to complete the passage below.

atom   **decreasing**   **element**
elements   **groups**   **increasing**
neutrons   **periods**   **protons**
rows   **similar**

In the Periodic Table of the _____, the elements are arranged in order of _____ atomic number. The atomic number of an _____ is the number of electrons or _____ in one atom of the element. Vertical columns are called _____ and contain elements which react in a _____ way. Horizontal _____ of elements are called _____ .

(8)

Name: Class: Date:

Science Companion © A Porter, M Wood, T Wood and Stanley Thornes (Publishers) Ltd 1995

5.6.1 Reactivity of metals

1 a) Why is the metal gold often found as the metal in the ground? (1)
   b) Why is iron never found naturally as the metal? (1)

2 When a piece of sodium is cut with a sharp knife the shinyness of the metal can be seen under the dull coating. The metal surface soon becomes dull again.
   a) Why does the metal surface become dull again? (1)
   b) Give the name of the chemical in the dull coating. (1)
   c) In order to prevent this corrosion how is sodium stored? (1)

3 Three metals were added to dilute hydrochloric acid. The diagram shows the results after one minute.

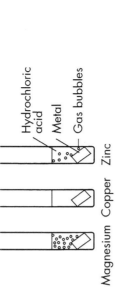

   a) Which metal was the most reactive (**A**, **B** or **C**)? (1)
   b) What is the name of the gas given off as it reacts with acid? (1)
   c) Which metal is the least reactive? (1)
   d) Name a metal which you think this least reactive metal might be. (1)
   e) Metal **D** was added to the same dilute hydrochloric acid. It was found to be more reactive than metal **B** but not as reactive as metal **C**. Draw a test tube which shows the result of adding metal **D** to the acid. (2)

4 Copper metal has replaced lead now for most water pipes in the home. Lead is more reactive than copper.
   a) Give two properties of lead which make it useful for water pipes. (2)
   b) Why is lead not used for modern water pipes? (1)
   c) Why is a less reactive metal like copper better for water pipes? (1)

Science Companion © A Porter, M Wood, T Wood and Stanley Thornes (Publishers) Ltd 1995

5.6.2 Reactivity of metals

1 a) Why is the metal gold often found as the metal in the ground? (1)
   b) Apart from being expensive, why is gold used in jewellery? (1)

2 When a piece of iron is cleaned and polished the shinyness of the metal can be seen. The metal surface can soon become rusty.
   a) What two things are needed to make the iron rust? (2)
   b) Give one way in which we can stop the iron from rusting. (1)

3 Three metals were added to dilute hydrochloric acid. The diagram shows the results after one minute.

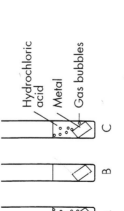

Magnesium   Copper   Zinc

   a) Which metal was the most reactive? (1)
   b) What is the name of the gas given off as it reacts with acid? (1)
   c) Which metal is the least reactive? (1)
   d) Name a metal which is more reactive than these three metals. (1)
   e) Metal **D** was added to the same dilute hydrochloric acid. It was found to be more reactive than metal **B** but not as reactive as metal **C**. Draw a test tube which shows the result of adding metal **D** to the acid. (2)

4 Copper metal has replaced lead now for most water pipes in the home. Lead is more reactive than copper.
   a) Give two properties of lead which make is useful for water pipes. (2)
   b) Why is lead not used for modern water pipes? (1)
   c) Why is a less reactive metal like copper better for water pipes? (1)

5.6.3 Reactivity of metals

Displacement reactions

1 When a metal is more reactive than another metal in a compound, there will be a reaction. The positions of the metals will swap over. We say one metal has displaced another.

e.g. IRON + LEAD OXIDE → LEAD + IRON OXIDE

a) Each of the four reactions shown below is a displacement reaction which will work. Use the idea above to predict what the names of the metals, solids and solutions are.

IRON OXIDE + ALUMINIUM → METAL **A** + SOLID **B**

METAL **C** + SOLUTION **D** → BATLIUM + IRON SULPHATE

METAL **E** + SOLUTION **F** → BATLIUM + ALUMINIUM BROMIDE

BATLIUM + SILVER OXIDE → METAL **G** + SOLID **H**

Metal **A** is _____

Solid **B** is _____

Metal **C** is _____

Solution **D** is _____

Metal **E** is _____

Solution **F** is _____

Metal **G** is _____

Solid **H** is _____ (8)

b) Now place the four metals – silver, iron, batlium and aluminium in the correct order of reactivity by completing the table below. (4)

REACTIVITY	NAME OF METAL
Most reactive	
2nd most reactive	
3rd most reactive	
Least reactive	

2 Use your knowledge of metals to judge whether the following statements are true or false. Tick the box which you think is right. (15)

Statements	True?	False?
1. Gold is classed as a reactive metal		
2. Silver is a good conductor of electricity		
3. Metals are good conductors of heat		
4. Gold corrodes more easily than silver		
5. Iron is commonly found naturally as the metal		
6. Ores are rocks which are rich in metals		
7. Ores are turned into oxides before extracting the metal		
8. Carbon can be used to turn iron oxide into iron in a furnace		
9. Iron is more reactive than carbon		
10. Copper reacts vigorously with water		

3 Complete the labels on the diagram of the blast furnace. The words to use are:
CARBON DIOXIDE, COKE, HOT AIR, IRON METAL, IRON ORE, LIMESTONE, SLAG (7)

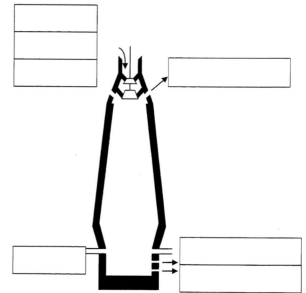

b) Use the seven labels to complete the table below in order to show which are raw materials and which are products of the blast furnace. Iron ore has been done for you. (6)

Raw Materials	Products
iron ore	

6.1.2 Solids, liquids and gases

1 Look at the following list of common materials. Place them in a table with the headings 'solid', 'liquid', 'gas' showing in which state you normally find them.

air   apple   ice   milk   paper   petrol   steam   vinegar (8)

2 Oil floats on top of water.
  a) What name is given to a pair of liquids which do not mix? Is it:
     insoluble   immiscible   insolvent? (1)
  b) Which liquid is the more dense, the liquid on top, or the liquid at the bottom? (1)
  c) Name a piece of laboratory equipment used to separate such a pair of liquids. (1)

3 What must you give ice particles in order to turn them into water particles? (1)

4

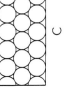

A   B   C

The diagrams represent particles in the three states of matter. The states are solid, liquid and gas. Say which state each letter represents. (3)

5 For each of the following substances say whether it is a solid, liquid or gas at 20°C.

Substance	Melting point (°C)	Boiling point (°C)
A	18	165
B	60	520
C	–116	–17
D	–116	17
E	–15	43

(5)

6.1.1 Solids, liquids and gases

1 Here are some descriptions of newly discovered materials. Say whether each is normally a solid, liquid or gas at 15°C.
  a) Porterite is a white plastic of definite shape. It melts at 125°C and boils at 400°C. (1)
  b) Batlium is less dense than air. It can be compressed very easily. (1)
  c) Woodite burns with a pale yellow flame. It pours easily from one container to another and settles into the bottom of its new container. A definite surface can be seen on it. (1)
  d) Minerite is more dense than air. When released it pours out of a vessel and forms a layer at floor level. It is easily compressed. It melts at –100°C and boils at –20°C. (1)
  e) Glenium boils at 70°C and melts at –1°C. It is green in colour. (1)

2 There are many 'breathable' fabrics on the market today which can be used in waterproof outdoor clothing. The idea of these is that they contain holes. The holes let vapour out, and stop liquid getting in. Explain how a fabric like this works. (2)

3 Water can exist as ice, water (liquid) and steam. What can you say about the energy of the water particles in the different states? (2)

4 What is the difference between a suspension and a solution? (3)

5 A fizzy drink is made by dissolving carbon dioxide in the water of the drink.
  a) Under what special conditions is the dissolving carried out? (1)
  b) When a fizzy drink is opened, why does it loose its fizziness if it is allowed to warm up slightly? (2)

6 What do you understand by each of the following words or phrases?
  a) immiscible liquids (1)
  b) solvent (1)
  c) solute (1)
  d) alcohol and water are completely miscible (2)

6.1.3 Solids, liquids and gases

1 Complete this word triangle. The clues are for words going across. The number tells you how many letters the word has.

CLUES

1. The symbol for the Celsius scale.
2. To drink coffee, the coffee is put _____ a cup or mug.
3. The solid state of steam.
4. To mix a solid and liquid using a glass rod.
5. The liquid state of steam.
6. A word to describe a solid when it is a liquid.
7. The units of temperature.
8. Solids are regular arrangements of these.
9. A solute did this to make a solution.
10. Lowering
11. A scale used to measure this in °C. (11)

2 Use the words in the list below to complete the following passage. Each word may be used once, more than once, or not at all.

dissolves   insoluble   solid
solids   soluble   solute   solution
solvent   suspension

An insoluble solid is one which stays as a _____ when it is added to a liquid. If the solid dissolves then we say it is _____ in the liquid. A liquid which dissolves _____ is called a _____. A solid which dissolves may be called a solute. When a _____ dissolves in a _____ it forms a _____. (7)

3 Study the following information and then answer the questions which follow.

Gore-tex fabric is the world's most advanced weather-proof fabric. Its success lies in the special membrane (2) made from an expanded form of Teflon – PTFE. This membrane contains 14 million holes per square millimetre. Each hole is 20 000 times smaller than a drop of water, but 700 times larger than a particle of water as a gas. This means that rainwater cannot get through so the wearer stays dry. But perspiration can evaporate and escape, so the material 'breathes' and is pleasant to wear. The outer fabric (1) and the liner (3) are selected depending on the use of the final garment, and are permanently bonded either side of the Gore-tex membrane.

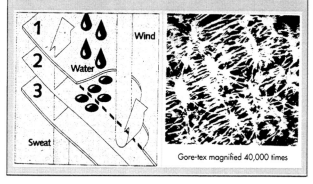

a) How much smaller is a hole in Gore-tex than a drop of water?

_____ (1)

b) Why is Gore-tex described as breathable?

_____
_____
_____ (1)

c) Why are an outer fabric and liner used?

_____
_____
_____ (2)

d) Why is a plastic garment not as comfortable to wear as Gore-tex?

_____
_____ (2)

Name: Class: Date:

Science Companion © A Porter, M Wood, T Wood and Stanley Thornes (Publishers) Ltd 1995

6.2.2 Atoms and temperature scales

1 Study the following thermometers very carefully. Watch the scales because they are all different! What temperature is each thermometer showing? (2)

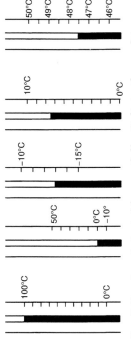

2 Study the following data table carefully. It lists the melting and boiling points of some common metals.

Metal	Melting point (°C)	Boiling point (°C)
Aluminium	660	2470
Copper	1084	2570
Gold	1064	3080
Iron	1540	2750
Lead	327	1740
Silver	961	2212
Tin	232	2270

a) Which metal has the highest melting point? What is this temperature? (2)

b) Which metal has the highest boiling point? What is this temperature? (2)

c) Work out the range of temperature over which each metal is a liquid. (7)

d) Name the metal which is a liquid over the smallest temperature range. (1)

3 Dalton said the following about atoms:
 (1) Everything is made up of atoms.
 (2) These atoms cannot be split, made or destroyed.

a) Which of these statements is now untrue? (1)

b) Explain why it is untrue. (1)

c) Atoms are very small but we can see them using a special type of microscope. Name this piece of equipment. (1)

6.2.1 Atoms and temperature scales

1 Why does it seem a good idea for the kelvin temperature scale to have its zero at −273°C? (2)

2 Dalton said the following about atoms:
 (1) Everything is made up of atoms.
 (2) These atoms cannot be split, made or destroyed.

a) Today, we know a lot more about atoms than Dalton. Which of his statements is now not true? Explain your answer. (2)

b) Name the special piece of apparatus which enables scientists today to see atoms. (1)

3 a) When a thermometer is placed in a beaker of boiling water why do you have to wait a few moments before taking the correct temperature of the water? (2)

b) Name the narrow tube inside a thermometer. (1)

c) A beaker of water is being heated over a bunsen on a tripod and gauze. Why should the thermometer be held in the body of the water to read the temperature and not placed on the bottom of the beaker? (2)

d) Suggest why mercury thermometers cannot be used outside the temperature range −39°C to 360°C. (2)

e) What type of thermometer would you use in the Antarctic, and why? (2)

4 Research, for a short talk in class, or a 200 word essay for homework, information about the work of one of the following: (6)

a) John Dalton (1766–1844)

b) Anders Celsius (1701–44)

c) William Thomson (Baron Kelvin of Largs 1824–1907, often known as Lord Kelvin)

d) Gabriel Daniel Fahrenheit (1686–1736).

6.2.3 Atoms and temperature scales

1 Study the following information on temperature scales and then complete the table which follows.

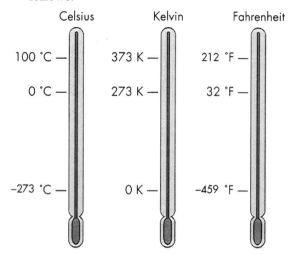

To convert temperatures from Fahrenheit to Celsius, take 32 from the Fahrenheit temperature and multiply by $\frac{5}{9}$.

To convert temperatures from Celsius to Kelvin simply add 273 on to the Celsius figure.

Complete the table below giving your answers to the nearest whole number.

	Celsius	Kelvin	Fahrenheit
a)			0
b)			61
c)			151
d)			462
e)		393	

(10)

2 Fill in the following horizontal words using the clues that follow. Then find the two words in sentence (i) using the letters from the numbered squares.

	1	2	3	4	5	6	7	8	9	10	11
a)											
b)											
c)											
d)											
e)											
f)											
g)											
h)											

CLUES
a) One hundredth of a degree. Also commonly used as the name for the Celsius temperature scale.
b) 70°F. Complete this F scale word.
c) Units of temperature.
d) The scale with the freezing point of water at 0.
e) We measure this {answer (e)} using the apparatus for answer (f).
f) See clue (e).
g) A red or blue dye is added to this liquid in these pieces of apparatus to measure temperature.
h) The only liquid metal at room temperature. (8)

Use letters from the squares above to complete this sentence:
i) Gabriel Fahrenheit was a

c3	d2	h6	f5	e7	b6

scientist who worked in

g5	f6	g7	d3	b2	a3	c1

for a time as a glassblower. (2)

Below is a bar chart showing the melting points of some common metals. The temperature units are kelvin. Study it carefully before answering the questions which follow.

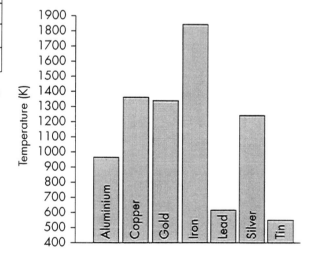

a) Place the seven metals in order of decreasing melting point with the highest first. _____
_____ (2)

b) Name the metals which melt between 927°C and 1127°C.
_____ (3)

Name: Class: Date:

6.3.2 Particles and changes of state

1 A drop of black ink is added to a bowl of cold water. A student watches and draws three sketches. The first at zero time, the next after five minutes, and the last ten minutes from time zero. The student labels the diagrams incorrectly. The student's sketches are shown below. Say which you think is which, and explain your answer. (6)

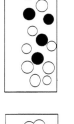
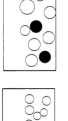
B   A   C

KEY ○ water particles
     ● black ink particles

2 Below are five words linked with 'the effect of heating a substance'.
   **boiling   condensation   freezing
   melting   sublimation**
   For each of the following statements choose one word from the list which best fits it.
   a) On heating, the solid goes straight to a gas without first being a liquid. (1)
   b) Steam particles reach a cold surface and turn into water. (1)
   c) On warming, ice turns into liquid water. (1)
   d) On warming, alcohol liquid turns into alcohol vapour whilst the temperature remains constant. (1)
   e) Water stays at a fixed temperature into ice. (2)

3 The element Morlium has just been discovered on the planet Mingle. It melts at 18°C and boils at 118°C. Draw its cooling curve as it is allowed to cool from 130°C to 0°C and label any special features. (5)

4 When a solid is heated the temperature rises. Explain why this temperature rise stops when the solid melts even though heat is being supplied all the time. (3)

6.3.1 Particles and changes of state

1 Imagine you and your classmates are playing the parts of solid particles. Describe what you would do as you were warmed up and changed into a liquid. (3)

2 Explain why washing dries on a line in the garden faster, when it is sunny and windy, than when it is dull and there is no wind. (3)

3 When ice is melted into water, why does the temperature remain the same even though heat is being supplied? (2)

4 How do particles move in the solid state? (1)

5 Explain why liquid particles diffuse more slowly than gas particles. (2)

6 Below is a diagram representing the diffusion of two gases **A** and **B** with some facts about each gas.

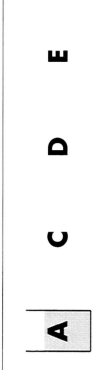

A   C   D   E   B

Beaker containing liquid **A** gives off gas **A**.

Beaker containing liquid **B** gives off gas **B**.

Particles of gas **A** are twice as heavy as those of gas **B**. Will the gases meet at **C**, **D** or **E**? Explain your answer. (2)

7 The following properties relate to solids, liquids and gases. Use one or more of these states of matter to match each statement.
   a) Particle arrangement is regular.
   b) Compressibility – none.
   c) Shape – spreads to fill the whole container.
   d) Particles are quite close. (6)

6.3.3 — Particles and changes of state

1 Wordsearch. Hidden among the 120 letters below are words all to do with 'the effect of heating a substance'. Circle the words as you find them, and tick them off on the list. Words go from left to right, top to bottom or diagonally top left to bottom right.

G	N	I	F	T	V	E	I	C	J
S	O	L	I	D	M	Z	X	O	S
U	P	F	R	E	E	Z	I	N	G
B	A	N	G	M	C	F	H	D	C
L	R	M	Y	A	H	H	M	E	O
I	T	E	I	E	S	E	J	N	O
M	I	L	K	R	Y	A	U	S	L
A	C	T	H	T	R	T	R	I	I
T	L	I	Q	U	I	D	E	N	N
I	E	N	R	E	H	O	F	G	G
O	S	G	C	O	O	L	I	N	G
N	H	B	O	I	L	I	N	G	A

The words are:
boiling
condensing
cooling
freezing
gas
heat
liquid
melting
particles
solid
sublimation (11)

Which word appears twice? _____ (1)

2 The triangle below links together the three states of matter. Five words are represented by letters. Write the five words in the spaces provided.

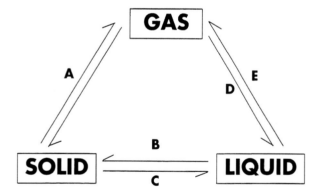

I think the missing words are:

A _____ (1)

B _____ (1)

C _____ (1)

D _____ (1)

E _____ (1)

3 The sets of apparatus shown at the top of the next column have been set up to study the diffusion of gases.
Apparatus **A** is for the diffusion of hydrogen and air.

Apparatus **B** is for the diffusion of carbon dioxide and air.

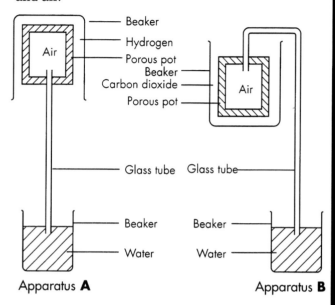

a) Why is it necessary to use two different designs of apparatus?

_____
_____
_____
_____ (2)

b) State, and explain, what will be seen in apparatus **A**.

_____
_____
_____
_____
_____
_____ (3)

c) State, and explain, what will be seen in apparatus **B**.

_____
_____
_____
_____
_____
_____ (3)

Name: Class: Date:

Science Companion © A Porter, M Wood, T Wood and Stanley Thornes (Publishers) Ltd 1995

6.4.1 Atoms, ions and molecules

1 From the diagrams below, choose the one which best shows the arrangement of particles in:

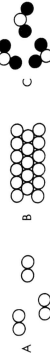

a) water (1)
b) nitrogen gas (1)
c) iron (1)
d) sodium chloride (1)
e) oxygen gas (1)

2 Give definitions for each of the following:
a) an element (1)
b) an atom (1)
c) a compound (1)
d) a mixture (1)
e) a molecule (1)
f) an ion (1)

3 Explain why atoms always have an overall charge of zero. (2)

4 In salt, the sodium ions are positive and the chloride ions are negative. Water molecules contain hydrogen and oxygen atoms. When salt is added to water, and dissolved, it is found that the oxygen atom in the water points to a sodium ion. Similarly, the hydrogen atoms in the water molecule point to a chloride ion. Suggest why this is so. (2)

5 Explain how we know that hydrochloric acid is a compound of hydrogen and chlorine and not a mixture of these two elements. (5)

Science Companion © A Porter, M Wood, T Wood and Stanley Thornes (Publishers) Ltd 1995

6.4.2 Atoms, ions and molecules

1 Below, are six words.

atom   compound   element   ion   mixture   molecule

In questions (a) to (f) below are six definitions (meanings). In your answer match each word with the correct definition (meaning).

a) a single pure substance that cannot be split into anything simpler. (1)
b) the smallest part of an element that can exist and still behave like the element. (1)
c) a substance which is made when two or more elements team up in fixed amounts. (1)
d) does not contain fixed amounts of each substance and the individual substances keep their own properties. (1)
e) the smallest part of a compound which can behave as the compound. (1)
f) an atom which has gained or lost electrons. (1)

2 In the following table, the values of some things have been replaced by the letters **A–E**. On your answer sheet write down the letter followed by the number which should be in the table at that point.

Particles in an atom	Charge on each	Mass of each in whole units
protons	A	D
neutrons	B	E
electrons	C	very small

3 The atom of carbon contains 6 protons and 6 electrons. Explain why the overall charge on this atom is zero. (3)

4 Sodium chloride is made up of positive ions of sodium and negative ions of chloride. The ions hold fixed places in the solid, but are free to move about in the liquid. Suggest why sodium chloride has a high melting point. (3)

5 Hydrogen is an explosive gas. Oxygen gas supports burning. Water is a compound of hydrogen and oxygen. Explain how we know water is a compound and NOT a mixture of hydrogen and oxygen. (3)

Science Companion © A Porter, M Wood, T Wood and Stanley Thornes (Publishers) Ltd 1995

6.4.3 Atoms, ions and molecules

1 In this puzzle there are seven clues. Write the answers to the clues into the grid on the right. When you have done this, transfer the letters to the grid below. If you get the right letters in the correct spaces, a sentence will appear.

CLUES

a) Practicals in a science lesson.

| 1a | 1b | 1c | 1d | 1e | 1f | 1g | 1h | 1i | 1j | 1k |

b) The amount of liquid in cm³ is its _____.

| 2a | 2b | 2c | 2d | 2e | 2f |

c) Short for one thousand grams.

| 3a | 3b | 3c | 3d |

d) Sand can be separated from salt water by this method.

| 4a | 4b | 4c | 4d | 4e | 4f | 4g | 4h | 4i |

e) To warm something up we add this.

| 5a | 5b | 5c | 5d |

f) Element number six in the Periodic Table which makes diamond and graphite.

| 6a | 6b | 6c | 6d | 6e | 6f |

g) The state of matter of which ice is an example.

| 7a | 7b | 7c | 7d | 7e |

(7)

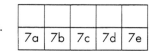

(18)
(1 mark per word)

Name: Class: Date:

Science Companion © A Porter, M Wood, T Wood and Stanley Thornes (Publishers) Ltd 1995

6.5.2 Radioactivity

1 The diagrams below show the structures of three possible isotopes of hydrogen.

Key:
x electron
p proton
n neutron

Isotope **A**   Isotope **B**   Isotope **C**

a) What do you understand by the word 'isotopes'? (3)
b) Name isotopes **B** and **C**. (2)
c) Isotopes can be written in a short-hand way, for example:

Mass number ⟶ A
                    X ⟵ Symbol of the element
Mass number ⟶ Z

Write isotope **C** in this way. (3)

2 Copy and complete the following table.

Isotope	Number of protons	Number of neutrons	Number of electrons
$^{4}_{2}$He			
$^{12}_{6}$C			
$^{41}_{19}$K			
$^{238}_{92}$U			

(2)
(2)
(2)
(2)

3 Whenever we take readings of radiation we must note the 'background radiation'. What is 'background radiation' and what causes it? (3)

4 Explain very simply how a nuclear reactor produces electricity by considering the following points:
a) What makes the heat? (2)
b) Where does the heat go? (1)
c) How is the heat used to make the electricity? (3)

6.5.1 Radioactivity

1 A Geiger-Müller tube was used to measure the radioactivity of a piece of radioactive metal. Below are the results in counts per minute after every hour.

Time (hours)	Counts per minute
0	844
1	603
2	575
3	422
4	301

During the experiment it was noted that the radioactivity was easily stopped by placing a piece of paper between the metal and the tube.

a) What type of radiation was the metal emitting? Explain your answer. (2)
b) What was the 'half-life' of the metal? (1)
c) Explain the meaning of the term 'half-life'. (2)
d) What should the counts per minute be after six hours? (1)

2 The isotope of carbon which contains eight neutrons is radioactive. It has a half-life of 5600 years. The atomic number of carbon is 6.

a) Explain what isotopes are. (2)
b) What is the mass number of this carbon isotope? (1)
c) If the radioactivity of a sample of carbon starts at 16 counts per minute, sketch a graph showing how the radioactivity changes over 22 400 years. (2)
d) A piece of woollen material was found to emit 25 counts per minute. Assuming it started with a count of 100 counts per minute when it was newly made, how old was the cloth? (2)
e) Why is this process less accurate for dating modern materials? (2)

3 Why are alpha, beta and gamma radiations often called ionising radiation? (2)

4 Upon what historic event is the film badge radiation detector based? (3)

5 The following questions all relate to the working of a nuclear reactor.
a) What is nuclear fission? (2)
b) How is the heat carried away? (1)
c) How is the reaction controlled? (1)
d) How are neutrons made more effective? (1)

6.5.3 Radioactivity

1. Find the missing words in rows (a) to (i) below using the clues. Then find the name of the scientist in column 4.

	1	2	3	4	5	6	7	8
a)								
b)								
c)								
d)								
e)								
f)								
g)								
h)								
i)								

   CLUES
   a) Radiation which needs a thin sheet of aluminium to stop it.
   b) The first name of the husband of a famous husband and wife team involved in the discovery of radioactivity. Her first name is your clue for (g). Their surname is your clue for (e).
   c) This material is needed to stop the most deadly radiation.
   d) A long word meaning need. It starts and ends with RE.
   e) See (b) above.
   f) Half-life is a measure of this feature of a radioactive substance.
   g) See (b) above.
   h) A nuclear reactor produces this.
   i) Radiation which can be stopped by a thin sheet of paper. (9)

   The name of the famous scientist in column 4 is _____. (1)

2. Below is a list of words.

   **alpha   bacteria   beta   boiling
   damage   gamma   germs
   hospitals   patients   sterilisation**

   Read the passage which follows very carefully, and use words from the list to complete the gaps. Each word may be used once, more than once, or not at all.

   Gamma radiation can be used to kill cancer cells. It can also be used to kill _____ or germs. When used to kill bacteria or _____ the method is called _____ This method is used for dressings and instruments in _____ which must be free from bacteria and germs before they are used on _____. Boiling in water is a common form of sterilisation. (8)

3. The following statements are about smoke detectors. Fill in the gaps.

   a) The type of radiation used in smoke detectors is _____ _____. (2)

   b) The radiation carries the _____ across an air gap. (1)

   c) In a fire _____ stop the radiation. (2)

   d) An _____ sounds when the circuit is broken in a fire. (1)

   e) Smoke detectors are fairly cheap and can save a lot of _____. (1)

4. The following questions are about alpha, beta and gamma radiation.

   a) Which of the three types of radiation are deflected by a magnetic field?

   _____ (2)

   b) Why is gamma radiation the most dangerous to people?

   _____
   _____
   _____
   _____
   _____
   _____ (3)

Name: **Class:**   **Date:**

6.6.2 Solvents, glues and hazards

1 Below are four hazard signs found on bottles of chemicals in a science laboratory. They are labelled **A**, **B**, **C** and **D**.

A   B   C   D
CORROSIVE   EXPLOSIVE   HIGHLY FLAMMABLE   TOXIC

Match these labels to the list of substances shown below. Suggest one label for each substance. On your answer paper write the letter **A**, **B**, **C** or **D** followed by the name of the substance.

a) concentrated sulphuric acid
b) gunpowder
c) rat poison
d) pure alcohol
e) concentrated nitric acid
f) petrol
g) arsenic   (7)

2 Explain why each of the following is a hazard.
a) Crossing the road without using a crossing and without looking. (2)
b) Heating chemicals in a laboratory without wearing safety goggles. (2)
c) Going back to a firework which was lit but has not gone off. (2)
d) Falling asleep on an inflatable mattress on a beach. (2)
e) Not fastening safety belts in a car. (2)

3 Why must glues be stored carefully away from young children? (2)

4 Suggest a hazard sign that could be designed by the local council to cover the possibility of a cloud of poisonous gas being ahead of road users. (3)

5 Pedestrian crossing places on roads are often governed by lights which look like traffic lights. Apart from flashing signs to say when it is safe to cross, why do these crossings have bleepers to sound and also often have paving slabs near with lumps? (3)

Science Companion © A Porter, M Wood, T Wood and Stanley Thornes (Publishers) Ltd 1995

6.6.1 Solvents, glues and hazards

1 There are many hazard signs on bottles at home and in science laboratories. Explain briefly, the meaning of each of the following words or phrases.
a) corrosive
b) explosive
c) harmful
d) highly flammable
e) irritant
f) oxidising
g) toxic   (7)

2 Explain why each of the following is a hazard.
a) Driving a car whilst under the influence of alcohol. (2)
b) Driving too close to the car in front, especially on wet roads. (2)
c) Riding a motorcycle without wearing a crash helmet. (2)
d) Running after a ball into the road. (2)
e) Leaving a long electric cable trailing from a kettle full of boiling water in a kitchen. (2)

3 A solute is a solid which will dissolve in a solvent. The solvent is a liquid. When dissolved, the solute and solvent produce a solution.

Summary: SOLUTE + SOLVENT ⟶ SOLUTION

Use the words below to make three summaries as shown above using named substances.

common salt   copper sulphate   potassium nitrate

water   (3)

4 In a science laboratory explain why:
a) we should wear safety goggles when heating things. (1)
b) we should never eat. (1)
c) all bags should be stored safely out of the way. (1)
d) when heating something in a test tube it should be pointed along the bench. (1)
e) there should be no running. (1)

Science Companion © A Porter, M Wood, T Wood and Stanley Thornes (Publishers) Ltd 1995

6.6.3 Solvents, glues and hazards

1 Alongside is a label for Domestos multi surface kitchen and bathroom cleaner. Study it carefully before answering the questions below.

a) Which parts of the body are liable to be irritated by this product?

_____ and _____ (2)

b) What should you do if you get any in your eyes? _____

_____ (2)

c) Which gas may be produced if this product is mixed with an acid?

_____ (1)

d) What does the product contain which should not be spilt on clothes, carpets and upholstery? _____ (1)

e) From the list of ingredients give the chemical name of a bleach. _____
_____ (1)

f) In use, you are given capfuls of the product to add to water. How many cm^3 of product does each capful hold?

(note 1 ml = 1 cm^3). _____ (1)

g) How much of the product is needed to disinfect a dishcloth? _____ (2)

h) From the label suggest the approximate volume of water which the manufacturers expect to find in 1 full bucket _____. (2)

Domestos MULTI SURFACE
KITCHEN & BATHROOM CLEANER

NEAT USAGE
Small surfaces (eg. work surfaces, food preparation areas, baths etc).
To disinfect dishcloths add 3 capfuls to 1 litre of hot water. Cap size 30 ml.

DILUTE USAGE
Large surfaces (eg. floors, walls, fridge interiors etc.) add 2 capfuls to 2.5 litres of water (1/2 bucket).

Domestos Multisurface contains amongst other ingredients:

| Less than 5% | Nonionic surfactant; Chlorine based bleaching agents (Sodium hypochlorine); Soap. |

CAUTION
This product contains bleach. Avoid contact with clothes, carpets and upholstery

- Irritant to Eyes and Skin.
- Keep out of reach of children.
- In case of contact with eyes, rinse immediately with plenty of water and seek medical advice.
- If swallowed seek medical advice immediately and show this container label.
- Warning DO NOT mix with other cleaning products or acid as they could give rise to dangerous fumes (Chlorine gas).

IRRITANT
750ml

UK Mainland: Lever Brothers Ltd, 3 St James's Road, Kingston-upon-Thames, Surrey, KT1 2BA England. Tel: 0800 010109

Island of Ireland: Lever Brothers (I) Ltd, Belgard Road, Tallaght, Dublin 24, Ireland. Tel: Dublin 572 222

2 Alongside is the label from a well-known make of liquid paper. Study the label carefully and then answer the questions below.

a) What is the hazard label warning about?

_____ (1)

b) Which part of the body should you 'avoid contact with'?

_____ (1)

c) What chemical which causes concern is in the product?

_____ (1)

Korrektur flüssigkeit: Gut schütteln. Sofort nach Gebrauch verschließen. Gesundheitsschädlich beim Einatmen und Verschlucken. Darf nicht in die Hünde von Kindern gelangen. Berührung mit den Augen vermeiden. Enthalt: 1,1,1-Trichloräthan: Unbrennbar.

Mindergiftig Harmful

Correction fluid. Shake well. Close cap tightly after use. Harmful if inhaled or swallowed. Keep out of reach of children. Avoid contact with eyes. Contains: 1,1,1-Trichloroethene: Non-flammable.

Name: **Class:** **Date:**

Science Companion © A Porter, M Wood, T Wood and Stanley Thornes (Publishers) Ltd 1995

7.1.2 Physical and chemical changes

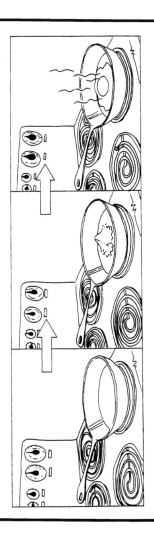

The three pictures above show the stages involved in frying an egg. These are: heating the pan, adding the fat and then the egg. The heating ring is at 150°C. Aluminium melts at 660°C and the fat melts at 65°C.

1. Why does the pan not melt on the heating ring? (1)
2. Why does the fat melt easily? (1)
3. When the fat melts is it a physical or chemical change? (1)
4. What will happen to the fat when the pan cools down? (1)
5. At what temperature will the fat change as the pan cools down? (1)
6. In what state is the raw egg? Explain how you know this. (2)
7. In what state is the cooked egg white? (1)
8. What is chemically changed, the pan, the fat or the egg? Explain your answer. (2)
9. Use the diagram below to answer these questions. Which letter best represents
   a) the melting fat
   b) the cooking egg
   c) the condensation on the kitchen window
   d) the cooling fat
   e) boiling? (5)

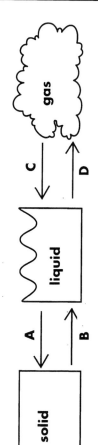

Science Companion © A Porter, M Wood, T Wood and Stanley Thornes (Publishers) Ltd 1995

7.1.1 Physical and chemical changes

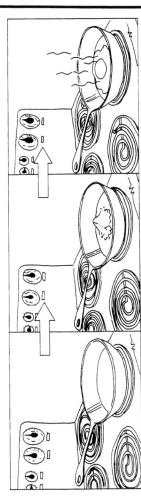

The three pictures above show the stages involved in frying an egg. These are: heating the pan, adding the fat and then the egg.

1. Describe what happens to the particles of the aluminium pan as the temperature gets higher. (1)
2. Why does the fat melt when the aluminium pan does not? (2)
3. When the fat melts, why does it spread out over the base of the pan? (1)
4. What will happen to the particles in the fat when the pan cools down again? (2)
5. Is the change in the fat physical or chemical? Explain your answer. (2)
6. What is the change in state as the egg cooks? (1)
7. Is the egg physically or chemically changed? Explain your answer. (2)
8. Choose which of the diagrams below best represents the particles in
   a) the aluminium pan
   b) the melted fat
   c) the raw egg
   d) the cooked egg (4)

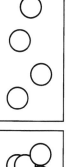 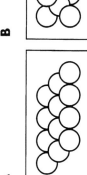

A   B   C

Science Companion © A Porter, M Wood, T Wood and Stanley Thornes (Publishers) Ltd 1995

7.1.3 Physical and chemical changes

Physical changes
Write the answers to the questions in the correct boxes in the grid. Rearrange the letters in the shaded boxes to complete the sentence below. (6)

Physical changes are _____

1. The process which happens to most solids when they are heated. (7 letters)
2. What steam will become when it condenses. (5)
3. Roads are this when the surface water freezes in winter. (3)
4. What liquids will do when they are heated. (4)
5. A word for a gas which evaporates from a liquid. (6)

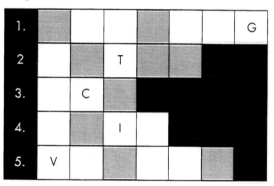

Chemical changes
Write the answers to the questions in the correct boxes in the grid. Rearrange the letters in the shaded boxes to complete the sentence below. (6)

Chemical changes are _____

6. This sulphate changes from blue to white when you heat it. (6 letters)
7. This is a raw material used to make glass. (4)
8. What you often need to do to a chemical to change it. (4)
9. A mixture of these materials is called an alloy. (6)
10. How to chemically change cake mixture. (6)

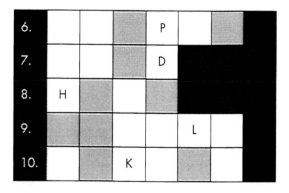

Supergreen
10 PHOTODEGRADABLE SWING BIN LINERS

RIM: 46 ins/1.17 m
DEPTH: 30 ins/76 cm

There are two types of plastic called thermosoftening and thermosetting plastics. One way in which they are different is in their reaction to heat. Complete the sentences below.

Thermosoftening plastics when heated will turn from a solid into a _____. This transformation is called _____. It is an example of a _____ change. These plastics are often collected from waste because they can be _____ (used again). This is done with the plastic used to make _____. Thermosetting plastics are different. They undergo a _____ change when they are heated and become new _____. The plastic used in the bin liners shown above also undergoes a chemical change. In this case it changes when exposed to _____. (8)

In the grid below try to find six physical changes and four ways to make a chemical change.

E	V	L	O	S	S	I	D
V	M	H	E	A	T	L	E
A	E	I	F	S	P	O	N
P	L	R	L	I	G	H	T
O	T	N	I	B	X	U	M
R	N	Y	O	L	U	K	I
A	Z	M	B	L	R	S	X
T	E	Z	E	E	R	F	P
E	S	N	E	D	N	O	C

Name: Class: Date:

7.2.2 Changes for the better from land and air

The pie chart opposite shows the uses to which sulphuric acid is put. The acid is made from sulphur which is found near volcanic areas. Most of the UK's sulphur comes from Poland where it occurs naturally. Sulphur is present in natural gas as hydrogen sulphide which can be converted to sulphur by burning it.

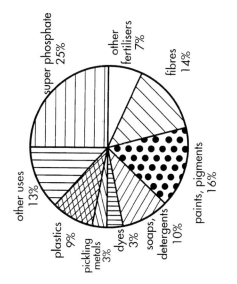

1. What test would show you that sulphuric acid was an acid? (2)
2. Sulphuric acid bottles always carry the hazard label for corrosive. Make a drawing of the hazard sign for corrosive. (1)
3. Why is there little sulphur found naturally in the UK? (1)
4. What gas in the air is needed to turn the sulphur into sulphur dioxide? (1)
5. Where does the UK's natural gas come from? (1)
6. Why is the manufacture of sulphuric acid often sited near a port? (2)
7. Why would the following manufacturers need sulphuric acid? (3)
    a) Fairy Liquid
    b) Dulux
    c) Dylon
8. List the eight uses shown in the pie chart in order of their importance. (1)
9. What type of pollution is associated with sulphur dioxide and how is it formed? (2)
10. Included in the other uses for sulphuric acid is its use in automobiles. How is sulphuric acid used in cars? (1)

Science Companion © A Porter, M Wood, T Wood and Stanley Thornes (Publishers) Ltd 1995

7.2.1 Changes for the better from land and air

The pie chart opposite shows the uses to which sulphuric acid is put. The acid is made from sulphur which is found near volcanic areas. Most of the UK's sulphur comes from Poland where it occurs naturally. Sulphur is present in natural gas as hydrogen sulphide which can be converted to sulphur by burning it.

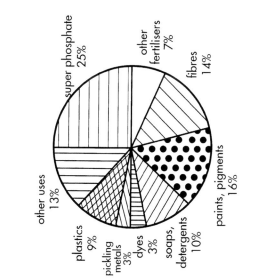

1. Sulphur is made into sulphur dioxide. Write a word equation for the reaction between sulphur and oxygen to make sulphur dioxide. (1)
2. Before becoming sulphuric acid, sulphur dioxide has to be turned into sulphur trioxide. Judging from the new name, what do you think has to be added? (1)
3. When hydrogen sulphide is burned in oxygen, it will form sulphur. What will happen to the hydrogen in oxygen? (1)
4. Write a word equation for the reaction in question 3. (2)
5. What would happen to the sulphur if too much oxygen was used in this reaction? (1)
6. If hydrogen sulphide is left in natural gas it could cause pollution. Explain how this can happen. (2)
7. Before building a plant to make sulphuric acid, a chemical company has to consider the following:
    a) easy access to ports
    b) easy access to railways and motorways
    c) have a town or city nearby
   Say why each of these considerations is important. (6)
8. Included in the other uses for sulphuric acid is its use in automobiles. How is sulphuric acid used in cars? (1)

Science Companion © A Porter, M Wood, T Wood and Stanley Thornes (Publishers) Ltd 1995

7.2.3 Changes for the better from land and air

1 Uses of Oxygen

The table below shows the uses to which the gas oxygen is put. Use the figures in the table to complete the pie chart below. Colour in and label the sections. Each section equals 5%. (5)

Use of oxygen	%
Steel making	55
Making chemicals	25
Other uses (eg. medical)	10
Cutting metal	5
Rockets and explosives	5

a) In steel making oxygen is used to get rid of unwanted carbon in the iron. What gas will be made when carbon reacts with oxygen?
_____

b) What use would oxygen have in a hospital?
_____

c) Acetylene (or ethyne) is used with oxygen in torches which can cut through metal. Why is the oxygen used instead of air?
_____

d) In the space shuttle oxygen is used with hydrogen to help launch it into space. What is the product of hydrogen burning in oxygen?
_____
(4)

2 Uses of Limestone

Limestone is a rock which was formed millions of years ago from sea shells and coral. The main chemical in limestone is calcium carbonate. This can be changed into calcium oxide by heating the limestone in a kiln. To get the high temperatures needed, the limestone is mixed with a fuel called coke (mainly carbon). The mixture is poured into the top of the kiln and air is blasted in from the bottom. The coke burns in the air, raising the temperature. The limestone is turned into quicklime (calcium oxide) and carbon dioxide. The blast of air removes the carbon dioxide as soon as it forms. Quicklime will turn into slaked lime when water is added to it. Slaked lime is called calcium hydroxide.

Use the words below to complete the flowchart of the production of slaked lime from limestone. You may need to use some words more than once. (6)

calcium carbonate, calcium hydroxide, calcium oxide, carbon, carbon dioxide, quicklime

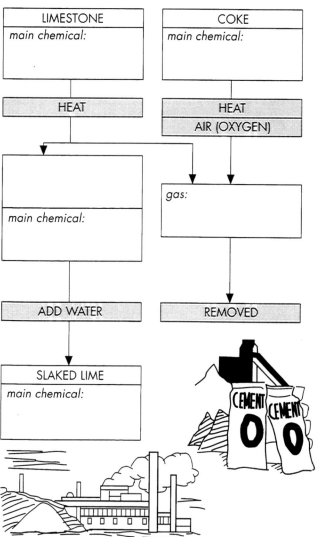

Name: **Class:**   **Date:**

7.3.2 Changes for the better from water and living things

Going for victory on a bicycle made from sea water

The frames on the new bike are the invention of Frank Kirk, an Essex engineer.

Magnesium is used for the frames instead of the steel tubing on traditional bikes.

When alloyed with aluminium, magnesium has the same strength as steel.

The density of steel is four times that of magnesium.

Magnesium is extracted from sea water. One cubic metre of sea water contains 1.3 kg of magnesium – enough for one frame.

Study the diagram above and then answer the questions below.

1. a) From what metal are conventional bicycles made? (1)
   b) What metal is used to make the new bicycles? (1)
   c) Why is tubing used to make bicycles instead of solid metal? (1)
   d) The new metal is taken out of sea water. What word in the article means 'taken out'? (1)
   e) The new metal has to be mixed with another metal before it is made into bicycles. What word in the article means 'metals mixed together'? (1)
   f) What other metal is needed to make the new bicycles? (1)
   g) Magnesium has a density of 1.7 g per cm³. What is the density of high tensile steel? (1)

2. The purple dye in red cabbage can be extracted using these steps:
   i Boil red cabbage leaves for 10 minutes
   ii Filter the juice from the leaves
   iii Evaporate the juice until dry

   Say which equipment you would need from the list opposite for each of the stages i to iii. (8)

   beaker, bunsen burner, evaporating dish, filter paper, funnel, gauze, heat mat, tripod.

7.3.1 Changes for the better from water and living things

Going for victory on a bicycle made from sea water

Metallurgy *Jonathan Theobold*

Some of the world's top cyclists will be riding the Tour de France next month on bicycle frames made from sea water. The frames are the invention of Frank Kirk, an Essex engineer who has abandoned steel tubing in favour of magnesium. The magnesium is extracted from sea water, with one cubic metre of water yielding 1.3 kilograms – enough material for one bicycle frame. Magnesium has only one quarter the density of high tensile steel but when alloyed with aluminium, its strength is just as good.

The new magnesium frame has a distinctive design. Freed from the constraints of tubing, its dimensions change with anticipated stresses. Kirk hopes to produce over 250 000 frames annually when his Chelmsford factory goes into full production next year.

Observer 20.5.1990

Read the newspaper article and then answer the questions below.

1. a) From what metal are conventional bicycles made? (1)
   b) Explain the word 'extracted' in line four. (1)
   c) How many cm³ are there in one cubic metre? (1)
   d) The density of sea water is approximately 1 g per cm³. How many kilograms are there in one cubic metre of sea water? (1)
   e) Magnesium has a density of 1.7 g per cm³. What is the density of high tensile steel? (1)
   f) Name another material which can be extracted from sea water. (1)
   g) Explain how the material you have chosen can be extracted from the sea. (1)

2. Listed below are the steps needed to extract the purple colour from red cabbage leaves. Put these steps in the correct order and describe what happens in each step. (8)

   boiling   chopping up   evaporating   filtering

7.3.3 Changes for the better from water and living things

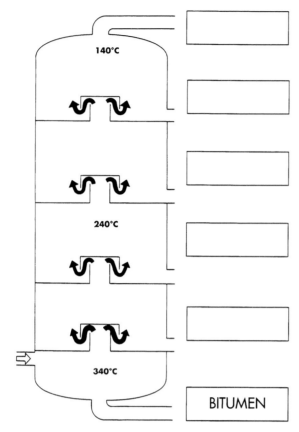

1 Look at the simplified diagram of the fractionating tower used at an oil refinery. Use the information in the table below to complete the labels on the diagram. (5)

FRACTION	BOILING RANGE
Bitumen	solid
Gases	below 25°C
Gas oil	200–400°C
Kerosene	174–275°C
Lube oil	over 300°C
Naphtha	20–200°C

2 Match the name of the fractions to their main uses in the diagram below. One has been done for you. (5)

BITUMEN — JET FUEL
GASES — GREASE
GAS OIL — CAMPING GAS
KEROSENE — PETROL FOR CARS
LUBE OIL — FUEL FOR TRAINS
NAPHTHA — TAR FOR ROADS

3 Write the answer to each question in the space provided. The shaded column should reveal the name of a heart drug which is found in foxgloves.

a.
b.
c.
d.
e.
f.
g.
h.
i.

a) What **S** comes after caustic and is used to turn fats into soap?
b) What **P** is a mould which can stop the growth of bacteria?
c) What **D** are extracted from many plants and can have an effect on the workings of the body?
d) What **A** is a headache pill and can be made from willow trees?
e) What **C** comes from seed heads and can be woven into fabrics?
f) What **S** can be made from natural plant materials and is used to wash our skin?
g) What **C** is a wrapping material made from extracted cellulose in plants?
h) What **M** is often made from plants and is used to treat or prevent diseases?
i) What **L** is an oil made from the flax plant? (10)

4 Look up the following words in a dictionary to try to find out what living thing they come from.

SILK _____

LATEX _____

COCHINEAL _____

LITMUS _____

CAMPHOR _____

(5)

Name: Class: Date:

Science Companion © A Porter, M Wood, T Wood and Stanley Thornes (Publishers) Ltd 1995

7.4.1 Chemicals from oil

FRACTION	BOILING POINT RANGE (°C)	NUMBER OF CARBON ATOMS IN A MOLECULE
Gases	below 25°C	1–4
Gasoline	20–200°C	4–12
Kerosene	174–275°C	11–15
Gas oil	200–400°C	15–19
Mineral oil	over 350°C	20–30
Fuel oil	over 400°C	30–40
Wax and grease	solid	41–50
Bitumen	solid	over 50

1. Crude oil is a mixture of different oils which are hydrocarbons. Which two elements are found in hydrocarbons? (2)
2. Explain how the fractions are obtained from crude oil. (3)
3. The gases are usually liquefied inside cylinders for use in portable heaters. They are not cooled down. How are they liquefied? (1)
4. Why do the fractions have a range of boiling points, rather than boil at a specific temperature? (1)
5. What is the link between the boiling point of the fraction and the number of carbon atoms in the molecule? (2)
6. The larger molecules are often cracked.
   a) What happens to the molecules when they are cracked? (1)
   b) What method is used to crack the molecules? (1)
   c) What is the reason for cracking the molecules? (1)
7. Small molecules which have a double bond inside them are very useful. They are called monomers.
   a) Why is the double bond useful? (1)
   b) What type of chemical can be made by reacting together these monomers? (1)
8. Give the names of two objects which are made from each of the following materials.
   a) Poly(ethene) *(polythene)*
   b) poly(chloroethene) *(PVC)*
   c) polystyrene (6)

Science Companion © A Porter, M Wood, T Wood and Stanley Thornes (Publishers) Ltd 1995

7.4.2 Chemicals from oil

FRACTION	BOILING POINT RANGE (°C)	NUMBER OF CARBON ATOMS IN A MOLECULE
Gases	below 25°C	1–4
Gasoline	20–200°C	4–12
Kerosene	174–275°C	11–15
Gas oil	200–400°C	15–19
Mineral oil	over 350°C	20–30
Fuel oil	over 400°C	30–40
Wax and grease	solid	41–50
Bitumen	solid	over 50

1. These fractions are obtained from crude oil by fractional distillation. Explain in your own words what happens to the oil in fractional distillation. (3)
2. What information in the table tells you that the first fraction, called Gases, is indeed made of gases? (1)
3. Which two fractions have *melting points* above 25°C? (2)
4. What happens to the boiling point as the number of carbon atoms in the molecules gets larger? (1)
5. How do the figures for the boiling points tell us that the fractions are not pure liquids but mixtures? (2)
6. All the fractions belong to a group of chemicals called hydrocarbons. As the name suggests, hydrocarbons are made from two elements. What are these two elements? (2)
7. Plastics are made from small, reactive molecules. Larger molecules have to be broken up into smaller ones to do this on a large scale.
   a) What name is given to the small, reactive molecules? (1)
   b) What is the process of breaking large molecules called? (1)
   c) Give one way in which this can be done. (1)
8. Give the names of two objects which are made from each of the following materials.
   a) Poly(ethene) *(polythene)*
   b) poly(chloroethene) *(PVC)*
   c) polystyrene (6)

Science Companion © A Porter, M Wood, T Wood and Stanley Thornes (Publishers) Ltd 1995

7.4.3 Chemicals from oil

1 Match the name of the monomer in the first column to the name of its polymer in the second column. Then match the name of the polymer to its use in the third column. One example has been done for you. (6)

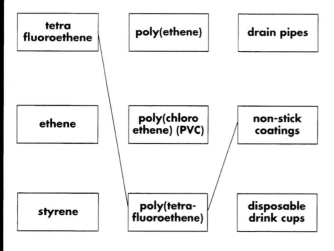

2 The table below shows the names of three groups of hydrocarbons (alkanes, alkenes and alkynes). The naming of the hydrocarbons follows a logical pattern based on the number of carbon atoms and the type of bond. Use this pattern to complete the table. (7)

3 Enter the answers to the following seven questions into the spiral grid below. The letter which ends one word, begins the next word and should be written in the shaded squares. (7)

1. Another word for crude oil (9 letters)
2. A small molecule which forms a repeating unit in a polymer (7 letters)
3. To make a plastic less flexible, makes it more _____ (5 letters)
4. A type of engine which uses fuel which is thicker than petrol (6 letters)
5. This means to move together smoothly, as when there is oil between two surfaces (11 letters)
6. Some frying pans have Teflon surfaces to make them _____ (two words 3–5)
7. The fraction of crude oil from which jet fuel is made (8 letters)

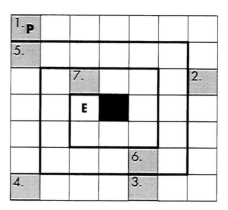

	ALKANES all single bonds	**ALKENES** one double bond	**ALKYNES** one triple bond
ONE CARBON	methane		
TWO CARBONS		ethene	ethyne
THREE CARBONS	propane		
FOUR CARBONS			butyne
FIVE CARBONS		pentene	

Name: Class: Date:

Science Companion © A Porter, M Wood, T Wood and Stanley Thornes (Publishers) Ltd 1995

7.5.1 Changing our food

1
a) When milk curdles it separates into a white lumpy solid and a watery liquid. What are these two parts called? (2)
b) Which part of curdled milk is used to make cheese? (1)
c) What is an enzyme? (2)
d) What is the name of the enzyme used to make cheese from milk? (1)

2
a) Why is yeast said to be related to mushrooms? (1)
b) Name two things that yeast needs to grow? (2)
c) Name the two products when yeast feeds on sugar. (2)
d) What is fermentation? (2)

3 To keep food fresh we usually put it in the fridge which is kept at a temperature between 2 and 5°C.

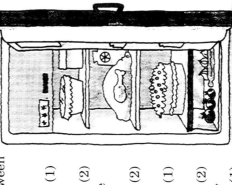

a) What effect does the cold temperature have on bacteria? (1)
b) What can happen to food outside the range of 2 to 5°C? (2)
c) What is wrong with the contents of the fridge in the picture? Explain your answer. (2)
d) What type of harmful bacteria is associated with chicken and eggs? (1)
e) Why can under-cooked chicken cause food poisoning? (2)
f) Why does cooking the chicken properly prevent food poisoning? (1)

4 There are many ways to prevent food from spoiling. For each of the three ways below explain how the process keeps the food from being spoiled by bacteria.
a) drying
b) canning
c) pickling (3)

7.5.2 Changing our food

1
a) What is the raw material used to make cheese? (1)
b) What are curds and whey? (2)
c) Why is rennet used in cheese making? (2)
d) What produces the blue veins in some cheeses (like Stilton)? (1)

2
a) What two things does yeast turn sugar into? (2)
b) Why is yeast used in the brewing industry? (2)
c) Why is yeast used in the bread making industry? (2)
Beer is brewed at a temperature between 21 and 24°C.
d) What happens if the temperature drops below this range? (1)
e) What happens if the temperature rises above this range? (1)

3 To keep food fresh we usually put it in the fridge which is kept at a temperature between 2 and 5°C.

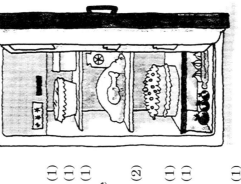

a) Why do bacteria grow more slowly in the fridge? (1)
b) How can food be damaged below 0°C? (1)
c) How can food be damaged above 5°C? (1)
d) What is wrong with the contents of the fridge shown in the picture? Explain your answer. (2)
e) Name a food which can carry the harmful bacteria called *Salmonella*. (1)
f) Why does salmonella make us ill? (1)
g) Why does cooking food thoroughly help stop *Salmonella* causing food poisoning. (1)

4 There are a number of ways of preventing bacteria from spoiling food. How can we keep bacteria from growing on the following foods?
a) baked beans
b) pickled onions
c) rice (3)

7.5.3 Changing our food

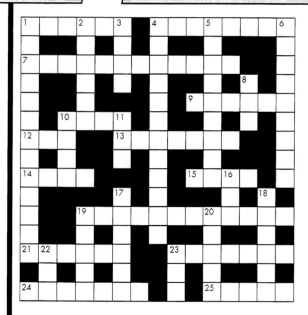

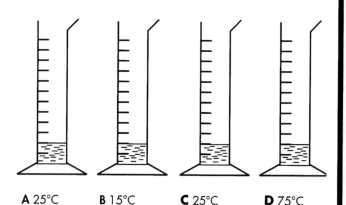

A 25°C   **B** 15°C   **C** 25°C   **D** 75°C

The diagram above shows the apparatus at the start of the following experiment:
A dough mixture was made from flour and water. A small amount of this dough was placed in cylinder **A**. Some yeast was mixed into the rest of the dough and this was added to cylinders **B**, **C** and **D**. The same amount of dough was added to each cylinder. Cylinders **A** and **C** were placed in a warm room. Cylinder **B** was placed in a cold room and cylinder **D** was heated to 75°C. All the cylinders were left for the one hour exactly.

Across
1. A solid food made from milk
4. These come from 3 down
7. A cold place to preserve food
9. This vegetable is the raw material for crisps
10. The temperature below which most food freezes
12. What could happen to you with food poisoning
13. This is made when beer or wine goes off
14. The ingredients for porridge
15. This is made by fermenting malted barley
19. The process of changing sugar into alcohol
21. This means 'can be eaten'
23. Exposure to the air can cause food to do this
24. Soaked in water like dried peas
25. The bad smell of rotten food

Down
1. A gas made when yeast mixes with sugar (2 words 6,7)
2. The place where many of our food laws are drawn up
3. These can sometimes carry salmonella
4. What milk separates into (what Miss Muffett ate) (3 words 5,3,4)
5. A sweet made from cocoa beans
6. Like salt, pepper or spices added to food for flavour
8. Putting food in one of these will keep it for a long time
10. A word used for the rind of a lemon or orange
11. The scientific word for 3 down
16. What we do with food
17. A cold way to preserve food
18. A toxic substance made by some bacteria
19. Plant tissue, healthy in a diet
20. These liquids (like lemon juice and vinegar) can preserve food
22. An insecticide which can end up poisoning food
23. Thick liquid extracted from olives

(30)

1 What was the temperature of the warm room?
_____ (1)

2 What was the temperature of the cold room?
_____ (1)

3 Give two things done in the experiment that made it a fair test.

a) _____
_____ (1)

b) _____
_____
_____ (1)

4 Which cylinder is the control experiment?
_____ (1)

5 On the diagram above mark clearly the positions of the dough after one hour. If there has been no change in the amount of dough then write 'no change' in the cylinder. (4)

6 What has made the dough in some of the cylinders rise up?
_____ (1)

Name: Class: Date:

7.6.2 Oxygen – the gas of life

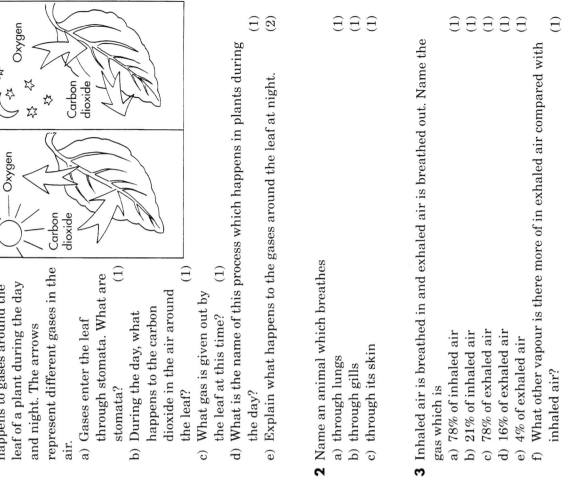

1 The diagram shows what happens to gases around the leaf of a plant during the day and night. The arrows represent different gases in the air.

a) Gases enter the leaf through stomata. What are stomata? (1)

b) During the day, what happens to the carbon dioxide in the air around the leaf? (1)

c) What gas is given out by the leaf at this time? (1)

d) What is the name of this process which happens in plants during the day? (1)

e) Explain what happens to the gases around the leaf at night. (2)

2 Name an animal which breathes

a) through lungs (1)
b) through gills (1)
c) through its skin (1)

3 Inhaled air is breathed in and exhaled air is breathed out. Name the gas which is

a) 78% of inhaled air (1)
b) 21% of inhaled air (1)
c) 78% of exhaled air (1)
d) 16% of exhaled air (1)
e) 4% of exhaled air (1)
f) What other vapour is there more of in exhaled air compared with inhaled air? (1)

7.6.1 Oxygen – the gas of life

1 The diagram shows what happens to gases around the leaf of a plant during the day and night. The arrows represent different gases in the air.

a) In your own words explain what happens to the gases in the air and in the leaf. (2)

b) How do the gases get into the leaf? (1)

c) What is the process of gas exchange called when there is daylight present? (1)

d) Respiration occurs in plants and is noticeable in darkness. What are the three products of respiration? (3)

2 Animals also have to respire in order to live. Explain how the following animals exchange gases with the air.

a) fish (1)
b) humans (1)
c) worms (1)

3 The mixture of gases in the air we breathe in is different from the gases we breathe out.

a) Which gas in the air do we use in respiration? (1)

b) What percentage of the air we breathe in is the gas in (a):
8%, 16%, 21%, 50% or 78%? (1)

c) What percentage of the air we breathe out is the gas in (a):
8%, 16%, 21%, 50% or 78%? (1)

d) Which gas takes up about 4% of the air we breathe out? (1)

e) We breathe out quite a lot of water vapour. Where does this come from? (1)

7.6.3 Oxygen – the gas of life

John Mayow (1640–79) was a lawyer by profession but enjoyed doing scientific experiments. He did experiments with air long before anyone knew what air really was. In his first experiment he stood a candle in a bowl of water and covered it with a bell jar. The candle continued to burn for a little while and then went out, but long before all the air in the jar was used up. Mayow could tell how much air in the jar was used up because the water level rose inside the jar. He noticed that about a fifth of the air had been used up.

Next he placed a mouse inside another jar. Rather cruelly he watched as the mouse slowly died inside the bell jar. The mouse like the candle flame, died long before all the air had been used up.

Mayow decided that the candle was only using the part of the air that was useful for burning. The mouse used only the part of the air that was useful for living. He began to wonder if they were both using the same part of the air.

1 a) The diagram below shows the first experiment which Mayow set up. Complete the second diagram to show what happened when the candle went out. (2)

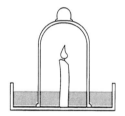 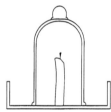

b) The paragraph opposite describes what would happen if the mouse and the candle were placed in the jar together. Complete the missing words.

Both the mouse and the candle use the gas _____. When this has been used the mouse will _____ and the candle will _____. The water will rise _____ of the way up the jar. The water will rise _____ than it did with just the mouse in. (5)

2 Enter the answers to questions 1 to 10 in the grid below and use the numbered letters to complete the sentence at the foot of the page. (18)

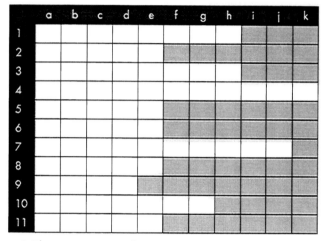

1 The major gas in the air
2 Unwanted (like carbon dioxide in our blood)
3 Got rid of (like carbon dioxide from our lungs)
4 The green pigment in plants
5 The organs we use to take in air
6 The organ we use to pump blood
7 Where food is digested
8 What fish use to take in oxygen
9 The percentage of carbon dioxide in exhaled air
10 Where food goes once swallowed
11 This carries gases round the body

6d	1g	8e	4g	7a	4e	2b	1c	7g	9b	5c							
												C	H	A	N	G	E	S

9a	10c	1e	3h				4d	3b	4i	8a	7i	1a		
		A	N	D										

				10f	6c	9d	11a	11c	7b		11e	8b	4f	3b	1b	3h	2e
I	N	T	O														

2a	10e	6e	3a	1d				3d	7h	3g	9d	5d	4i	
			A	N	D									

Name:　　　　　　　　　　　　　　　　　　　　**Class:**　　　　　　**Date:**

Science Companion © A Porter, M Wood, T Wood and Stanley Thornes (Publishers) Ltd 1995

7.7.1 Oxidation

1 Methane is a compound made of carbon and hydrogen. It is the main gas in North Sea gas and in a plentiful supply of air, it is flammable.
  a) When methane burns in air, what gas does it react with? (1)
  b) What compound is the carbon oxidised to? (1)
  c) What compound is the hydrogen oxidised to? (1)
  d) If there is not a plentiful supply of air, the methane will not oxidise completely. What compound will the carbon be oxidised to? (1)
  e) Why is it important that gas fires are well ventilated so that the fumes are removed from a room? (1)

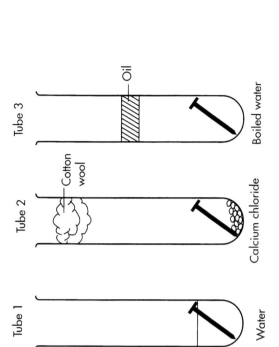

Tube 1 — Water
Tube 2 — Calcium chloride / Cotton wool
Tube 3 — Boiled water / Oil
Tube 4 — Sodium chloride solution

2 The diagram above shows an experiment which was carried out on some iron nails.
  a) For each tube 1 to 4 write down whether the nail was exposed to water, air or salt. (4)
  b) For each tube 1 to 4 describe the appearance of the nail after one week (4)
  c) What two substances are necessary for iron to rust? (2)
  d) Give two reasons why cars rust more in winter than in summer. (2)
  e) Give three ways the iron nail could be protected from oxidation. (3)

7.7.2 Oxidation

1 a) For something to be oxidised, what gas does it need to react with? (1)
  b) Why would a camping gas burner not light on the Moon? (1)
  c) Oxy-acetylene torches have got two gas cylinders supplying the flame. Why might an oxy-acetylene torch work on the Moon? (2)
  d) What would be the name of the compound formed when the following elements are oxidised?
    i) carbon
    ii) magnesium
    iii) hydrogen (3)

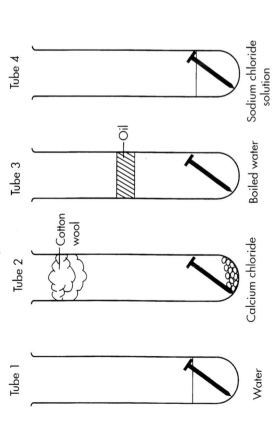

Tube 1 — Water
Tube 2 — Calcium chloride / Cotton wool
Tube 3 — Boiled water / Oil

2 The diagram above shown an experiment which was carried out on some iron nails.
  a) How would you know that the iron nail had gone rusty? (1)
  b) In which tubes was the nail exposed to oxygen? (2)
  c) In which tubes was the nail exposed to water? (2)
  d) Why did the nail rust in tube 1? (2)
  e) Why did the nail not go rusty in tubes 2 and 3? (2)
  f) What chemical, put on roads in winter, can increase the rusting? (1)
  g) Give three ways the iron nail could be protected from oxidation. (3)

7.7.3 Oxidation

1 Say what happens to the following foods when they are oxidised by oxygen from the air.

a) wine _____

b) a peeled apple _____
_____ (2)

Exhaust chemicals: Oxides of nitrogen, Water, Carbon monoxide, Carbon dioxide, Sulphur dioxide, Unburnt hydrocarbons, Lead compound

2 Complete the fire triangle above by writing in the three sides, the three things which are needed to make a fire burn. (3)

3 Match the metals and non metals to the properties in the middle with lines. Two have been done for you. (4)

- malleable
- oxidised to acids
- oxidised to base
- brittle
- solids at room temperature
- insulators of electricity

most metals are — most non-metals are

4 The picture at the top of the page shows the kind of chemicals which can be found in the exhaust fumes of a car. They are not in any order.

a) The fuel in a car is made from hydrocarbons. What word tells you that some of these are not oxidised?

_____ (1)

b) Underline the two chemicals below which are found in hydrocarbons.

nitrogen   hydrogen   carbon
sulphur   lead (2)

c) Name the three chemicals in the picture which are made when hydrocarbons burn.

_____
_____
_____ (3)

d) Which of the oxides comes from a gas in the air (besides oxygen)?

_____ (1)

e) Which of the oxides comes from impurities found naturally in the fuel?

_____ (1)

f) What environmental damage can these last two oxides cause?

_____ (1)

g) What test would show that carbon dioxide was present in the exhaust?

_____ (1)

h) What would be the result?

_____ (1)

Name: **Class:**   **Date:**

Science Companion © A Porter, M Wood, T Wood and Stanley Thornes (Publishers) Ltd 1995

7.8.2 Chemical energy

1 When methane reacts with oxygen in the air, heat is given out.

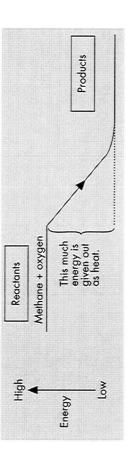

a) What do we call the type of reaction where heat is given out? (1)
b) What are the two products of burning methane? (2)
c) What happens to the temperature as the methane burns? (1)
d) The difference between the energy of the reactants and products is given the symbol ΔH. What does this stand for? (2)
e) Which of the following is likely to be the value for ΔH for this reaction? (1)

$+252\ \text{kJmol}^{-1}$ $0\ \text{kJmol}^{-1}$ $-252\ \text{kJmol}^{-1}$

2 The diagram alongside shows what happens when a solution of copper chloride is electrolysed.

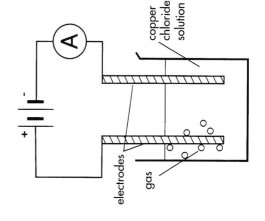

a) Why is this reaction classed as endothermic? (1)
b) What is the energy supply? (1)
c) What type of energy does this supply? (1)
d) Which has more chemical energy, the reactants or products? (1)
e) What two elements does the copper chloride split into? (2)
f) Which of these two elements collects around the negative electrode? (1)
g) When these two elements combine, will the reaction be exothermic or endothermic? (1)

7.8.1 Chemical energy

1 When methane reacts with oxygen, there is an exothermic reaction.

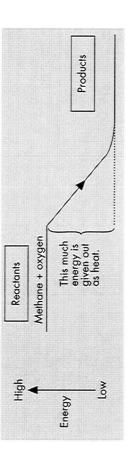

a) What are the products of the reaction? (2)
b) What happens to the temperature in this reaction? (1)
c) Which have more chemical energy, the reactants or products? (1)
d) What is the difference between the energy of the reactants and products known as? (1)
e) What is the symbol for this difference in energy? (1)

2 Describe what happens to the energy and the temperature in an endothermic reaction. (2)

3 The diagram alongside shows what happens when a solution of copper chloride is electrolysed.

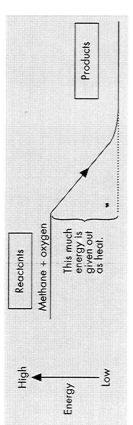

a) Why is this reaction classed as endothermic? (1)
b) In what form is the energy supplied? (1)
c) Which has more chemical energy, the reactants or products? (1)
d) What two elements does the copper chloride split into? (2)
e) Which of these two elements collects around the negative electrode? (1)
f) When these two elements combine, will the reaction be exothermic or endothermic? (1)

7.8.3 Chemical energy

1 A student investigated the temperature change when dissolving chemicals in water. Complete the last column in the table of results below. (4)

Temperature of water	Chemical added	Temperature of solution	Exothermic or endothermic?
18°C	A	10°C	
17°C	B	19°C	
18°C	C	−2°C	
19°C	D	36°C	

2 Complete the following paragraph about chemical energy changes. (7)

Some reactions involve an i_____ in the temperature. This can be measured with a t_____. In these reactions h_____ is given out to the surroundings. These type of reactions are called e_____ reactions. The opposite to these reactions are e_____ reactions. Here the surroundings have to supply e_____ to the reaction. So that this can happen, the t_____ of the reaction falls.

3 Match the reactions in the left hand column to the energy changes in the right hand column. Use one of the changes twice. Choose the most important change for each reaction. (5)

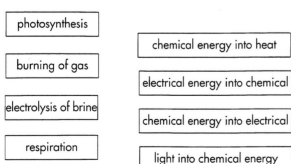

4 The diagram above shows how aluminium is made by passing electricity through liquefied aluminium oxide. This is the word equation for the reaction.

aluminium oxide → aluminium + oxygen

Add the words in the box below to the labels on the diagram. (3)

> **ALUMINIUM METAL**
> **OXYGEN GAS**
> **ALUMINIUM OXIDE**

Delete the incorrect words in each of the following sentences.

a) The positive electrode is **A/B** (1)

b) Energy is **supplied/given out** (1)

c) The reaction in the word equation is **exothermic/endothermic** (1)

d) The aluminium oxide has **more/less** energy than the products (1)

e) ΔH for the reaction in the word equation is **positive/negative** (1)

f) When aluminium and oxygen react the temperature will **increase/decrease** (1)

Name: **Class:** **Date:**

Science Companion © A Porter, M Wood, T Wood and Stanley Thornes (Publishers) Ltd 1995

7.9.1 — Rates of reaction

1 a) How does the temperature effect the rate of a chemical reaction? (2)
   b) Explain the effect in terms of particles. (2)

2 The illustration shows a carving on a church. It shows signs of acid corrosion. It has corroded more in the last fifty years than in the previous two hundred years.

   a) What chemical causes natural rainwater to be slightly acidic? (1)
   b) Why has the concentration of acid in rain increased in the last fifty years? (2)
   c) Explain why increasing the concentration of a solution should increase the rate of a chemical reaction. (2)
   d) How could you measure the changing acidity of the rainwater? (1)

3 Explain why potatoes cook faster when they are chopped into smaller pieces. (2)

4 What effect do catalysts have on a chemical reaction and how do they save energy in industrial reactions? (3)

Science Companion © A Porter, M Wood, T Wood and Stanley Thornes (Publishers) Ltd 1995

7.9.2 — Rates of reaction

1 a) How does the rate of a chemical reaction change with the temperature? (2)
   b) How does the rate of decay in a chicken change when it is placed in a kitchen, a fridge and a freezer? (2)

2 The illustration shows a carving on a church. It shows signs of acid corrosion. It has corroded more in the last fifty years than in the previous two hundred years.

   a) Carbon dioxide and sulphur dioxide both make rain acidic. Which is naturally present in rainwater? (1)
   b) Why has the concentration of acid in rain increased in the last fifty years? (2)
   c) What effect does increasing the concentration have on the rate of reaction? (1)
   d) How could you measure the changing acidity of the rainwater? (1)

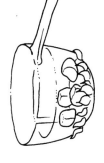

3 Explain why potatoes cook faster when they are chopped into smaller pieces. (2)

4 Give three ways you could increase the rate at which hydrochloric acid will react with magnesium. (3)

Science Companion © A Porter, M Wood, T Wood and Stanley Thornes (Publishers) Ltd 1995

7.9.3 Rates of reaction

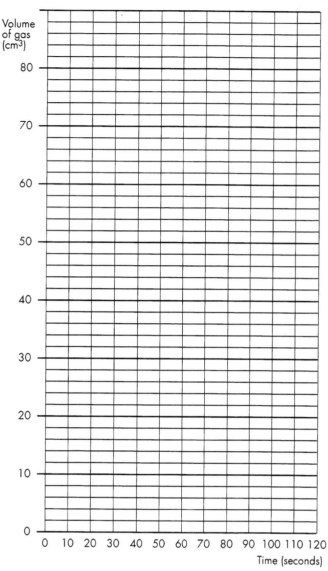

1 A student measured the volume of gas produced every 10 seconds when marble chips (calcium carbonate) reacted with dilute hydrochloric acid.
The marble chips were about 1 mm in diameter and the temperature was 20°C. There was more acid than marble chips.

a) Plot the results shown at the bottom of the page on to the above graph. Label it **A**. (4)

b) Sketch another curve on the graph which shows another experiment with the same dilute acid and the same sized marble chips, but half the amount of chips were used. Label it **B**. (2)

2 Use some of the words in the box below to complete the paragraph about catalysts. (6)

| chemically, decrease, increase, large, physically, small |

A catalyst is usually used to _____ the rate of a chemical reaction. It is used in _____ amounts. Catalysts are unchanged _____ during the reaction but may be changed _____.

3 Add the missing words from these five sentences to the grid below. (5)

R – Diluting solutions will _____ the rate of a reaction.

A – Changing the temperature _____ the rate of a reaction.

T – The rate can be measured by _____ a reaction.

E – Increasing the temperature gives particles more _____.

S – Having a smaller surface area will make a reaction _____.

R						
A						
T						
E						
S						

Time (in seconds)	0	10	20	30	40	50	60	70	80	90	100	110	120
Volume of gas (in cm³)	0	17	32	43	53	61	68	73	77	79	80	80	80

Name: Class: Date:

Science Companion © A Porter, M Wood, T Wood and Stanley Thornes (Publishers) Ltd 1995

8.1.2 The weather and weather forecasting

1 Below are five weather symbols (**A** to **E**) which may be seen on TV weather forecasts.

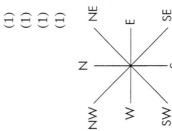

For each of the weather forecasts given below say which symbol you would use.

a) A warm summer day with a cloudless sky. (1)
b) Heavy rain. (1)
c) Thunderstorms very likely. (1)
d) Cloudy, no sunshine, dull with no rain. (1)
e) A cold, clear, frosty night. (1)

2 Alongside is a simple compass face. It shows 8 different points of the compass. Give the names for each of the points shown. (8)

```
      N    NE
 NW    |   /
    \  |  /
 W ---+--- E
    /  |  \
 SW    |   SE
       S
```

3 a) From which direction do the winds blow for most of the time in the UK? (1)
b) These winds bring our weather in over an ocean. Name this ocean. (1)
c) In which direction do the areas of weather usually go after they leave the UK? (1)

4 Areas of low pressure bring cloud and rain. The air in them swirls around. What causes the air to swirl? (1)

5 Weather fronts are narrow bands of rain which come with areas of low pressure. Draw the symbols used on weather maps for each of the three types of front. The types are:
a) warm (1)
b) cold (1)
c) occluded (1)

8.1.1 The weather and weather forecasting

1 Study the following weather data for 1990 and then answer the questions which follow.

	JANUARY	FEBRUARY	MARCH	APRIL
Mean temperature for central England	6.6°C	8.3°C	8.8°C	8.6°C
Monthly rainfall total for England and Wales	102.5 mm	100 mm	15 mm	30 mm
Monthly sunshine total for England and Wales	50 hrs	78 hrs	130 hrs	240 hrs

a) Which month had the highest mean temperature? (1)
b) Which was the wettest month? (1)
c) Which month had the least sunshine? (1)
d) What was the total rainfall for the four months? (1)
e) What was the total number of hours of sunshine for the four months? (1)
f) What was the average number of hours of sunshine per day in April? (1)
g) What was the average rainfall per day in April? (1)

2 Below is a list of six words or phrases to do with the weather.

anticyclone   depression   isobar   occluded front   warm front   cold front

For each of the weather statements which follow select one word from the list which best fits the description.

a) Drizzly rain at the leading edge of warm air. (1)
b) Heavy rain is expected. (1)
c) A warm front is followed immediately by a cold front. (1)
d) An area of low pressure. (1)
e) An area of high pressure. (1)
f) A line on a map joining places with the same air pressure. (1)

3 a) Why do winds blow away from the centre of an anticyclone? (1)
b) What sort of weather is usually found beneath stratus clouds in a depression? (1)
c) For the UK, which direction does most of the weather blow in from? (1)
d) In which direction do winds circulate around a depression? (1)
e) Suggest two methods of gathering weather data currently used by the MET office. (2)

8.1.3 The weather and weather forecasting

1 Study carefully these weather figures for Edinburgh and Jersey in March 1990. Then answer the questions which follow.

Edinburgh

Date	4	5	6	7	8	9	10
Rain (mm)	0.25	1.50	1.25	0.50	0.75	0.40	0.75
Maximum temperature (°C)	5	8	10	11	12	9	6
Sunshine (hours)	2.1	4.5	1.8	0.2	0.5	0.0	2.3

Jersey

Date	4	5	6	7	8	9	10
Rain (mm)	1.50	0.00	0.00	0.00	0.00	0.00	0.00
Maximum temperature (°C)	9	10	11	12	11	14	13
Sunshine (hours)	6.6	4.9	9.4	1.3	0.3	9.0	5.0

a) What was the total number of hours of sunshine for the week in Edinburgh? _____ hrs (2)

b) What was the total number of hours of sunshine for the week in Jersey? _____ hrs (2)

c) Use the graph paper below to draw a bar chart of the sunshine in Edinburgh (one colour) and Jersey (second colour). (14)

d) On which date did Edinburgh receive more sunshine than Jersey? _____ (1)

e) What was the average maximum temperature in Edinburgh during the week? _____ °C (2)

f) What was the average maximum temperature in Jersey during the week? _____ °C (2)

g) What was the total rainfall for the week in Edinburgh? (1) _____ mm (2)

h) What was the total rainfall for the week in Jersey? _____ mm (1)

2 Find the seven horizontal words using the clues (**a**) to (**g**) below and then find the name of the type of cloud in column 5.

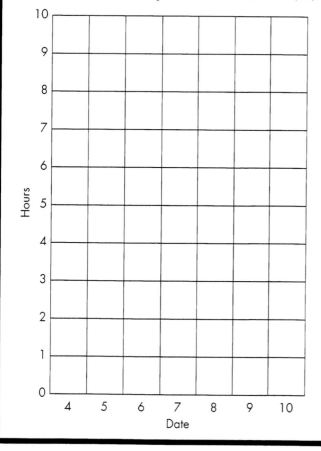

CLUES

a) Air _____ is the name given to the weight of air above us.

b) This is an area of high pressure.

c) This is an area of low pressure.

d) These winds blow from the south-west in the UK.

e) There are three types of weather _____.

f) The first high cloud which shows that a warm front is approaching.

g) Line on a map joining places with the same air pressure. (7)

The cloud in column 5 is _____ (2)

Name: **Class:**   **Date:**

8.2.1 Weather science

1 Study the diagram alongside about energy from the Sun and then answer the questions which follow.
   a) What percentage of energy from the Sun
   i) reaches the surface of the Earth? (1)
   ii) is reflected back into space? (1)
   iii) is stored by the air around the Earth? (1)
   b) In what form does energy from the Sun travel through space? (2)
   c) Why are thermometers in weather stations kept in the shade? (2)
   d) How do we know that energy from the Sun is not reaching Earth by conduction? (2)

2 a) Why does heat transfer due to conduction work better in a solid than in a liquid? (2)
   b) Why is heat transferred by convection in gases and liquids but not in solids? (2)

3 a) What is fog? (1)
   b) Explain the differences between fog, mist and smog. (3)
   c) The diagram alongside shows the formation of frontal fog. Explain how the fog forms. (2)

4 The following questions concern thunderstorms.
   a) What type of cloud is a thundercloud? (1)
   b) What are thermals? (1)
   c) What is the cause of static electricity in thunderclouds? (2)
   d) What is thunder? (2)

8.2.2 Weather science

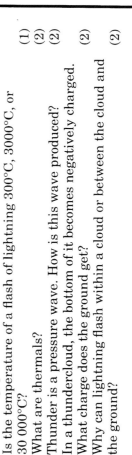

1 Study the diagram alongside about energy from the sun and then answer the questions which follow.
   a) What percentage of energy from the Sun
   i) reaches the surface of the Earth? (1)
   ii) is reflected back into space? (1)
   iii) is stored by the air around the Earth? (1)
   b) Why are thermometers in weather stations kept in the shade? (2)

2 Heat energy is moved around in three ways. These are: **conduction   convection   radiation**
   Which method is best described by each of the following?
   a) Energy from the Sun travels through the vacuum of space where there are no particles. The energy travel in waves. (1)
   b) Hotter material rises because it is less dense than cooler material. Energy is moved by these currents caused by different temperatures. (1)
   c) The vibration of particles in a material carries energy by this method. (1)

3 Use the diagram alongside to help answer the questions below.
   a) Which is denser, warm or cold air? (1)
   b) What effect does the cold air have on the water in the warm moist air? (1)
   c) What is fog? (2)
   d) What is smog? (2)
   e) What name is often given to thin fog where visibility is more than 1 km? (2)

4 The following questions are about thunderstorms.
   a) Is the temperature of a flash of lightning 300°C, 3000°C, or 30 000°C? (1)
   b) What are thermals? (2)
   c) Thunder is a pressure wave. How is this wave produced? (2)
   d) In a thundercloud, the bottom of it becomes negatively charged. What charge does the ground get? (2)
   e) Why can lightning flash within a cloud or between the cloud and the ground? (2)

8.2.3 Weather science

1 Read the following information and then answer the questions which follow.

> **Reflection of radiated heat**
> When heat energy is absorbed by an object, its molecules vibrate more and it gets hotter. Black surfaces absorb radiation the best. White or silvery surfaces are very poor at absorbing radiation and reflect most radiated heat energy which reaches them. The polar icecaps reflect most of the heat radiation they receive because they are white. The best absorbers of radiation are also the best emitters.

a) Why does a black teeshirt feel hotter in sunshine than a white one? (2)

b) Which part of the passage helps to explain why substances expand when they get hotter? (2)

c) A radiator not only heats a room by radiating heat. It also uses conduction and convection. Explain how it heats by these last two methods. (2)

2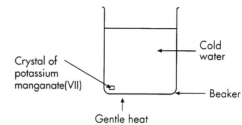

a) Explain what you would see happening in the beaker. Draw some arrows on the diagram to help your answer. (1)

b) In a second beaker, the crystal has been placed on the bottom in the middle. Draw arrows on the diagram below to show any convection current which would be seen. (2)

3 Why do birds sometimes fluff up their feathers in winter? (2)

Name: **Class:**   **Date:**

Science Companion © A Porter, M Wood, T Wood and Stanley Thornes (Publishers) Ltd 1995

8.3.2 Climate, farming and catastrophies

1. Study the diagrams alongside which show the way heat from the Sun reaches the Earth. Explain why the equator is hotter than the North Pole. (4)

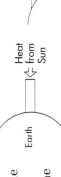

2. a) What is weather? (2)
   b) Over how many years is a weather pattern measured before a climate can be described. (1)

3. Study the map alongside. It shows the rainfall in the UK. Answer the following questions.

   a) Copy the compass bearing from the top left-hand corner of the map. Complete the three other points of the compass. (3)
   b) How much rain does most of Wales receive in a year? (1)
   c) Where does the North Atlantic Drift current start? (1)
   d) Is this current warm or cold? (1)
   e) Most of eastern England and Scotland is in a 'rain shadow'. Explain in detail what is meant by a 'rain shadow'. (5)

4. The diagram below shows how hot, moist air, at the Equator, rises and cools, causing heavy rain.

   a) What causes the air at the Equator to rise? (3)
   b) This air comes back down to the surface some 2000 miles away from the Equator. At this point there are many deserts. Why? (2)
   c) Why is air pressure low at the equator and high 2000 miles away? (2)

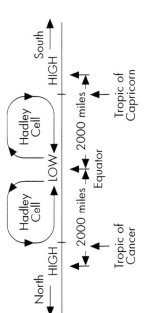

8.3.1 Climate, farming and catastrophies

1. What is the difference between weather and climate? (2)

2. There are at least six factors which influence climate. List four of them. (4)

3. Explain why the heat from the Sun is more concentrated at the equator than at the poles. Use diagrams to help you answer. (4)

4. Study the diagram of the Hadley Cells and answer the questions which follow.

   a) The heat from the Sun causes the air to rise at the equator. Why is there heavy rain around the equator? (2)
   b) The air descends 2000 miles away from the equator. Why are there deserts in this area? (2)
   c) There are places in deserts where nomadic people find water. What does the word 'nomadic' mean, and what name is given to the places where they find water? (2)
   d) What problem is produced by the heavy rain at the equator? (1)

5. The climate in the UK is greatly influenced by the North Atlantic Drift.
   a) Where does this current start? (1)
   b) Is this current warm or cold? (1)
   c) The prevailing wind also comes from that direction. Does this make the wind wet or dry? (1)

6. Why does the tilt of the Earth on its axis bring about seasonal changes? (2)

7. a) What causes tides? (2)
   b) What is a Spring Tide? (1)

8.3.3 Climate, farming and catastrophies

1 Study the Beaufort Scale for wind which is shown below.

Beaufort scale number	Wind	Effects
0	Calm	Smoke rises vertically
1	Light air	Smoke drifts in wind
2	Light breeze	Wind felt on face, leaves rustle, weather vanes move
3	Gentle breeze	Light flag blows, leaves and small twigs move
4	Moderate breeze	Small branches move, dust and loose paper blow
5	Fresh breeze	Small trees sway, waves seen on ponds, etc.
6	Strong breeze	Large branches move, telephone wires whistle
7	Moderate gale	Hard to walk into the wind, trees start to sway
8	Fresh gale	Very hard to walk into wind, twigs break off trees
9	Strong gale	Structural damage probable, slates and chimney pots lost
10	Whole gale	Trees uprooted, serious damage to buildings
11	Storm	Very rare inland, causes widespread damage
12	Hurricane	Major disaster

Now use the two squares which follow to draw simple pictures to illustrate the effects of the winds at two different values on the Scale. (8)

2 The west of the UK is warmer in the winter than the east.

a) Which current keeps the west warm?

_____ (1)

b) Where does this current come from?

_____ (1)

c) Why doesn't it affect the east?

_____
_____
_____ (1)

3 When 15 mm of rain falls in 3 hours, weather experts describe the conditions as a flash flood. Why does rain like this cause a flood?

_____
_____
_____
_____
_____
_____
_____
_____
_____ (2)

4 Use the clues below to find the 13 horizontal words in the grid below. Then find the flood prevention device in column 5.

	1	2	3	4	5	6	7	8	9
a)									
b)									
c)									
d)									
e)									
f)									
g)									
h)									
i)									
j)									
k)									
l)									
m)									

(13)

CLUES
a) The day by day changes in the air.
b) Eastern England is in a rain ————.
c) Wind of 7–10 on the Beaufort Scale.
d) 30 years of (a) in a pattern.
e) Wind of 2–6 on the Beaufort Scale.
f) Provides energy for the Earth.
g) A small tidal wave or a dull person.
h) (i) and (j) in the order (j) (h) (i) name a current which influences the UK.
k) Catastrophic wind developed from a cyclone.
l) Caused mainly by the pull of the moon on the sea.
m) A strong wind sometimes called a twister.

The flood protection device in column 5 is

_____ (1)

Name: **Class:**   **Date:**

8.4.2 — Air masses and Charles' law

1 Five air masses are given below:

arctic   polar maritime   polar continental   tropical continental   tropical maritime

Match the five names to the five numbers given to the air masses on the map below. (5)

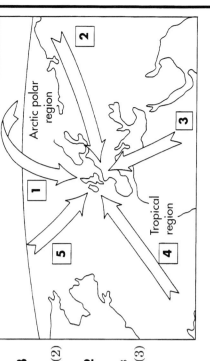

2 Why is air mass **3**
  a) hot
  b) dry? (2)

3 Why is air mass **2**
  a) cold
  b) not as moist as air mass **5**? (3)

4 Charles' law may be written as:
$$V_1/T_1 = V_2/T_2$$

  a) What do the symbols V_1, V_2, T_1, and T_2 each represent? (4)
  b) The units of temperature are in kelvins. How are Celsius units changed into kelvins? (1)

5 a) On graph paper work out an x-axis to run from $-300°C$ to $+100°C$, (400°C in all). The y-axis should be at 0°C and go from 0 cm³ to 100 cm³. Using the table of results below, plot two lines **A** and **B** on the same axes. (13)

Temperature in °C	0	20	40	60	80
line A cm³	50	57	65	72	79
line B cm³	71	76	81	87	92

  b) Which line best fits Charles' law and why? (2)

8.4.3 Air masses and Charles' law

1 Study the information about the main air masses that affect the British Isles, and then answer the questions which follow:

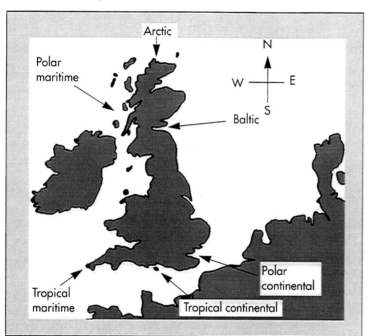

All our weather except thunderstorms is 'imported', because we are a small land area. Air which has come over the large land area of the continent has lost most of its moisture and so is usually dry when it reaches us. Maritime air has travelled over the sea for large distances and so is usually wet and brings rain. Air masses from the north (the polar region) are very cold, whilst in those from the south (towards the equator where it is hot) the air is warm.

The air masses are named after the places from which they come. The most common air mass to affect us is the polar maritime, which starts in the north-west and then swings round to the west or south-west.

a) What is an air mass?

_____ (1)

b) Here are four weather conditions:

  A hot, sunny, dry, dusty
  B cool, sunny periods, showers
  C cold, dry, bright, sunny
  D warm, humid, cloudy, light rain

From the four air masses given below, pick a letter, for the weather conditions you would expect, from the list above.

  i) polar continental _____
  ii) tropical continental _____
  iii) tropical maritime _____
  iv) polar maritime _____ (4)

2 Below is a graph of some results from an experiment to demonstrate Charles' law.

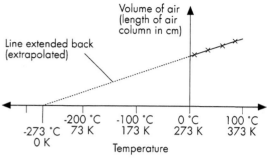

a) In this experiment the graph is extrapolated. What does this mean?

_____
_____
_____
_____ (2)

b) The graph is a straight line. What does this show?

_____
_____
_____
_____ (2)

3 In experiments about the gas laws there are four variables to consider. They are mass, pressure, temperature and volume. In experiments about Charles' law, which of these four variables must be kept constant?

_____ (2)

4 Use Charles' law to calculate the missing figures in the following table.

	Initial volume (cm³)	Initial temperature (K)	Final volume (cm³)	Final temperature (K)
a)		500	60	300
b)	140		88.2	63
c)	75	84		32
d)	600	475	320	
e)		275	85	125
f)	12		60	564
g)	520	220		240
h)	147	1489	8	
i)		90	50	150

(9)

8.5.2 Weathering and landscaping

1 Study the diagram alongside. It shows how water/ice can erode rocks. Explain what happens. (4)

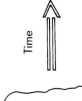

Water/ice

Time

2 How do trees, plants and grass stop heavy rain washing soil away? (1)

3 Read the following passage and then answer the questions which follow.

> Acid rain is caused by burning fossil fuels. The acid eats away the stonework of many buildings. Cathedrals which have stood for hundreds of years have been damaged more in the last 100 years than in the previous 400 years.
> Cathedrals are often made from sandstone or limestone. These materials are natural, soft and easy to carve into statues and other decorative features.

a) What stones are used to make many of the Cathedrals in the UK? (2)
b) This building stone is natural. What does this mean? (1)
c) Why is this used where statues and other decorations are needed? (2)
d) Why have the last 100 years seen a big increase in the damage caused by acid rain? (2)

4 The diagram alongside shows how a river might flow across a large area of flat land.

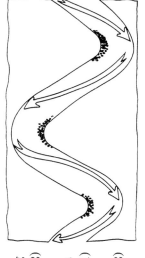

a) Why is this river not forming a deep canyon? (2)
b) What name is given to the way a river moves in a series of bends across flat land? (1)
c) How does this river change its shape over many years? (2)

5 Niagara Falls are some 50 metres high. Why is there a 50 metre deep plunge pool immediately under the waterfall? (3)

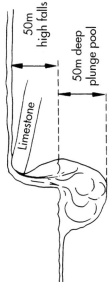

8.5.1 Weathering and landscaping

1 Study the following information about Monument Valley, and then answer the questions which follow.

> Monument Valley cuts across the border between Utah and Arizona in the south-western United States of America. There are two types of feature arising from the plain. These are buttes and mesas.
> The land is made of alternate layers of hard and soft rock. Over millions of years, the action of wind and water have worn away much of the soft rock, producing mesas, which are rocky, table-like structures. Further erosions cut the mesas into the narrow columns called buttes. The scenery has formed the backdrop to many famous western films. Today it is the home of the Navajo Indians.

a) In which country is Monument Valley? (1)
b) Whereabouts is it situated in that country? (1)
c) Over which two states does it stretch? (2)
d) Which native Indian tribe now live in the valley? (1)
e) What are mesas, and how were they formed? (3)
f) What is a butte? (1)
g) What type of films are often made in the valley? (1)

2 Why is water/ice a good eroder? (2)

3 How can soil be prevented from erosion by rain? (2)

4 Niagara Falls are some 50 metres high. Why is there a 50 metre deep plunge pool immediately under the waterfall? (2)

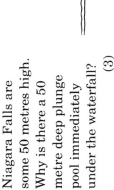

5 Acid rain is the cause of a lot of weathering on cathedrals in the UK.

a) What is the cause of acid rain? (1)
b) What gas in the rain causes this large amount of acidity? (1)
c) Why has this type of pollution only been seen over the last 100–150 years? (1)
d) Name **one** type of soft stone used to build cathedrals. (1)

8.5.3 Weathering and landscaping

1 Study the following information about the River Nile, and then answer the questions.

> The River Nile in Egypt used to carry a lot of mud, called silt, with it. In 1970 the Aswan High Dam was completed and the quantity of mud in the river fell.
> Before entering Egypt, the Nile flows through the Ethiopian Highlands. At Egypt's southern boundary it is held behind the Aswan High Dam, and forms Lake Nasser which is almost 500 km long.
> Below are the silt readings (in parts per million) taken 100 km downstream of the Aswan High Dam site in 1889 and 1989.

	Jan	Feb	Mar	Ap	May	Ju	Jul	Aug	Sept	Oct	Nov	Dec
1889	60	50	40	45	45	90	700	2700	2400	900	130	75
1989	45	46	45	50	50	48	47	45	45	46	47	46

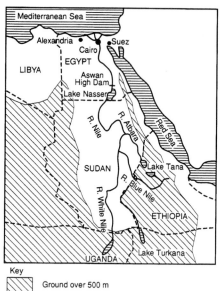

a) Why did Egypt get a lot of flooding along the banks of the River Nile before 1970?

_____
_____
_____
_____
_____
_____ (2)

b) What was the advantage of this flooding?

_____
_____
_____ (2)

c) Why do Egyptian farmers have to buy fertiliser today?

_____
_____
_____ (1)

d) The River Nile has a very large delta, built up over millions of years. What caused the delta to form?

_____
_____
_____ (2)

2 Use clues (a) to (j) to complete the horizontal words in the grid below and then find the word in column 6.

	1	2	3	4	5	6	7	8	9	10	11	12	13
a)													
b)													
c)													
d)													
e)													
f)													
g)													
h)													
i)													
j)													

CLUES
a) Moving air which is said to blow.
b) The sea into which the River Nile flows.
c) This type of rain eats away our buildings.
d) The most popular natural building stone in Britain.
e) Rock like this stays. It is not eroded.
f) A pile of rock at the bottom of a rock face.
g) The process of being eroded.
h) The Grand Canyon is in this state in America.
i) The most famous river in Egypt.
j) World famous waterfalls on the USA/Canada border. (10)

The word in column 6 is _____ (1)

Name: **Class:** **Date:**

8.6.1 The water cycle

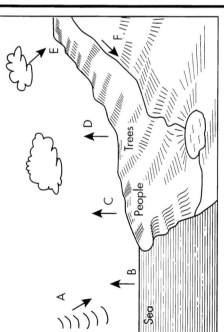

1 Alongside is a sketch of the water cycle. What is happening at each of the letters **A–F**? (6)

2 Name the process, which is not shown on the sketch, by which water sinks slowly through the rocks. (1)

3 What percentage of the Earth's water is actually fresh, and not unsalty water? Is it **3%, 13%, 23%, 33%** or **43%**? (1)

4 Where is most of this fresh water stored? (1)

5 What name is given to the level at which water settles underground? (1)

6 When water is extracted from a river, what is the purpose of the first screening tank? (1)

7 Why are aluminium sulphate and calcium hydroxide added to water at a treatment works? (1)

8 Why is chlorine added to our water supply? (1)

9 Where is local water usually stored before it is used? (1)

10 At the start of a sewage treatment plant there is usually an Archimedes' screw. How does this screw carry water uphill? (2)

11 The Primary sedimentation tank at a sewage treatment plant removes solid matter. This solid matter is removed as sludge after treatment. Give one use to which it is put? (1)

12 What destroys the waste matter in a biological filter at a sewage treatment plant? (1)

13 What is a sewerage system? (1)

14 Name **one** of the water-borne diseases which was spread before sewerage systems were built. (1)

8.6.2 The water cycle

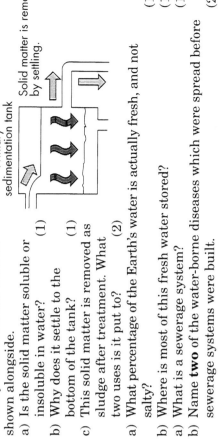

1 A student was asked to make a quick sketch to show the water cycle. The sketch is shown alongside.
   a) What is happening at each of the letters **A – F**? (6)
   b) Why does process **E** change with the height of the mountains? (1)
   c) What is the energy source for the water cycle? (1)

2 At a water treatment works, water is removed from lakes, rivers or boreholes, purified and then sent to homes, offices and factories for use.
   a) What job is done by the first screening tank? (1)
   b) Why are aluminium sulphate and calcium hydroxide added to the water? (1)
   c) Why is chlorine added to our water supply? (1)

3 In a sewage treatment plant solid matter is removed by settling in the Primary sedimentation tank as shown alongside.
   a) Is the solid matter soluble or insoluble in water? (1)
   b) Why does it settle to the bottom of the tank? (1)
   c) This solid matter is removed as sludge after treatment. What two uses is it put to? (2)

4 a) What percentage of the Earth's water is actually fresh, and not salty? (1)
   b) Where is most of this fresh water stored? (1)

5 a) What is a sewerage system? (1)
   b) Name **two** of the water-borne diseases which were spread before sewerage systems were built. (2)

8.6.3 The water cycle

1 Study the following information and then answer the questions which follow.

WORLD NEWS   27.10.94

A new mineral water has been launched on to the world market today. It comes from a tiny spring near Stirling in Scotland. 'Glenwater' is formed from centuries of Scottish rains which have percolated through the rocks to form an underground source of natural mineral water with a taste and purity which is certain to become known worldwide. It is purified as it is formed by the rain filtering through the rocks. This removes solid impurities and dissolves minerals from the rocks which give 'Glenwater' its lovely taste. Years of heavy rain have produced a high water table leading to a forceful spring which should last for ever even with sales expected at 1 million litres per year. The slightly acid water (pH 7.3) contains many healthy minerals.

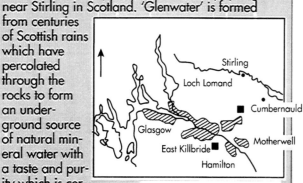

MINERAL ANALYSIS

Mineral	mg per litre
calcium	23
magnesium	6
potassium	0.2
sodium	6.7
hydrogen carbonate	81
sulphate	13
nitrate	1
chloride	9
fluoride	0.06
silicates	5.8
dry residue at 190°C	96
pH	7.3

a) What is the serious science error which the newspaper reporter has made?

_____ (1)

b) What is the main metallic element present?

_____ (1)

c) Which mineral is present in the smallest amount?

_____ (1)

d) What benefit is this mineral to humans?

_____ (1)

e) How are unwanted solids removed from the water?

_____ (2)

2 Study the following information and then answer the questions which follow.

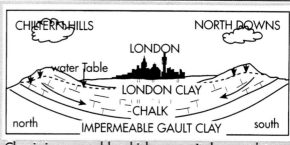

Clay is impermeable which means it does not let water through, and London sits on clay. Beneath the clay is a chalk syncline running from the Chiltern Hills to the North Downs. Rain on the Hills and Downs drains through the permeable chalk until it reaches the water table. The water table is higher under the Hills and Downs than below London. This means that the water below London is under pressure and up comes the water. This type of situation is known as an artesian well. The chalk syncline is an example of an artesian basin.

a) What is a water table?

_____ (1)

b) Why is the water under London at a high pressure?

_____ (1)

c) From the diagram, what do you think a syncline is?

_____ (1)

Name: **Class:**   **Date:**

Science Companion © A Porter, M Wood, T Wood and Stanley Thornes (Publishers) Ltd 1995

8.7.1 The rock cycle

1 Clay is made from many disc-like sedimentary particles. It forms sedimentary rocks called mudstone and shale.
  a) Where does the pressure come from to turn clay into the metamorphic rock slate? (1)
  b) What is the difference between the way the discs are arranged in sedimentary and metamorphic rocks? (2)
  c) How does the structure of slate make it suitable as a roofing material? (2)

2 a) How are sedimentary rocks formed? (2)
  b) If you could dig through layers of sedimentary rocks, where would you find the oldest ones? (1)

3 a) What is the meaning of the word 'metamorphosis'? (2)
  b) What two forces are generally involved in the creation of metamorphic rocks from other types of rock? (2)

4 a) Why do sedimentary rocks often contain fossils? (2)
  b) Why do metamorphic rocks not contain fossils? (2)

5 Igneous rocks are formed from molten magma.
  a) What is magma? (2)
  b) Igneous rocks can be made of small or large crystals. What conditions bring about each size of crystal? (4)
  c) What is magma called when it reaches the Earth's surface? (1)
  d) Why can the rock called pumice often float on water? (2)

6 a) To which sedimentary rock is marble related? (1)
  b) How do the crystals of marble compare with those of its sedimentary relative? (2)
  c) How do these crystals influence the properties of marble compared with those of its sedimentary relative? (2)

Science Companion © A Porter, M Wood, T Wood and Stanley Thornes (Publishers) Ltd 1995

8.7.2 The rock cycle

1 a) Where does the pressure come from to turn clay into the metamorphic rock slate? (1)
  b) What is the difference between the way the discs are arranged in sedimentary and metamorphic rocks? (2)
  c) How does the structure of slate make it suitable as a roofing material? (2)

clay → slate

2 Sedimentary rocks are formed from pieces of other rocks called sediments. The sediments are carried by rivers to places where they can settle.
  a) Name two possible types of place where rocks can settle. (2)
  b) Are old or new rocks deeper down in sedimentary layers? (1)
  c) What type of rock is formed when muddy sediments settle? (1)
  d) What type of rock is formed when sand settles? (1)
  e) Why are fossils found in sedimentary rocks? (2)

3 Igneous rocks are formed from molten magma.
  a) What is magma? (2)
  b) Granite rock is formed by the slow cooling of magma. The magma is undisturbed. Are granite crystals large or small? (1)
  c) What name is given to magma which reaches the Earth's surface? (2)
  d) Is cooling faster at the surface or in the body of the Earth? (2)
  e) Are crystals formed at the surface smaller or large than granite crystals? (1)
  f) Pumice is porous because it contains a lot of holes. How have these holes been formed? (2)

4 a) Which two forces may be used to make metamorphic rocks? (2)
  b) From what chemical is limestone made? (2)
  c) Marble is made from limestone. How are the crystals changed as limestone becomes marble? (2)
  d) What new properties do the crystals give marble compared with limestone? (2)

Science Companion © A Porter, M Wood, T Wood and Stanley Thornes (Publishers) Ltd 1995

8.7.3 The rock cycle

1 Study the following three paragraphs on rock formation and then answer the questions which follow.

> **A** The hot sun beating down on warm, shallow seas means that lots of water evaporates. Calcium carbonate which has dissolved in the water is then forced to precipitate as crystals and settle on the sea bed. This process is continuous, and may last millions of years. So, along with deposits of shells from dead animals, limestone rocks many thousands of metres thick are created.

> **B** During volcanic eruptions, the magma pours out as lava and cools, releasing gases, turning into rocks such as basalt. This rock is often highly resistant to erosion and can result in the volcanic rock remaining long after the surrounding cone of ashes has eroded.

> **C** Marble is made from limestone which has been subjected to heat and pressure. The original sedimentary rock has been turned into a dense crystalline structure which can take lovely carvings and a high polish. Mudstone, another sedimentary rock, under similar conditions becomes slate which can be split into thin sheets.

a) What type of rock is being made in

   A _____

   B _____

   C _____ ? (3)

b) In **A**, what causes the water to evaporate?

   _____ (2)

c) What mineral makes limestone?

   _____ (1)

d) How long may the process in **A** last?

   _____ (1)

e) In **B**, is basalt formed inside the Earth or on the surface?

   _____ (1)

f) Is basalt a hard or soft rock? _____ (1)

g) Explain why you think your answer to part (f) is correct. _____

   _____

   _____ (1)

h) Is an ash cone hard or soft? _____ (1)

i) Explain why you think your answer to part (h) is correct.

   _____

   _____ (1)

j) How is marble made from limestone?

   _____ (2)

k) What sort of new structure is in marble which was not in limestone?

   _____ (2)

l) From what named rock is slate made?

   _____ (1)

2 Below are the names of eight rocks or minerals.

china clay   clay   coal   copper

granite   iron ore   limestone

tungsten

Fit them into the following table in the place where each fits best. Each rock of mineral may only be used once. (8)

Rocks and minerals	Uses
	pottery, paper
	plumbing, electrical cables
	iron and steel
	cement, chemicals
	drill tips
	bricks, cement
	building stone
	fuels, chemicals

(8)

Name: **Class:**   **Date:**

8.8.1 Earthquakes and volcanoes

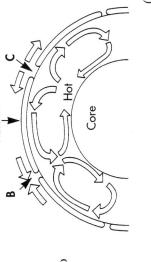

1. The Earth's crust is divided into large areas called plates. Below the crust is a mantle which in turn floats on top of the core. The following questions relate to the diagram alongside.
   a) In which direction is plate **A** moving? (1)
   b) Why are there convection currents in the mantle? (1)
   c) What is happening to the plates at point **B**? (1)
   d) What is happening to the plates at point **C**? (1)
   e) Plates are moving and creating the Mid-Atlantic ridge. Which is typical of the Mid-Atlantic ridge, point **B** or point **C**? (1)

2. San Francisco lies on the San Andreas Fault.
   a) What is the San Andreas Fault? (1)
   b) Why do a lot of fires start during and after an earthquake which takes place in a large city like San Francisco? (1)
   c) There was a strong wind in 1906, and no wind in 1989. What effect did this have on the two occasions? (1)

3. Mt. Etna in Sicily is an example of an acid lava volcano. It is formed where one of the Earth's plates slide under another.
   a) As the crust descends into the mantle, what happens to it and why? (2)
   b) Thick molten magma is produced. Is this slow or fast moving. Explain your answer. (2)
   c) What is the name of the chamber inside a volcano within which molten rock is stored before an eruption? (1)
   d) Why are the sides of an acid lava volcano steep? (2)

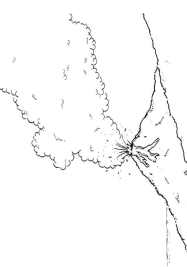

Science Companion © A Porter, M Wood, T Wood and Stanley Thornes (Publishers) Ltd 1995

8.8.2 Earthquakes and volcanoes

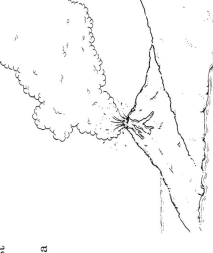

1. The Earth's crust is divided into large areas called plates. Below the crust is a mantle which in turn floats on top of the core. The following questions relate to the diagram alongside.
   a) The convection currents in the mantle move the plates above them. In which direction is plate **A** moving? (1)
   b) What is happening to the plates at point **B**? (1)
   c) What is happening to the plates at point **C**? (1)
   d) What causes part of the mantle to rise, and part to fall? (1)

2. The San Andreas Fault in California lies almost underneath the City of San Francisco. The city was destroyed by a major earthquake in 1906.
   a) Fires caused the most damage, how did they start? (1)
   b) When there is an earthquake, why is there a problem putting out the fires? (1)
   c) In 1989 another earthquake struck the city. The modern skyscrapers stood. Suggest how engineers have tried to make skyscrapers 'quake-proof'. (2)
   d) There was a strong wind in 1906, and no wind in 1989. What effect did this have on the two occasions? (2)

3. An acid lava volcano like Mt. Etna in Sicily is formed where one of the Earth's plates slides under another.
   a) What happens to the crust as it descends into the mantle? (1)
   b) This produces acid magma which is thick. Does this magma move quickly or slowly? Explain your answer. (2)
   c) When magma like this leaves the crater of a volcano it forms steep-sided cones. Why? (1)
   d) What name is given to magma when it is on the surface of the Earth? (1)

Science Companion © A Porter, M Wood, T Wood and Stanley Thornes (Publishers) Ltd 1995

8.8.3 Earthquakes and volcanoes

1 Study the following information about the Pacific Ocean, and then answer the questions which follow.

10° north of the Equator, in the Pacific Ocean, south of Mexico, two plates are moving apart. Sea water seeps down into the gap, gets hot and rises in springs which are rich in minerals. These springs or vents support strange life forms 8,500 feet below sea level. Among the life forms are giant tube worms which love the highly acidic water which contains little salt, but a lot of hydrogen sulphide. Minerals including iron, zinc and copper are brought up from deep inside the Earth, by water which reaches a temperature of 380°C under the enormous pressure at this great depth. Tube worms die within weeks when earthquakes close the mineral-rich vents.

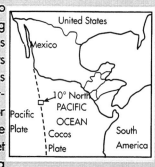

a) Name the two plates which are moving apart in this area of the Pacific Ocean.

_____ (2)

b) How can water reach 380°C and still be a liquid?

_____
_____ (1)

c) Name three metals which are present in the water from the vents.

_____ (3)

d) What proof is suggested in the notes above, that tube worms live on the very hot mineral-rich water?

_____
_____ (1)

e) Suggest which chemical causes the water from the vents to be highly acidic.

_____ (1)

2 Study the following information about Mount Pinatubo, and then answer the questions which follow.

On 15 June 1991, 5,770 feet high Mount Pinatubo in the Philippines erupted. Sulphur dioxide was thrown 25 miles up into the atmosphere where along with moisture it created a thin cloud which circled the Earth in 21 days. Satellite data showed that 2% of sunlight would be deflected from the Earth, lowering temperatures a little. Ash accompanied the sulphur dioxide. The ash contained bits of old volcano, pumice, sulphur-rich anhydrite and crystals of hornblende. The ash covered many square miles of the Philippines and poses a constant threat turning into an unstable slurry every time it rains.

a) Name the volcano which erupted in the Philippines on 15 June 1991.

_____ (1)

b) How long did it take for the high thin cloud from the volcanic eruption to circle the Earth?

_____ (1)

c) How did scientists measure the effect of this cloud on our climate?

_____
_____ (1)

d) Name four things found in the ash.

_____
_____
_____ (4)

e) What is a slurry?

_____
_____ (2)

f) Name the main gas produced in the eruption.

_____ (1)

g) What type of rain is this gas known to produce?

_____ (1)

Name: **Class:**   **Date:**

Science Companion © A Porter, M Wood, T Wood and Stanley Thornes (Publishers) Ltd 1995

9.1.2 — How things work

1 Read the following passage about live-wire linesmen who work for the National Grid.

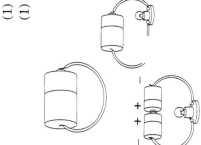

The cables which hang between pylons can carry electricity from power stations to homes and factories across the country. It is the job of the live-wire linesmen to inspect the cables while the current still flows through them.

They wear a special suit which has thin wires of copper, coated with silver, woven into the fabric. They ride a trolley which is suspended from the cables. The trolley is lowered on to a cable from a helicopter. The ropes from the helicopter to the trolley are made from a strong plastic.

a) What is the National Grid? (1)
b) From what type of material are the cables made? (1)
c) Why is this type of material chosen? (1)
d) If the linesmen are electrocuted, why does the electricity not go through their bodies? (2)
e) From what type of material are the helicopter ropes made? (1)
f) Why is this type of material chosen? (1)
g) From what type of material are the discs which separate the cable and the pylon made? (1)
h) Why is this type of material chosen? (1)

2 There are three circuits shown alongside.

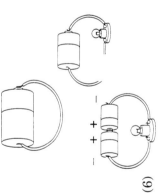

a) Why is the first circuit called a short circuit? (1)
b) Why is a short circuit bad for the cell? (1)
c) Explain why the bulbs in the other two circuits will not be alight. (2)
d) What can be done in these circuits in order to make the bulb light? (2)

9.1.1 — How things work

1 Read the following passage about live-wire linesmen who work for the National Grid.

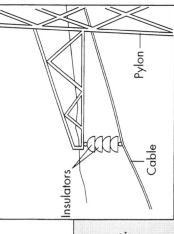

The cables which hang between pylons can carry electricity from power stations to homes and factories across the country. It is the job of the live-wire linesmen to inspect the cables while the current still flows through them.

They wear a special suit which has thin wires of copper, coated with silver, woven into the fabric. They ride a trolley which is suspended from the cables. The trolley is lowered on to a cable from a helicopter. The ropes from the helicopter to the trolley are made from a strong plastic.

a) What is the National Grid? (1)
b) Explain what the terms 'live-wire' means. (1)
c) Why does the suit have metals woven into it? (1)
d) What would happen if the men wore ordinary work suits while they were working on the cable? (2)
e) Why are the ropes from the helicopter made from plastic rather than steel? (1)
f) Why would someone touching the base of a pylon not receive an electric shock? (2)
g) Why are birds which land on overhead cables not electrocuted? (1)

2 Here are three rules when using cells in electric circuits. For each one say why it is an important rule. (6)

a) Never connect one end of a cell directly to the other.
b) Do not leave any gaps in the circuit.
c) Do not put cells in a circuit facing each other.

9.1.3 How things work

1 The symbols for a cell, a bulb and a switch are:

Use these symbols to complete the following circuits.

a) One bulb which will light when one switch is pressed. (3)

b) Two cells connected to three bulbs. (5)

2 The bulb in the circuit below is said to glow with NORMAL brightness.

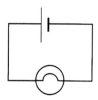

Say how bright the bulbs in the following circuits would be. Write one of the words, BRIGHT, NORMAL, DULL, OFF in the box below each circuit. (6)

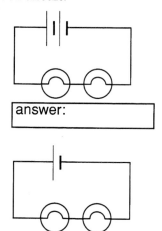
answer:

answer:

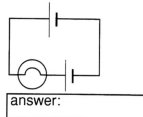
answer:

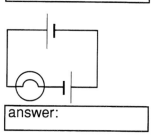
answer:

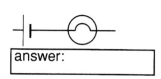
answer:

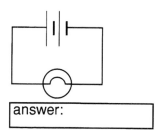
answer:

3 Put the following household objects into the correct side of the table below.

steel fork   jumper   key   magnet
newspaper   rubber   scissors
spanner   tap   teacup
glass window (11)

INSULATORS of electricity	CONDUCTORS of electricity

Name: **Class:**   **Date:**

Science Companion © A Porter, M Wood, T Wood and Stanley Thornes (Publishers) Ltd 1995

9.2.1 Go with the flow

1 Explain what you understand by the following terms.
 a) electric current
 b) series circuit
 c) parallel circuit
 d) ammeter
 e) resistor (5)

2 The circuit below shows how to measure the voltage across a resistor and the current going through it. The number of cells can be changed.

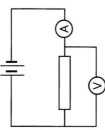

VOLTAGE	CURRENT
4 V	0.4 A
2 V	0.2 A
1 V	0.1 A

 a) Draw the four symbols used in the circuit and name each one. (4)
 b) What conclusion can you draw from these results? (3)

3 In the circuit shown below,

$$\text{resistance (ohms)} = \frac{\text{voltage (volts)}}{\text{current (amperes)}}$$

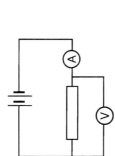

The cell used was 6 volts.

 a) What is the resistance if the current was 2 amperes? (1)
 b) What is the current if the resistance was set at 6 ohms? (1)
 c) What voltage would be needed to produce a current of 1 ampere through a resistance of 3 ohms? (1)

9.2.2 Go with the flow

1 Here are five descriptions. What word does each sentence describe?
 a) A device which measures the flow of electricity in a circuit.
 b) A circuit where the electricity flows through each bulb in turn.
 c) The flow of electric charges in a circuit.
 d) A circuit where each bulb is connected directly to the power source.
 e) Something which slows down the flow of electricity. (5)

2 The circuit below shows how to measure the voltage across a resistor and the current going through it. The number of cells can be changed.

VOLTAGE	CURRENT
4 V	0.4 A
2 V	0.2 A
1 V	0.1 A

 a) Copy the diagram and label the symbols using the following words **AMMETER, CELL, RESISTOR, VOLTMETER.** (4)
 The experiment gave the results shown in the table above.
 b) What current would flow with a 3 volt supply? (1)
 c) What voltage would be needed for a current of 0.5 ampere? (1)

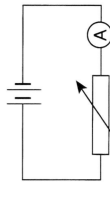

$$\text{resistance (ohms)} = \frac{\text{voltage (volts)}}{\text{current (amperes)}}$$

 d) What is the value, in ohms, of the resistor in the circuit? (1)

3 a) What is meant by a short circuit? (1)
 b) It is very important when building electric circuits to avoid including a short circuit. What damage to the cell does a short circuit do? (1)
 c) Will the bulb in this circuit light up? (1)

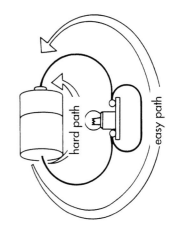

9.2.3 Go with the flow

1 Put the missing words from the passage below into the crossword grid. (16)

1.		2.		3.		4.		5.			6.
7.				8.					
		9.					10.		
					11.				
12.		13.							
		14.							

If a material will let electricity flow through it, it is called a (*14 across*). Water from the (*12 across*) will conduct electricity. The flow of electricity is caused by electric (*1 down*) when they (*10 across*). In wires this produces an electric (*1 across*) which is measured in a unit called an (*7 across*). D.C. stands for direct current and (*13 down*) stands for alternating current. Electric charges which are the same will (*2 down*) each other.

The electric light bulb was invented by Thomas (*3 down*). He (*4 down*) different materials for the coil inside the bulb, called the (*5 down*). If the coil becomes (*9 across*), the bulb will not work. Electricity can be fatal. Heat from an electric (*5 across*) comes off the glowing (*6 down*). You have to (*8 across*) the casing to stop electricity going through you, otherwise you could (*11 across*) when you touch it.

2 The passage below contains words in bold. In each case, cross out the word which is incorrect.

Anything that **slows down/speeds up** the current in a circuit is called a resistor. The current could be measured with **a voltmeter/an ammeter**. The bigger the resistance in a circuit, the **smaller/larger** the current would be. A resistor that can be altered is called a **variable/constant** resistor. Resistance is measured in units called **amperes/ohms**. (5)

3 Match the symbol to the name of the electrical apparatus. (5)

 BULB

 SWITCH

 AMMETER

 RESISTOR

 CELL

4 The bulb in the circuit below is said to glow with normal brightness.

Say how bright the bulbs in the circuits below would be. Write one of the words BRIGHT, NORMAL, DULL, OFF in the box below each circuit. (4)

a)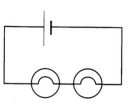
answer:

b)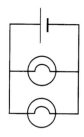
answer:

c)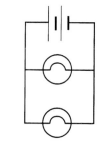
answer:

d)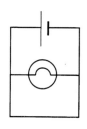
answer:

Name: **Class:** **Date:**

Science Companion © A Porter, M Wood, T Wood and Stanley Thornes (Publishers) Ltd 1995

9.3.2 Electricity in your home

1 This is a copy of an electricity bill for one quarter (3 months). It shows how many units of electricity the household used and the price for each unit.

Tariff	Meter Reading Present	Meter Reading Previous	Units Used	Unit Price	Total Charge
D1	37400	36900	?	8p	?

a) How many units of electricity did they use? (1)
b) What is the price of each unit? (1)
c) What is the total charge? (1)

2 For one unit of electricity you can run an iron for one hour, or you can have a light bulb on for 10 hours. This is because they use electricity at different rates. (This is known as the power and it is given on each appliance in watts or W for short).

APPLIANCE	POWER
Iron	1000 W
light bulb	100 W
hair dryer	500 W
television	75 W
microwave oven	700 W
electric fire	2000 W

a) List the six appliances in order from the most to the least expensive to run. (6)
b) How many hours could you run a hair dryer for one unit? (1)
c) How many units would be used if an electric fire was on for two hours? (1)

3 A plug contains three useful safety devices. The earth wire, the flex grip and the fuse.

a) What colours are used for the earth wire? (1)
b) Why is the flex grip an important safety device? (1)
c) What size fuses would you need for a lamp and an iron? (2)

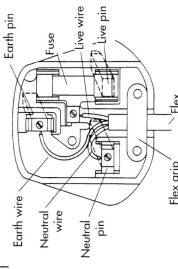

9.3.1 Electricity in your home

1 This is a copy of an electricity bill for one quarter (3 months). It shows how many units of electricity the household used and the price for each unit.

Tariff	Meter Reading Present	Meter Reading Previous	Units Used	Unit Price	Total Charge
D1	37385	36854	?	7.67p	?

a) How many units of electricity did they use? (1)
b) What is the price of each unit? (1)
c) What is the total charge (to the nearest penny) (1)

2 Electricity units are measured in kilowatt hours. The number of units used depends on two things – the time something is left plugged in (in hours) and the power (kilowatts). A two kilowatt electric fire left on for 4 hours will use $4 \times 2 = 8$ units. How many units will the following items use?

a) An iron (1 kilowatt) left on for half an hour. (1)
b) A bulb (100 watt) left on for 5 hours. (1)
c) A hair dryer ($\frac{1}{2}$ kilowatt) left on for 30 minutes. (1)
d) A T.V. (77 watt) left on for 6 hours. (1)
e) To the nearest penny, how much would it cost to run the four appliances above? (Use the unit price in question 1) (4)

3 A plug contains three useful safety devices. The earth wire, the flex grip and the fuse. In your own words explain what each of these safety devices does. (4)

9.3.3 Electricity in your home

1 The steps below are part of the story of the electricity to your home. The steps are in the wrong order. Put the steps into the correct order in the boxes **A–E**.

- ❏ Electricity passes through the National Grid at 440 000 V
- ❏ Electricity enters homes through the meter
- ❏ Fuel is burned in a power station
- ❏ Electricity is transformed to 240 V supply
- ❏ Water boils into steam to drive turbines (5)

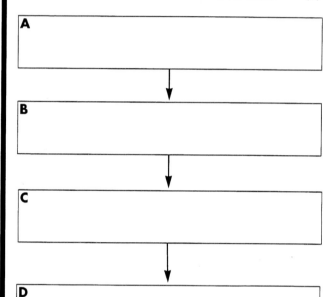

2 Complete the sentences by writing the answers in the rows **a–d**. If the answers are correct, there should be a word reading down the first column.

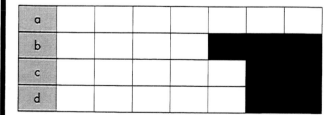

a) Electric wires are covered in _____ to insulate them.
b) The brown wire is the _____ wire.
c) Kilowatt hours are the _____ of electricity on a meter.
d) The earth wire is yellow and _____. (4)

3 Below is a diagram of a standard electrical plug.

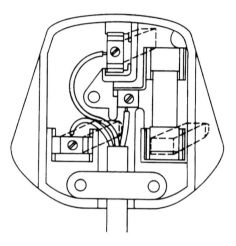

a) Colour in the three wires coming out of the flex. You will need a blue, brown, green and yellow pencil or felt-tip. (3)
b) Label the diagram with the words Earth wire, Live wire, Neutral wire, Fuse. (4)

4 The answers to the clues **a–d** will fit into the grid below. If they are correct, they will spell out an important part of a plug in the first column.

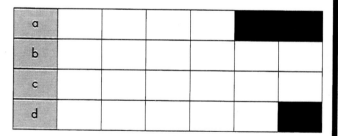

a) The electricity cable between the plug and the television.
b) To disconnect the power supply from a lamp.
c) The hole in the wall which supplies electricity.
d) The safety wire in the plug. (4)

Name: **Class:**   **Date:**

Science Companion © A Porter, M Wood, T Wood and Stanley Thornes (Publishers) Ltd 1995

9.4.1 The life of Michael Faraday

Read the following extract about Michael Faraday.

> Faraday is best known for his work on electricity. It was not until he was forty that he began a famous series of experiments. He connected a coil of wire to what we now call an ammeter. He noticed that when a magnet was pushed in and out of the coil, the needle of the ammeter moved. We use this principle to generate electricity today. When a coil of wire is moved through a magnetic field, an electric current is made. It did not take long for Faraday to realise that the reverse must be true. If an electric current was passed through a coil of wire, it could be used to make a magnet turn round. It was this discovery which led him to invent the first electric motor.

1. Explain in your own words how a bicycle dynamo produces electricity for the front and rear lights. (3)

2. What does an ammeter do? (1)

3. Name three household appliances which use an electric motor. (3)

4. Research. You will need to find a text book in which there are references to Michael Faraday.
   a) When was Faraday born and when did he die? (2)
   b) Where was he born? (1)
   c) What was his father's occupation? (1)
   d) Which scientist did Faraday work for at the Royal Institute? (1)
   e) How did Faraday get a job at the Royal Institute? (1)
   f) When did he start the annual Christmas Lectures at the Royal Institute? (1)
   g) In the field of chemistry, Faraday discovered many new materials. Name one of them. (1)

9.4.2 The life of Michael Faraday

Read the following extract about Michael Faraday.

> Faraday is best known for his work on electricity. It was not until he was forty that he began a famous series of experiments. He connected a coil of wire to what we now call an ammeter. He noticed that when a magnet was pushed in and out of the coil, the needle of the ammeter moved. We use this principle to generate electricity today. When a coil of wire is moved through a magnetic field, an electric current is made. It did not take long for Faraday to realise that the reverse must be true. If an electric current was passed through a coil of wire, it could be used to make a magnet turn round. It was this discovery which led him to invent the first electric motor.

1. A bicycle dynamo is one common use of Faraday's work.
   a) Where does the dynamo get the energy to make electricity? (1)
   b) The dynamo has a coil of wires. What is made to turn inside these wires to make electricity? (1)
   c) What is the electricity made by the dyanamo used for? (1)
   d) What does an ammeter do? (1)

2. Electric motors can be found in (a) spin dryers (b) hover lawn motors and (c) fans.
   For each of these appliances, say what the motor is used for. (3)

3. Research. You will need to find a text book in which there are references to Michael Faraday.
   a) When was Faraday born and when did he die? (2)
   b) Where was he born? (1)
   c) What was his father's occupation? (1)
   d) Which scientist did Faraday work for at the Royal Institute? (1)
   e) How did Faraday get a job at the Royal Institute? (1)
   f) When did he start the annual Christmas Lectures at the Royal Institute? (1)
   g) Name one of the new materials which Faraday discovered. (1)

9.4.3 The life of Michael Faraday

1 Here is an illustration of an electric vehicle.

a) What is the vehicle used for?

_____ (1)

b) Where is the electrical energy stored?

_____ (1)

c) Why do some people consider electric vehicles more environmentally friendly?

_____ (1)

d) What advantages does an electric vehicle have over one powered by diesel in this case?

_____ (1)

e) In early mornings in winter, electric motors have a particular advantage over diesel engines. What is it?

_____
_____ (1)

f) What two disadvantages do electric motors have over diesel engines?

_____
_____ (2)

g) Name two other vehicles which use electric motors.

_____
_____ (2)

```
L E E T S S S E L N I A T S B
F A R D A V A M T R O Y A L B
D A Y L F Y P S E R Y N A V E
M I C H A E L O N T Y C R T N
O V Y E R R U S G L K R U Y Z
T D E Z A H R N A S L T V R E
O T A V D P R T M M I I S U N
R T L V A M I I N T O K L C E
S U R Y Y U T O S K C L A R B
I N S T T H U N A M M E T E R
Z N O T G N I W E N K S M M R
```

2 The answers to the clues below can be found in the grid above. Circle each answer as you find it.

a) The main character in the title (2 words).

_____

b) His tutor in London (2 words).

_____

c) His town of birth.

_____

d) His county of birth.

_____

e) His father's job.

_____

f) The place in London where he worked (2 words)

_____

g) A material he discovered – found in petrol.

_____

h) Another material he discovered – used to make cutlery.

_____

i) An electric machine for driving vehicles.

_____

j) An iron bar which attracts other iron objects.

_____

k) Wire wound round in a circle.

_____

l) This is used to measure electric current.

_____

m) A poisonous metal used in early electricity experiments.

_____ (16)

Name: **Class:**   **Date:**

Science Companion © A Porter, M Wood, T Wood and Stanley Thornes (Publishers) Ltd 1995

9.5.1 Electromagnetism

1
a) The movement of what cause an electric current in a metal wire? (1)
b) What are a.c. and d.c. short for? (2)
c) Explain the difference between a.c. and d.c.? (2)

2
a) Why are the compasses pointing in the same direction? (1)
b) In which direction are they pointing? (1)
c) Why is there no flow of electricity in the circuit as shown in the diagram? (1)
d) What causes an electric current to flow in the circuit when the switch is turned on? (1)
e) Say what happens to the six compasses when the switch is turned on. (2)
f) What causes the change in the compasses? (1)

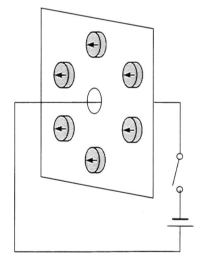

3
a) From what metal are magnets usually made? (1)
b) When do two magnets repel each other when pushed together? (2)
c) When do two magnets attract each other when pushed together? (1)

4 Explain what electromagnets do in
a) loudspeakers (2)
b) doorbells (2)

9.5.2 Electromagnetism

1
a) What does d.c. stand for? (1)
b) How does the current flow in a d.c. circuit? (1)
c) What does a.c. stand for? (1)
d) How does the current flow in an a.c. circuit? (1)

2
a) How are compasses made? (2)
b) In which direction do compasses usually point? (1)
c) What causes them to point in this direction? (1)
d) What would happen to the needle of a compass as the south pole of a bar magnet was brough closer to the compass? (2)

3
a) From what metal are most magnets usually made? (1)
b) What would happen if the pairs of magnets below were pushed together? (4)

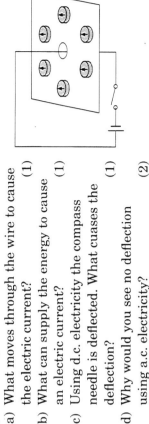

i) | N | S | | N | S |
ii) | | S | | N | |

iii) | S | N | | S | N |
iv) | | | | N | S |

4 A compass is held close to a wire. An electric current can be made to pass through the wire.
a) What moves through the wire to cause the electric current? (1)
b) What can supply the energy to cause an electric current? (1)
c) Using d.c. electricity the compass needle is deflected. What cuases the deflection? (1)
d) Why would you see no deflection using a.c. electricity? (2)

9.5.3 Electromagnets

1 Put the answers to the clues in the grid. You should reveal two words in the shaded column. (18)

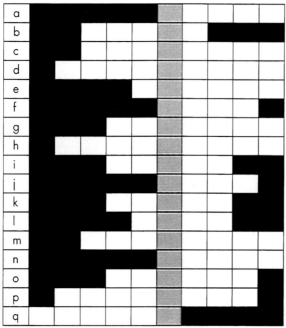

a) The opposite pole to the north.
b) The unit of frequency, symbol Hz.
c) An iron bar which attracts objects made from iron.
d) These move through metals when they conduct electricity.
e) The sort of current which a dry cell supplies.
f) Wire twisted round a cylinder.
g) Part of a radio which uses an electromagnet.
h) Something which does not conduct electricity.
i) A magnetic _____ forms between the north and south poles of a magnet.
j) This can become charged when you pull it through your hair.
k) A material which conducts electricity.
l) A metal used as the core of an electromagnet.
m) The kind of charge which is made when electrons are removed from a material.
n) These supply electricity for radios, torches, etc.
o) This turns electrical appliances on and off.
p) The opposite charge to m).
q) Coal and gas are _____ sources for generating electricity.

The words in the shaded column are
_____

2 a) Put the following labels on the diagram below.

Car battery,   Coil of wire,
Electromagnet,   Ignition switch,
Soft iron core (5)

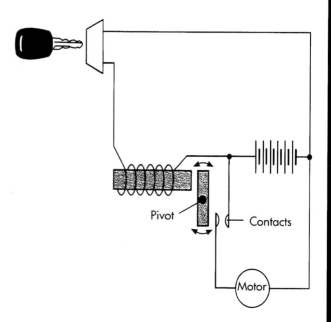

b) Fill in the missing words.

To start, the car motor needs energy supplied by

the _____. The motor cannot start until

the _____ in the circuit come together.

The motor circuit can be completed when the

_____ pushes the contacts. An

electromagnet is used to move the _____.

The electromagnet is made from a _____ of

wire wrapped around a core made of _____.

When the _____ switch is turned, it

operates the electromagnet. (7)

Name: Class: Date:

9.6.2 Numbers and words

1 a) What number is the decimal system based on? (1)
   b) Why do we use this number for counting? (1)

2 What is an abacus and how is it used? (2)

3 a) How are banking details stored on credit and cash cards? (1)
   b) Why is your money usually safe is someone steals your cash card? (2)

4 a) What is this set of lines on items of shopping called? (1)

   b) What do shops use to scan the lines? (1)
   c) Give two advantages this system gives to the supermarket. (2)
   d) What advantage does the shopper get from this system? (1)

5 a) How many symbols do we use in our alphabet? (1)
   b) Why is the Chinese alphabet much larger than ours? (1)
   c) Who used hieroglyphics as their alphabet? (1)

A B C D E F
G H I J K L

English alphabet   Chinese pictograms   Hieroglyphics

Science Companion © A Porter, M Wood, T Wood and Stanley Thornes (Publishers) Ltd 1995

9.6.1 Numbers and words

1 South American Incas used to use knots tied in string in order to keep track of numbers they needed to remember.
   a) Why might they have needed to remembers some numbers? (1)
   b) Give two disadvantages of using knots in string to remember important numbers. (2)

2 a) Who is credited with inventing the first true computer? (1)
   b) He invented it before the days of electricity, so how did it work? (1)

3 a) How are banking details stored on credit and cash cards? (1)
   b) Why is your money usually safe if someone steals your cash card? (2)

4 a) What is this set of lines on items of shopping called? (1)

   b) What do shops use to scan the lines? (1)
   c) Give two advantages this system gives to the supermarket. (2)
   d) What advantage does the shopper get from this system? (1)

5 What are hieroglyphics and who used them in their written language? (2)

Science Companion © A Porter, M Wood, T Wood and Stanley Thornes (Publishers) Ltd 1995

9.6.3 Numbers and words

1 Each clue has a missing word. Put this word in the correct space in the grid below. (17)

Across

3 A _____ might use a code made up of letters and numbers.
5 A Chinese pictogram uses a simple _____ to represent a word.
7 Scientists have been able to move an individual _____ to make the smallest writing ever.
8 Johannes _____ invented the first movable type in the 1450s.
11 _____ are machines which print letters by pressing them against a ribbon soaked in ink.
13 To use a cash card you need to know your _____ number.
14 A _____ _____ holds financial information on a magnetic strip set in its plastic.
15 In Roman numerals I represents the number _____.

Down

1 Pictures have been used in the _____ to represent words.
2 Magnetic tape is made from the metal we call _____.
4 The chips inside electronic devices are called _____.
5 Another word for a finger is a _____.
6 To store information on a plastic card you need to have _____ tape.
9 We can see words and numbers using our _____.
10 Most supermarkets use a laser to scan the _____ _____ on an item of shopping.
12 Since decimalisation 100 _____ make one pound.
13 Three letters which tell you to look on the next page.

2 Each clue is a number. Change the number into Roman numerals and put it into the grid. (7)

1 down – 103
2 across – 14
2 down – 11
3 down – 8
4 across – 3
5 across – 54
6 across – 4

3 The table below shows the symbols which the ancient Mayan civilisation of South America used for their numbers.

a) Complete the table by drawing in the missing symbols. (5)

Number	Symbol
1	•
2	••
3	
4	••••
5	
6	•̲
7	
8	
9	••••̲
10	̳
11	

b) Write the answer to the following sums as Mayan symbols.

i) ••• × •• =
ii) •̲ + •••• =
iii) ̳ − ••̲ =
iv) ••••̲ ÷ ••• =   (4)

4 By shading the right sections in the grids alongside show how a calculator displays the numbers 3 and 5. (2)

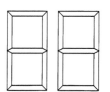

Name: Class: Date:

Science Companion © A Porter, M Wood, T Wood and Stanley Thornes (Publishers) Ltd 1995

9.7.2 Microelectronics

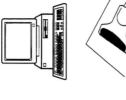 

1 a) What name is given to a miniaturised circuit where all the components are etched into a slice of silicon? (1)
   b) What does ROM stand for? (1)
   c) What does RAM stand for? (1)
   d) Which of the two memory chips (RAM or ROM) only stores information temporarily? (1)

2 a) What does a word processor do? (1)
   b) What does a scanner do? (1)
   c) What does a sampler do? (1)

3 Calculators have become very small and yet they can still do many thousands of sums in a second.
   a) What do calculators do? (1)
   b) How are numbers entered into the calculator? (1)
   c) What element are the chips inside the calculator made from? (1)
   d) What is the output of the calculator? (1)
   e) Where are liquid crystals used in the calculator? (1)

4 A microprocessor interprets information as a series of electronic pulses (a series of ons and offs).
   a) What name is given to the code of ons and offs? (1)
   b) This code is made up of just two characters. What are they? (2)
   c) How many characters are there in the decimal system? (1)

9.7.1 Microelectronics

1 Look at the table alongside which shows how to convert decimal numbers into binary numbers. In binary you add up the number of 1's, 2's, 4's, 8's, etc. that appear in the number.
e.g. in binary 111 = one 4, one 2 and one 1 = 4 + 2 + 1 = 7.

DECIMAL	BINARY				
	16's	8's	4's	2's	1's
1					1
2				1	0
3				1	1
4			1	0	0
5			1	0	1
6			1	1	0
7			1	1	1
8		1	0	0	0

   a) What decimal numbers do these binary numbers represent? (5)
      i) 101   ii) 1001
      iii) 1100   iv) 1111
      v) 10101
   b) What is the binary number for these decimal numbers? (5)
      i) 6   ii) 11
      iii) 18   iv) 20
      v) 31

2 Explain in your own words what the following phrases mean.
   a) integrated circuit (1)
   b) ROM (1)
   c) RAM (1)

3 Say how the following can be stored in a computer
   a) words (1)
   b) pictures (1)
   c) sounds (1)

4 Calculators have become very small and yet they can still do many thousands of sums in a second.
   a) How are numbers entered into the calculator? (1)
   b) What element are the chips inside the calculator made from? (1)
   c) What is the output of the calculator? (1)
   d) Where are liquid crystals used in the calculator? (1)

9.7.3 Microelectronics

1 Binary information has been used to send coded messages into outer space. You can see the principle with this message in binary.

0111001000011100101001110

There are 25 digits and 25 squares in the grid alongside. Take the first five numbers of the message. Where there is a zero leave a square in the first line of the grid blank. Where there is a one, then shade in the square on the grid. Use the next five numbers to fill in the second line. Continue until you have completed the squares to reveal another number. (5)

2 Complete the words in the passage below and then look for them in the wordsearch grid. (15)

A microprocessor contains a miniaturised

_____ circuit made from _____.

The circuit contains small electronic switches.

Each one, called a _____, lets a _____

of electricity through the circuit. The pulses form

a stream of on and offs called _____ code.

The memory chips come in two versions – RAM

(stands for random _____ _____) and

ROM (_____ _____ memory).

You can input words into a computer through the

_____ and see the _____ on a

monitor screen. You can _____ a

photograph and use a _____ to capture a

sound. The computer can output its information

to a _____ after which it is usually called a

hard _____.

P	Y	S	I	E	S	C	A	N	Y
T	R	A	N	S	I	S	T	O	R
D	A	M	T	L	L	U	C	E	O
R	N	P	E	U	I	F	A	L	M
A	I	L	G	P	C	D	H	X	E
O	B	E	R	T	O	N	L	Y	M
B	T	R	A	P	N	C	G	A	C
Y	R	E	T	N	I	R	P	A	O
E	S	S	E	C	C	A	N	K	P
K	J	E	D	I	S	P	L	A	Y

3 Unscramble the following words. The clues given for each word should help.
a) TERMCOUP _____ (1)
(a machine that handles words, pictures, numbers and sounds, electronically)
b) LOLUTACCAR _____ (1)
(This can help with difficult maths)
c) KOWTERN _____ (1)
(One way in which computers can 'talk' to each other)
d) TEAFROWS _____ (1)
(What computers need in order to work)
e) MORYME _____ (1)
(Where information is stored in a computer)

4 Underline one of the words in brackets which best matches the meaning of the word in capitals.
a) INTEGRATED –
   (small) (whole) (separate)
b) MICROPROCESSOR –
   (chip) (cell) (display)
c) TRANSISTOR –
   (code) (gate) (tape)
d) DATA –
   (robot) (pulse) (information)
e) DIGITAL –
   (numbered) (continuous) (ten) (5)

Name: Class: Date:

9.8.2 Logic gates

1 Sensors can detect changes in the environment.
  a) What changes in the environment will cause the following sensors to work? (3)
     i) push switch
     ii) temperature sensor
     iii) light sensor
  b) Which of these sensors could be used in:
     i) a night light over a front door (1)
     ii) a thermostat to turn off heating in a room (1)
     iii) a touch sensitive sensor? (1)

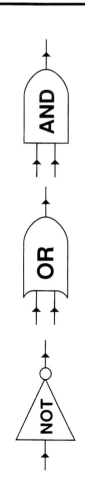

push switch

temperature sensor

light sensor

2 The diagrams below show three logic gates commonly used in microelectronics.
  a) How many inputs does the NOT gate have? (1)
  b) If the input to the NOT gate is ON, what is the output? (1)
  c) How many inputs does the OR gate have? (1)
  d) There is only one combination of inputs to the OR gate which will produce an output which is OFF. What is it? (1)
  e) How many inputs does the AND gate have? (1)
  f) There is only one combination of inputs to the AND gate which will produce an output which is ON. What is it? (1)

3 Which of the three logic gates would you need in the following circuits?
  a) A frost alarm in a greenhouse. A bell must sound when the temperature sensor drops below 0°C. (1)
  b) An alarm to sound when a bath is full of hot water. A temperature sensor can be fitted near to the top of the bath. A push switch must also be able to activate the alarm so the circuit can be tested. (1)
  c) A doorbell which only works during the day. The light sensor stops the bell sounding at night. (1)

Science Companion © A Porter, M Wood, T Wood and Stanley Thornes (Publishers) Ltd 1995

9.8.3 Logic gates

1 There are three sensors shown below. For each one say how it can be switched on. (3)

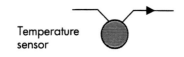
Temperature sensor

_____

Light sensor

_____

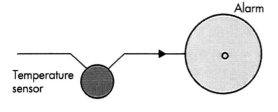
Push switch

_____

2 The four circuits below show sensors connected to outputs (alarms, bells, lamps, relay switches). Most of the circuits use a logic gate (NOT, OR, AND). For each circuit say what must be done to switch on the output. Also give one practical use for each circuit. (8)

[Temperature sensor → Alarm]

How to switch on alarm:

_____

Use for circuit:

_____

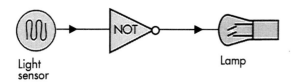

How to switch on lamp: _____

_____

Use for circuit: _____

_____

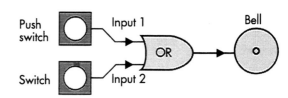

How to switch on bell: _____

_____

Use for circuit: _____

_____

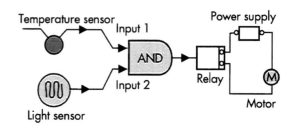

How to switch on relay: _____

_____

Use for circuit: _____

_____

3 Complete the tables below showing the outputs of NOT, OR and AND gates. (9)

a) NOT

Input 1	Output
ON	OFF
OFF	

b) OR

Input 1	Input 2	Output
ON	ON	
ON	OFF	
OFF	ON	
OFF	OFF	

c) AND

Input 1	Input 2	Output
ON	ON	
ON	OFF	
OFF	ON	
OFF	OFF	

Name: Class: Date:

10.1.1 Fuels

1
- a) What is meant by a non-renewable fuel? (1)
- b) Why are some fuels classed as fossil fuels? (1)
- c) In your own words explain how fossil fuels were formed. (3)
- d) Name three fossil fuels. (1)

2
- a) What is petroleum a mixture of? (1)
- b) What method is used to split petroleum into its separate parts? (1)
- c) On what property does this method depend? (1)

3
- a) What is the main use to which coal is put? (1)
- b) Why are imprints of long extinct plants found in some pieces of coal? (1)
- c) What pollution problem was caused when a lot of the houses in towns and cities used to burn coal? (1)
- d) How did the Clean Air Act of 1956 help solve the pollution problem? (1)
- e) What advantage does coke have over coal as a solid fuel? (1)

4
- a) What is the only product when hydrogen burns? (1)
- b) Why does this make hydrogen a clean fuel? (1)
- c) What disadvantage is there in using hydrogen as a fuel? (1)

5 Nuclear fuels are radioactive metals.
- a) Name a metal used in a nuclear power station. (1)
- b) What happens to atoms of this metal inside the reactor? (1)
- c) How is the heat from a radioactive metal turned into electricity? (1)

10.1.2 Fuels

1
- a) Did it take hundreds, thousands or millions of years for fossil fuels to form? (1)
- b) Why does this mean that they are classed as non-renewable fuels? (1)
- c) Which three of the following fuels are fossil fuels? (3)

   wood, oil, hydrogen, coal, natural gas, nuclear

- d) Which of these fossil fuels is a solid? (1)
- e) When this solid fuel burns it makes a lot of smoke. Why was it necessary to introduce the Clean Air Act in 1956? (1)
- f) The solid fuel can be turned into a smokeless fuel by heating it in the absence of air. What is this smokeless fuel called? (1)

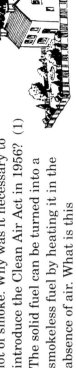

2 Gas and oil are often found together under ground.
- a) From where does the UK get most of its natural gas? (1)
- b) How is gas transported across the country? (1)
- c) What stops the oil and gas underground from rising to the surface? (2)
- d) From what were oil and gas formed? (2)
- e) Oil has to be purified to make it more useful. How is this done? (1)

3 Hydrogen could be a fuel of the future. It burns in air.
- a) What gas in air does hydrogen react with when it burns? (1)
- b) What common substance is formed when hydrogen burns in air? (1)
- c) Why is hydrogen often called a clean fuel? (1)
- d) The space shuttle Challenger used hydrogen to help get it into space. How did this cause a disaster in 1986. (1)
- e) Airships used to use hydrogen to lift them off the ground. Why do they now use helium? (1)

10.1.3 Fuels

1 Use the words below to complete the passage.

> animals, carbon, carbon dioxide, impermeable, layers, oxygen, rocks, sea

How oil and gas were formed

Dead plants and _____ which were living in the _____ drifted down to the bottom. They became covered by _____ of sand and mud. This stopped the gas _____ getting to them. As they decayed the _____ in their bodies could not turn into _____ _____ and they became oil and gas. Over millions of years the _____ changed as the Earth's crust moved. The oil and gas rose to the surface. Where they met a layer of _____ rock, the oil and gas became trapped. (8)

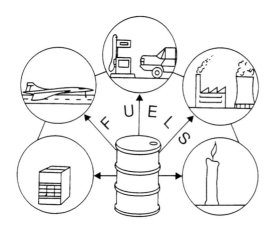

2 The diagram above shows five fuels which can be made from petroleum. Complete the table below to show what these fuels are and what they are used for. (10)

Name of fuel	What it is used for

3 Enter the missing words in the sentences below into the grid. This should reveal a hidden word in the shaded column. (6)

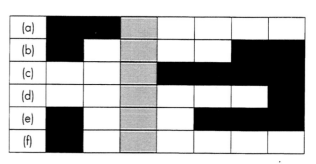

a) Substances used for heating, when they are burned, are called _____.
b) Many power stations use a solid fuel called _____ to generate electricity.
c) _____ is supplied to homes through pipelines.
d) If you want to separate a mixture of liquids, you must _____ them.
e) Petroleum is also known as crude _____.
f) Many fuels were formed from dead _____.

4 The pie chart below is divided into sections each worth 5%. The table shows approximate uses of fuels in the world. Use the figures in the table to complete the pie chart. (6)

Type of fuel	How much is used in the world
COAL	35
OIL	43
GAS	20
OTHERS	2

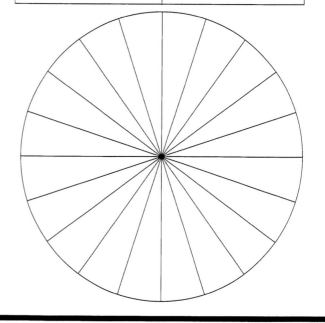

Name: Class: Date:

10.2.1 Making use of energy

1 Use the diagram below to help you explain how energy from the Sun ends up as energy to heat a room. (4)

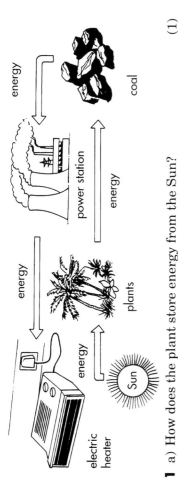

2 The battery in a car stores energy.
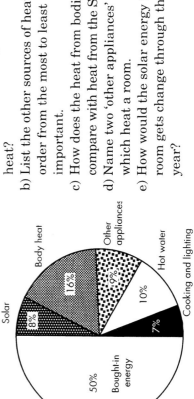
  a) Which metal is used in the battery? (1)
  b) Which acid is used in the battery? (1)
  c) What kind of energy is stored in the battery? (1)
  d) What is this stored energy used for in starting the car? (1)
  e) Where does the energy come from to recharge the battery? (1)

3 The pie chart shows the ways in which an average room gets heated.
  a) Half the heat in a room comes from fuels paid for by the occupiers. Name two types of energy which could do this. (2)
  b) List the other five energy sources in order from most to least important. (1)
  c) If there were more people in the room, what effect would this have on the bought-in energy? (1)
  d) Why would large, south-facing windows help reduce the cost of heating a room? (1)
  e) Why do lights warm up a room? (1)

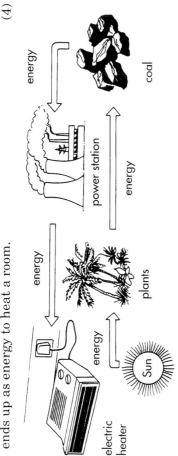

10.2.2 Making use of energy

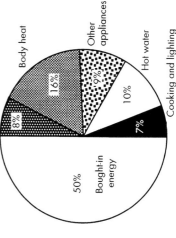

1 a) How does the plant store energy from the Sun? (1)
  b) How does the energy stored in the plant become energy in coal? (1)
  c) What form of energy is stored in coal? (kinetic, thermal, electrical or chemical) (1)
  d) How is energy in coal converted into electrical energy? (1)
  e) Into what form of energy is the electrical energy in the fire converted? (1)

2 The toy plane uses stored energy in order to work.
  a) Where is the energy stored? (1)
  b) How is the energy put into the plane? (1)
  c) What is the stored energy used for? (1)

3 The pie chart shows the ways in which an average room gets heated.
  a) How does the room get half of its heat? (1)
  b) List the other sources of heat in order from the most to least important. (1)
  c) How does the heat from bodies compare with heat from the Sun? (2)
  d) Name two 'other appliances' which heat a room. (2)
  e) How would the solar energy the room gets change through the year? (1)

10.2.3 Making use of energy

1 Draw lines to match the type of energy to its use. (5)

ENERGY	USE
ELECTRICAL	HUMAN BODY
SOLAR	SUN-BEDS
ULTRA VIOLET LIGHT	TV REMOTE CONTROL
INFRA-RED LIGHT	TELEVISION
FOOD	PLANTS

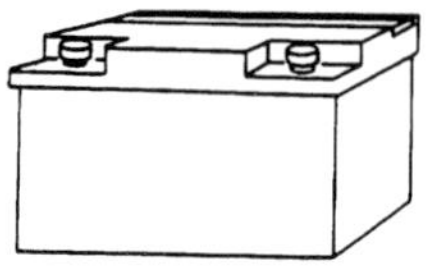

Lead–acid battery

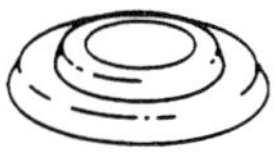

Mercury cell

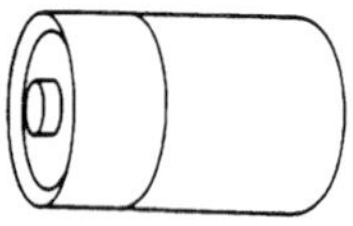

Zinc–carbon cell

Alkaline cell

Nickel–cadmium cell

2 Which of the cells in the diagram above,

a) is a single rechargeable cell?

_____ (1)

b) is used in watches?

_____ (1)

c) is used in a car?

_____ (1)

d) lasts longer than normal dry cells?

_____ (1)

3 The missing words in the nine sentences will fit into the word spiral. the last letter of one word is also the first letter of the next word. The start letter of each word is shown by the shaded square. (9)

1. Transistor _____ collect invisible signals through an aerial. (6 letters)
2. Energy from the Sun is called _____ energy. (5 letters)
3. You can _____ some batteries and use them again. (8 letters)
4. An _____ fire can heat a room without a flame. (8 letters)
5. A solid fuel made from plants is _____. (4 letters)
6. A form of energy our eyes can see is _____. (5 letters)
7. Burning gas _____ chemical energy into heat energy. (10 letters)
8. You can store energy when you _____ an elastic band. (7 letters)
9. Fuels burn to produce _____ energy. (4 letters)

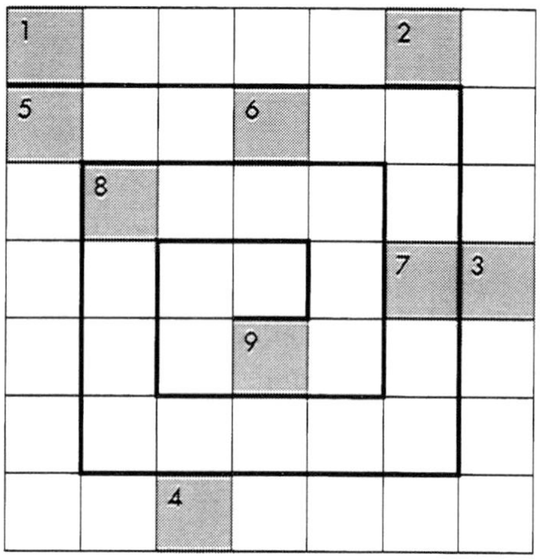

4 Use the words chemical, electrical or heat to complete the sentences below.

e.g. a paraffin heat changes chemical energy into heat energy.

a) A digital thermometer changes _____ energy into electrical energy. (1)

b) An electric fire changes _____ energy into _____ energy. (2)

c) A dry cell changes _____ energy into _____ energy. (2)

d) Recharging a cell changes _____ energy into _____ energy. (2)

Name: Class: Date:

10.3.2 On the move

1 Energy can be stored in each of these devices. Which of these devices releases the energy in the form of,
   a) heat (1)
   b) electricity (1)
   c) movement? (1)

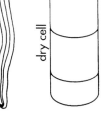
elastic band

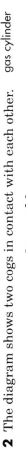
dry cell

 gas cylinder

2 The diagram shows two cogs in contact with each other.

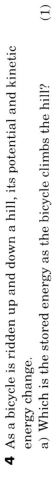

   a) In which direction is the first cog turned? (1)
   b) In which direction will cog **A** turn? (1)
   c) If the first cog turns 60 times a minute, how fast will cog **A** turn? (1)
   d) If cog **A** was larger, how would that change its speed? (1)

3 The picture shows two bicycles in different gears. The pedals on both bicycles are turned at the same speed.

   a) Where does the energy come from to turn the pedals? (2)
   b) How is the energy transferred from the pedals to the back wheel? (2)
   c) In which bicycle would the back wheel be turning the faster? (1)

4 As a bicycle is ridden up and down a hill, its potential and kinetic energy change.
   a) Which is the stored energy as the bicycle climbs the hill? (1)
   b) Which is the energy of movement? (1)
   c) When does the bicycle have the most potential energy? (1)

Science Companion © A Porter, M Wood, T Wood and Stanley Thornes (Publishers) Ltd 1995

10.3.1 On the move

1 The following items can have energy stored in them. How is this energy stored? (3)

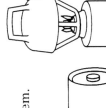

a) elastic band   b) dry cell   c) natural gas

2 a) Explain how the movement of the cog transfers energy to the cog labelled **A**. (2)
   b) In what way does the movement of these two cogs differ? (1)
   c) How does the movement of cogs **B** and **C** compare with the cog which is turning them? (2)

3 The picture shows two bicycles in different gears. The pedals on both bicycles are turned at the same speed.

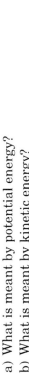

   a) Which back wheel will turn faster? (1)
   b) Which bicycle is in a high gear? (1)
   c) Which of the two bicycles is in the better gear for going uphill? (1)

4 a) What is meant by potential energy? (1)
   b) What is meant by kinetic energy? (1)
   c) As a child climbs the ladder of a slide, what happens to her potential energy? (1)
   d) When is the potential energy transferred to kinetic energy on the slide? (1)

Science Companion © A Porter, M Wood, T Wood and Stanley Thornes (Publishers) Ltd 1995

10.3.3 — On the move

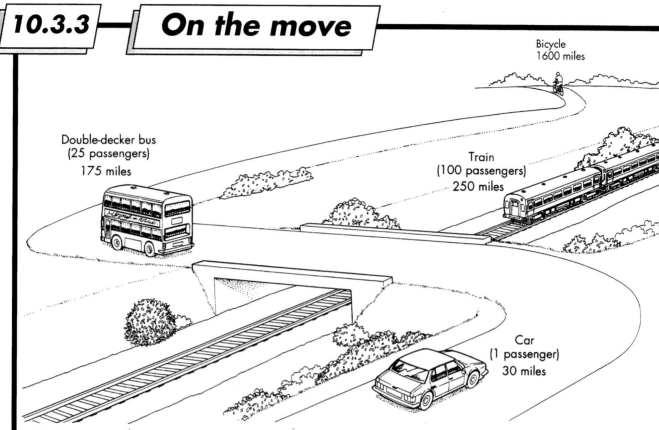

1 The drawing above shows you how far you could go for about £2 worth of fuel per person. For the car this is about one gallon (5 litres) of petrol. It also shows you the number of people you can transport this distance.

a) Complete the table below. (4)

Transport	No. of passengers	Distance for £2 of fuel per person
bicycle		
bus		
car		
train		

b) List the four forms of transport from the most efficient to the least. (4)

most efficient: _____

_____

_____

least efficient: _____

c) What fuels do the forms of transport need?

bicycle: _____ (1)

bus/train _____ (1)

car: _____ (1)

2 The prices below are for four energy sources – gas, coal, electricity and fuel oil. The cheapest is fuel oil and electricity is four times more expensive than gas.

a) Complete the table by putting the energy source next to the correct price. (4)

Energy source	Price per unit
	22.0p
	13.5p
	16.5p
	88.0p

b) For each energy source say how it can be used to transport things or people.

Coal: _____ (1)

Fuel oil: _____ (1)

Gas: _____ (1)

Electricity: _____ (1)

c) Why do you think electricity is so much more expensive than the other sources?

_____

_____ (1)

Name: **Class:**   **Date:**

Science Companion © A Porter, M Wood, T Wood and Stanley Thornes (Publishers) Ltd 1995

10.4.2 Energy conservation

1

The Manx Electricity Authority

This is responsible for the provision and maintenance of a public supply of electricity in the Isle of Man. Fossil fuels need to be imported into the island. The Authority has three diesel generating stations at Douglas, Peel and Ramsey, and also has a small hydroelectric station. The total plant capacity is 97.15 MW. The island's population of 70 000 is served by a network of approximately 850 substations.

a) What area does the Manx Electricity Authority serve? (1)
b) What fuel is used in three of the power stations? (1)
c) The fuel has to be imported. What does this mean? (1)
d) What costs are saved by having the power stations on the coast? (1)
e) What is the alternative method for producing electricity? (1)
f) How are the generators powered in this type of station? (1)
g) As the island is surrounded by water, what other power source might be used to generate electricity? (1)

2 In order to conserve supplies of fossil fuels, scientists are looking at other ways of generating electricity. Copy the table below and put the words in the second column in the correct order. (4)

Energy source	Method of making electricity
WIND	GEOTHERMAL
LAKES	SOLAR
HOT ROCKS	HYDROELECTRIC
SUN	WIND TURBINES

3 Match the methods of insulating a house to the correct part of the house. (4)

Insulation	Part of house
Cavity insulation	The roof
Double glazing	Walls
Carpet underlay	Windows
Loft insulation	The floor

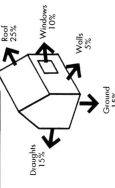

10.4.1 Energy conservation

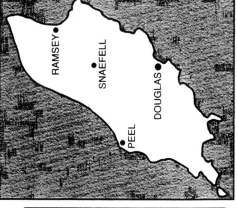

1

The Manx Electricity Authority

This is responsible for the provision and maintenance of a public supply of electricity in the Isle of Man. Fossil fuels need to be imported into the island. The Authority has three diesel generating stations at Douglas, Peel and Ramsey, and also has a small hydroelectric station. The total plant capacity is 97.15 MW. The island's population of 70 000 is served by a network of approximately 850 substations.

a) What are the problems in using fossil fuels in a small island? (1)
b) Which fuel is used to generate most electricity in the Isle of Man? (1)
c) Which fossil fuel is the source of this fuel? (1)
d) Why is it cheaper to have power stations on the coast? (1)
e) What is the alternative to using fossil fuels? (1)
f) In your own words explain how electricity is generated by this alternative method. (2)

2 In order to conserve supplies of fossil fuels, scientists are looking at other ways of generating electricity.
a) Explain how wind power could be used to generate electricity. (2)
b) What heat source does geothermal energy use? (1)
c) How can sunlight be converted directly into electricity? (1)

3 Heat can be lost from a house in many ways. Explain how you could stop heat escaping through
a) windows
b) walls
c) the roof
d) the floor (4)

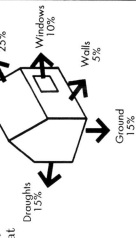

10.4.3 Energy conservation

1 The diagram below shows how double glazing keeps heat from being lost. The gap between the two panes of glass can be changed. Single glazing has a gap of zero mm.

a) On the grid below plot a line graph. Along the bottom put the gap between the panes. Up the side put the heat loss. Join the points with a smooth curve. (8)

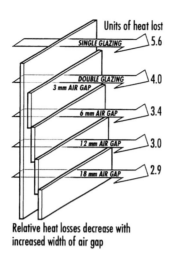

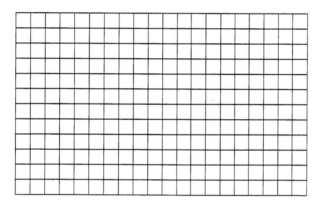

b) What is link between the air gap and the amount of heat lost?

_____

_____ (1)

c) In what way can double glazing help the environment?

_____

_____ (1)

d) Apart from heat insulation, what other advantage can double glazing offer?

_____ (1)

e) What disadvantage is there in fitting double glazing?

_____ (1)

2 The diagram below shows how energy can be extracted from the ground using hot rocks. Holes are drilled into the Earth's crust. Explosive charges are detonated at the bottom of each hole to connect them.

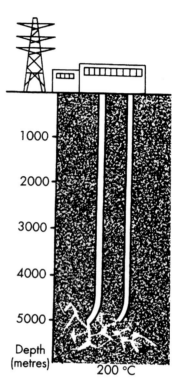

a) How deep are the holes?

_____ (1)

b) How do the explosions connect the two holes?

_____ (1)

c) Cold water is pumped down one hole. What happens to it at the bottom of the hole?

_____

_____ (2)

d) How does this enable electricity to be generated?

_____ (1)

e) What name is given to this type of energy?

_____ (1)

f) Give one advantage of using this method to generate electricity.

_____ (1)

g) Why would these power stations not last forever?

_____ (1)

Name: Class: Date:

10.5.2 Human life and activity – the energy we need

1 Name two ways that food is used by the body. (2)

2 Place the foods below into the three food groups: carbohydrates, proteins and fats. (6)

cheese   cream   flour   meat   potatoes   oil

3 Finish the following sentences:
  a) Foods which we need to grow are in the group called _____ (2)
  b) Foods which we need for energy are in the group called _____ (2)
  c) Foods which provide energy and warmth are in the group called _____ (2)

4 Look at the food label below and answer the questions.

NUTRITION INFORMATION		
	TYPICAL COMPOSITION	
	per biscuit	per 100 g
ENERGY	142 kJ	1984 kJ
	34 kcal	470 kcal
PROTEIN	0.5 g	6.7 g
CARBOHYDRATE	5.3 g	74.2 g
FAT	1.1 g	15.7 g

  a) What is the food? (1)
  b) How much energy is there in 100 g of the food? (1)
  c) Is the food mostly fat, protein or carbohydrate? (1)
  d) What do the letters kcal and kJ stand for? (2)
  e) Do a survey of food labels and make up a booklet showing energy values of foods. (5)

5 Look at the snacks below.

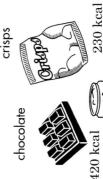

chocolate   crisps   cola
420 kcal   230 kcal 110 kcal

  a) Which snack contains the most energy? (1)
  b) Which snack contains the most sugar? (1)
  c) Which snack contains the most fat? (1)
  d) Why should we eat fewer snacks like these? (2)
  e) Suggest a healthier snack food. (1)

10.5.3 Human life and activity – the energy we need

1 Read the table which shows the nutritional value of some foods, then answer the questions which follow.

FOOD	% PROTEIN	% FAT	% CARBOHYDRATE	ENERGY per 100 g
Brown bread	8.0	2.0	52.0	1042
Weetabix	11.0	2.0	77.0	1470
Rice	2.6	1.1	31.3	586
Peas	5.0	0	8.0	210
Yogurt	4.7	1.0	4.6	193
Chicken	30.0	7.0	0	770
Fish fingers	21.0	5.0	3.0	800
Baked beans	6.0	0	17.0	390
Chips	6.0	38.0	49.0	2940
Bananas	1.1	0.3	19.2	337
Cola	0	0	7.0	110
Sausages	12.0	25.0	13.0	1530

a) Write the percentage of fat, carbohydrate and protein in each food below.

Bread
- Protein _____ %
- Fat _____ %
- Carbohydrate _____ %

Fish fingers
- Protein _____ %
- Fat _____ %
- Carbohydrate _____ %

Yogurt
- Protein _____ %
- Fat _____ %
- Carbohydrate _____ %

Cola
- Protein _____ %
- Fat _____ %
- Carbohydrate _____ %

Chips
- Protein _____ %
- Fat _____ %
- Carbohydrate _____ %

(5)

b) Work out the energy values, in kJ of each meal below.

200 g sausages _____ kJ

100 g chips _____ kJ

100 g baked beans _____ kJ

200 g chicken _____ kJ

100 g rice _____ kJ

100 g peas _____ kJ

Meal 1. _____ kJ   Meal 2. _____ kJ

(8)

c) Write a sentence about the nutritional value of each meal.

Meal 1. _____
_____
_____
_____
_____
_____ (1)

Meal 2. _____
_____
_____
_____
_____
_____ (1)

d) Which meal provides the most energy?
_____ (1)

e) Which meal is healthier and why?
_____
_____
_____ (1)

Name: **Class:**   **Date:**

11.1.1 — What is a force?

1 Forces are measured in units called newtons. Alongside is a 100 g mass which gives a reading of 1 newton when hung on a newtonmeter.

For each of the masses given below, say how many newtons would be measured if it were hung on a newtonmeter in your school laboratory.

a) 200 g
b) 1200 g
c) 10 g
d) 6 g
e) 0.1 g   (5)

2 a) Name the force which pulls downwards on all masses which are hung on a newtonmeter in your school laboratory. (1)
b) What name is given to the downwards pull of this force on a mass? (1)
c) In outerspace, people have mass, but are without what? (1)

3 A papergirl puts her large bag of papers down on the ground whilst pushing the Saturday paper sections through a letterbox.
a) What force acts downwards on the bag of papers as it sits on the ground? (1)
b) What force acts upwards on the bag of papers as it sits on the ground? (1)
c) After a minute, the bag is still in exactly the same place. What can you say about the forces acting on the bag of papers? (1)
d) One week later she returns. It has rained heavily all week. The bag sinks into the mud as she puts the sections of the paper through the letterbox. What can you say about the forces on the bag of papers this time? (2)

4 Imagine you are sitting in a bus at 60 mph as it cruises down the quiet motorway on a Sunday morning.
a) Why do you feel no forces acting on you? (1)
b) A dog runs across the motorway, and the bus brakes to avoid the dog. Which way will you move, and why? (2)
c) The bus accelerates back to 60 mph. What can you feel? (2)

5 When a tank fires a shell, the tank is seen to rock after the shell has been fired. Why is this? (2)

11.1.2 — What is a force?

1 Forces are measured in units called newtons. Alongside a a 100 g mass hangs on newtonmeter and this gives a reading of 1 newton (1N).

For each of the masses given below, say how many newtons would be measured if it were hung on a newtonmeter in your school laboratory.

a) 200 g
b) 700 g
c) 500 g
d) 800 g
e) 1000 g   (5)

Now you are given the newtonmeter reading. Say what mass you would hang on to get this reading in your school laboratory.

f) 4 N
g) 8 N
h) 0.5 N
i) 0.1 N
j) 1.1 N   (5)

2 A papergirl puts her large bag of papers down on the ground whilst pushing the Saturday paper sections through a letterbox.
a) What force acts downwards on the bag of papers as it sits on the ground? (1)
b) What force acts upwards on the bag of papers as it sits on the ground? (1)
c) After a minute, the bag is still in exactly the same place. What can you say about the forces acting on the bag of papers? (1)
d) Next Saturday she returns and does exactly the same thing. The ground is now very soft and muddy as it has rained all week. The bag of papers sinks into the mud as she puts the paper sections through the letterbox. Why does the bag of papers sink into the mud? (3) (Think of your answers to parts a), b) and c).

3 You are sitting in a bus at 60 mph as it cruises down the quiet motorway on a Sunday morning. You feel comfortable and it seems as if no forces are acting on you, but they are. The forces are constant. You only notice them when you change speed.
a) A dog runs across the motorway, and the bus brakes to avoid the dog. Which way will you move, and why? (2)
b) The bus accelerates back to 60 mph. What can you feel? (2)

11.1.3 — What is a force?

1 Jane's group collected a set of results after adding known masses to a hook on the bottom of a spring. The increase in the length of a spring from its unstretched length is called the extension.

a) Use the table of results alongside to plot a graph on the graph paper below which shows the extension of a spring as different masses are added.

Extension (cm)	Mass added (g)
0	0
5	10
10	20
15	30
20	40
25	50
30	60
35	70
40	80

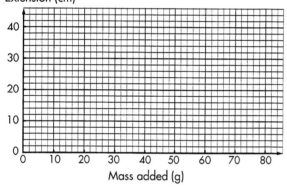

(4)

b) What will be the extension for the following masses?
　i) 0 g　_____ cm³　(1)
　ii) 25 g　_____ cm³　(1)
　iii) 45 g　_____ cm³　(1)
　iv) 75 g　_____ cm³　(1)

2 Use the six clues (a) to (f) to find the horizontal words in the grid below and then find the word in column 5 with help from clue (g)

	1	2	3	4	5	6	7	8
a)	■							
b)		■						■
c)								■
d)								■
e)	■							
f)								

CLUES
a) Forces are measured in these units.
b) The limit of this on the motorway is 70 mph.
c) Masses are pulled downwards by this force.
d) An object can be moving and still have balanced forces acting on it. You only feel a force when the balance of the forces _____.
e) The opposite of pulling.
f) Forces act in pairs. This is the opposite force to action.　(6)

Now the clue for the word in column 5.
g) The force with which gravity pulls down on an object is _____　(1)

3 Forces are measured in units called newtons. If you hang a 100 g mass on a newtonmeter, then the meter will read 1 newton (1 N). A mass of 1 kilogram (1000 g) will give a reading of 10 newtons (10 N).

Use the above to help complete the following table.

Mass in grams	Newtonmeter reading in newtons
300	
	6
2000	
	40
	1.1
12	
8	
	0.56
	3.7
274	

(10)

Name:　　**Class:**　　**Date:**

11.2.1 Going down and slowing down

1 The human spine contains 33 vertebra which are separated by cartilage discs.
 a) You have been lying down all night. Why are you shorter after standing up for an hour? (2)
 b) Why do astronauts on long spaceflights become taller by as much as 5 cm? (2)

2 a) What is mass? (1)
 b) In what units is mass measured? (1)
 c) On Earth, gravity pulls down on mass with a force. What is the name of this pulling force? (1)
 d) What are the correct units for measuring weight? (1)
 e) In space, astronauts are described as 'weightless'. Why are they 'weightless' rather than 'massless'? (2)
 f) Why is gravity on the Moon less than that on the Earth? (3)

3 Explain how a parachute works in slowing down an object falling through the air. (2)

4 Study the diagrams alongside.
A student set up this equipment to measure resistance to movement through a thick liquid.

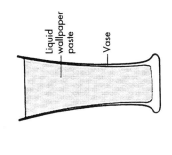

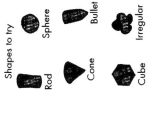

Of the six shapes tried
 a) Which two do you think moved the fastest, and why? (3)
 b) Which one would probably move the slowest, and why? (2)

11.2.2 Going down and slowing down

1 You have set up the equipment as shown alongside. You are trying to show that liquids give resistance to things travelling through them.
 a) Why does the vase need to be clear glass or plastic rather than pot? (1)
 b) Why should all the plasticine sections be the same size before you mould them into their special shapes? (2)
 c) Why use liquid wallpaper paste rather than plain water? (2)
 d) Why should the objects under test be released gently at the liquid's surface rather than be dropped in? (2)
 e) Which of the six shapes would you expect to be the two fastest, and why? (3)
 f) Which one would probably be the slowest, and why? (2)

2 a) What is mass? (1)
 b) In what units is mass measured? (1)
 c) On Earth, gravity pulls down on mass with a force. What is the name of this pulling force? (1)
 d) What are the correct units for measuring weight? (1)
 e) In space, astronauts are described as 'weightless'. Why are they 'weightless' rather than 'massless'? (2)

3 Jet fighter aircraft are made of a metal called titanium which has a very high melting point. These planes fly very quickly, about four times the speed of a Jumbo Jet. Why do you think they are not made from Aluminium like a Jumbo Jet? (2)

11.2.3 Going down and slowing down

1 A mass of 10 kg weighs 100 N on Earth. The table alongside shows the weight in newtons of 10 kg on each of the planets in the solar system.

Planet	Weight of 10 kg (N)
Mercury	40
Venus	70
Earth	100
Mars	40
Jupiter	230
Saturn	90
Uranus	80
Neptune	110
Pluto	40

a) Use the graph paper below to draw a bar chart to illustrate the figures in the table above.

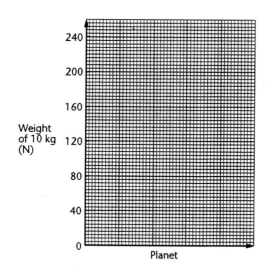

(9)

b) On which planets would an astronaut weigh more than on Earth?
_____ (2)

c) On which planets would a 1 kg mass weight 4 N?
_____ (3)

d) On which two planets is the gravity almost the same as on Earth?
_____ (2)

e) Explain your answer to part (d)
_____
_____
_____ (3)

f) On which planet would a mass of 1 kg have a weight of 7 N?
_____ (1)

2 Study the newtonmeters **A–J**. Using the table to Question 1, give the names of **one** planet on which the newtonmeter is being used.

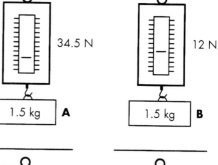

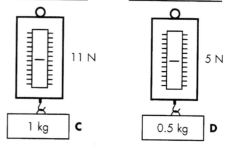

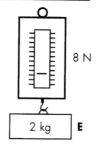 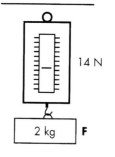

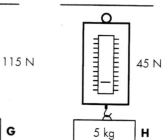

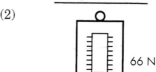 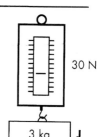

(10)

Name: **Class:** **Date:**

Science Companion © A Porter, M Wood, T Wood and Stanley Thornes (Publishers) Ltd 1995

11.3.1 Starting and stopping

1 The diagram below shows the braking and thinking distances, in metres and in car lengths, in dry conditions for the average family car and driver.

At 30 m.p.h. (48 km/h)
Thinking distance 9.2 m | Braking distance 13.7 m | Overall stopping distance 22.9 m

At 50 m.p.h. (80 km/h)
Thinking distance 15.2 m | Braking distance 38.1 m | Overall stopping distance 53.3 m

At 70 m.p.h. (113 km/h)
Thinking distance 21.3 m | Braking distance 74.7 m | Overall stopping distance 96.0 m

a) What is the thinking distance at 30 mph? (1)
b) What is the thinking distance at 50 mph? (1)
c) How much further is the thinking distance at 50 mph than at 30 mph? (2)
d) How much further is the braking distance at 50 mph than at 30 mph? (2)
e) How much further is the overall stopping distance at 50 mph than 30 mph? (2)
f) If braking distances are doubled in wet weather, what will be the wet weather braking distances for each of the three speeds? (3)
g) Now find the wet weather overall stopping distances for each of the three speeds. (3)

2 When bicycle brakes are used forcefully the bicycle stops quickly but the brakes get hot. Why is heat produced? (2)

3 Alongside is a diagram of a human hip joint.
a) What is the lubricant for this joint? (1)
b) From what material is the slippery surface of the bone in the joint made? (1)
c) What happens to joints in osteoarthritis? (2)

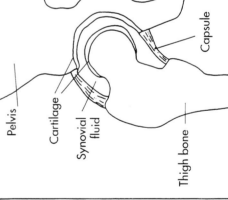

Pelvis
Cartilage
Synovial fluid
Thigh bone
Capsule

Science Companion © A Porter, M Wood, T Wood and Stanley Thornes (Publishers) Ltd 1995

11.3.2 Starting and stopping

1 The diagram below shows the braking and thinking distances, in metres and in car lengths, in dry conditions for the average family car and driver.

At 30 m.p.h. (48 km/h)
Thinking distance 9.2 m | Braking distance 13.7 m | Overall stopping distance 22.9 m

At 50 m.p.h. (80 km/h)
Thinking distance 15.2 m | Braking distance 38.1 m | Overall stopping distance 53.3 m

At 70 m.p.h. (113 km/h)
Thinking distance 21.3 m | Braking distance 74.7 m | Overall stopping distance 96.0 m

a) Give the speed in km/h for (i) 30 mph, (ii) 50 mph, (iii) 70 mph. (3)
b) What are the thinking and braking distances at 30 mph? (2)
c) In wet conditions, braking distances are doubled. What would be the wet weather braking distance at 30 mph? (2)
d) Using your answers to (b) and (c), what is the overall stopping distance at 30 mph when the road is wet? (2)
e) What are the thinking, and braking distances at 50 mph? (2)

2 What is the name of the lubricant used to reduce friction in car engines? (2)

3 Alongside is a diagram of a human hip joint. This is a synovial joint which has special features to reduce friction.
a) What is the lubricant for this joint? (2)
b) From what is the slippery surface of the bone in the joint made? (2)

4 One of the many spinoffs from space technology is the new material **polytetrafluoroethene**. It is very slippery.
a) For what purpose is it used as a coating on frying pans? (2)
b) Look at the long name of this material. Which one of the following is its common short 'name'? Is it POLYTET, PTFE, POLY, or PFE? (1)

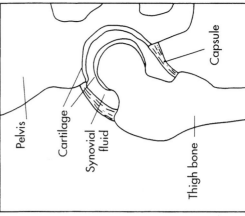

Pelvis
Cartilage
Synovial fluid
Thigh bone
Capsule

Science Companion © A Porter, M Wood, T Wood and Stanley Thornes (Publishers) Ltd 1995

11.3.3 Starting and stopping

1 Use the words below to complete the following sentences about friction. You may use a word more than once.

**doubled   further   lubricant
microscope   oil   rain   rough
surface**

No oil   Oil

Friction can be reduced by using a _____. One such substance is _____. If you study the surfaces through a _____ you can see they are _____. Friction is reduced because _____ smooths out the sliding of one _____ over the other. _____ reduces the friction on the roads. The distance it takes a car to stop is more than _____ in wet weather. Cars need to keep _____ from one another in the wet to prevent accidents. (9)

2 Use clues (a) to (n) to find the words to complete the spaces below. As an additional clue, the word downwards in column 4 is the name of the painful disease of the joints suffered by many elderly people.

CLUES
a) A liquid used as a lubricant in car engines.
b) The top layer of a road.
c) The part of the leg just below the hip.
d) Short for polytetrafluoroethene the non-stick coating on pans.
e) There are many of these in our skeleton. Nickname of Dr. McCoy in *Star Trek*.
f) On a bicycle and car these help it to stop.
g) Tyres need a good _____ on the road to help a vehicle to stop.
h) An alloy of iron used to make car bodies.
i) Friction makes this. A bunsen burner supplies this.
j) A force which stops one surface sliding over another.
k) A car can do this when it loses its grip on a slippery road surface.
l) The covering on some joints which provides a slippery surface for movement.
m) A natural joint may be replaced by an _____ one.
n) The hip is this type of joint.

The word in column 4 is
_____ (15)

3 There is a coded message below. Use the squares and their letters from question 2 above to place a letter in a blank space. There are no M and W letters in question 2 so these have been done for you.

.											M			
5a	1l	5i		6c	3h		8b		2k	—	7n	4e	4c	8c

											.
4b	5b	3l	5m	4f	1l	6h		9m	10j	5k		7e	7n

										W
5m	6b	5g	6j	4h	5j	9j	5e		8j	3h		11n	7n	—

M												
—	1m	5f	4k	6n	8l		4n	4l	4e	6g	1d	5l	6n	3g

5k	4m	7b	3j	6m	9b	5b	10m	7j

(6)

Name: Class: Date:

Science Companion © A Porter, M Wood, T Wood and Stanley Thornes (Publishers) Ltd 1995

11.4.2 Sinking and floating

1 1 g of water has a volume of 1 cm³ so its density is 1 g/cm³. Look at these two metal containers of identical volume.

$$\text{Density} = \frac{\text{mass of air + metal}}{\text{volume}} \text{ g/cm}^3 \qquad \text{Density} = \frac{\text{mass of water + metal}}{\text{volume}} \text{ g/cm}^3$$

a) Air is less dense than water. In which container will the density be lower? (1)
b) Which one of the words below completes the sentence which follows?

greater   less   more   similar

'The volume of air has _____ mass than the same volume of water. (1)
c) What is in the ballast tanks of a submarine when it is floating? (1)
d) What is in the ballast tanks of a submarine when it is sinking? (1)
e) Suggest how a submarine might keep to a fixed depth in the water. (2)

2 Below are two diagrams showing a modern airship.

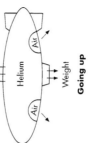

a) What is the purpose of the helium gas? (1)
b) In older airships, hydrogen was used instead of helium. What was the big disadvantage of hydrogen? (1)
c) Why does the airship go up when air is forced out of the ballonets? (2)
d) Why does the airship go down when air is sucked into the ballonets? (2)
e) Why is the envelope of an airship made from fabric rather than metal? (2)

3 Calculate the density of each of the following elements using the formula alongside.

$$\text{Density} = \frac{\text{mass}}{\text{volume}}$$

ELEMENT	aluminium	calcium	gold	lead	silver	tin
MASS (g)	27	15	38.6	22.6	10.5	73
VOLUME (cm³)	10	10	20	20	1	10

(6)

11.4.1 Sinking and floating

1 Look at these two metal containers of identical volume.

$$\text{Density} = \frac{\text{mass of air + metal}}{\text{volume}} \text{ g/cm}^3 \qquad \text{Density} = \frac{\text{mass of water + metal}}{\text{volume}} \text{ g/cm}^3$$

The volume and mass of metal are the same in both cases.
a) Which container has the lower density? (1)
b) How does this explain why a submarine floats when its ballast tanks are full of air, and sinks when they are full of water? (4)
c) What does it mean when we say that the density of pure water is 1 g/cm³? (1)

2 Look at the object alongside which is floating in a beaker of water.

a) What name is given to the water pushed away as the object floats? (1)
b) For a floating object, what is special about this weight of water which it pushes away? (1)

3 When airships became popular in the 1930s, their envelopes were filled with hydrogen. Today airships have their envelopes filled with helium.
a) What is the purpose of the hydrogen or helium gas? (1)
b) Why do you think hydrogen was used in the 1930s rather than helium? (1)
c) What was the problem with using hydrogen? (1)
d) What advantage has helium over hydrogen? (1)
e) What disadvantage does helium have compared with hydrogen? (1)
f) What name is given to the tanks of air inside the envelope of an airship? (1)
g) What is the purpose of these tanks of air? (4)
h) Why is the envelope of an airship made from fabric rather than metal? (2)

11.4.3 Sinking and floating

1 Hot-air ballooning is very popular nowadays. Study the following information and then complete the sentences which follow.

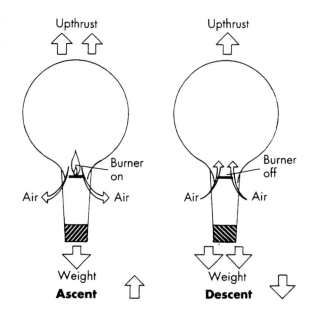

FACTS
- The burner uses propane gas as a fuel.
- The burner heats the air in the envelope to about 100°C.
- Hot air takes up more room than cold air.
- When the burner cuts out, the air in the envelope starts to cool.
- Cold air takes up less room than hot air.
- Cold air is more dense than hot air.

Use the words from the list below to complete the following sentences about hot-air balloons. The words may be used once, or more than once.

decreases   falls   greater
increases   into   less   out

As the air in the balloon is heated its volume _____. This _____ the total weight of the balloon, because some of the air is forced _____ of the balloon. Upthrust is now _____ than weight and so the balloon rises.

When the burner is switched off the air in the envelope cools and takes up _____ volume. Air is sucked _____ the balloon which _____ the total weight and the balloon _____. (8)

2 Density = $\dfrac{\text{mass}}{\text{volume}}$

a) Calculate the density of the following elements. (8)

ELEMENT	MASS (g)	VOLUME (cm³)	DENSITY (g/cm³)
diamond	35	10	
copper	890	100	
magnesium	34	20	
mercury	68	5	
potassium	86	100	
sulphur	63	30	
titanium	90	20	
iron	63.2	8	

b) Which of the elements in the above table is
 i) less dense than water.
 _____ (1)
 ii) less dense than sulphur
 _____
 and
 _____ (2)
 iii) the most dense?
 _____ (1)

3 Below is the Plimsoll Line which is drawn on the side of a ship. The lines show the loading limits for different seas and seasons.

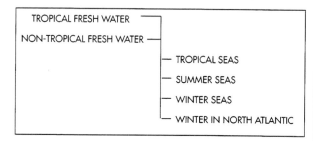

a) What happens to the ship as it is loaded with more goods at the dock-side?
 _____ (1)

b) Why are there different lines for the different types of water?
 _____
 _____ (1)

c) In which water will the ship be furthest out of the water for a fixed loading?
 _____ (1)

Name: Class: Date:

Science Companion © A Porter, M Wood, T Wood and Stanley Thornes (Publishers) Ltd 1995

11.5.2 Some forces in action

1 A car has broken down in your road. It will not start and you are asked to give a push.

a) Which of the following will you find the easiest
   i) push the car on your own
   ii) push with the help of 1 friend
   iii) push with help from 2 friends? (1)

b) Explain your answer to part (a). (2)

c) When a car is moving and the driver wants it to stop, what does the driver do? (1)

d) In general, small cars have small engines. Why do car engines get larger as cars get bigger? (2)

e) Using your answer to part (d), suggest one way of producing a really fast small car. (1)

2 Alongside is one possible way of making a simple model hovercraft.

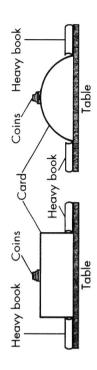

a) What do you need to do to the balloon before fixing it to the model? (1)

b) What will make the hovercraft float? (1)

c) A small hole is drilled in one side of the tube joining the balloon to the wood. What will happen? (2)

3 Below are two ideas for model bridges.

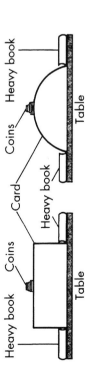

a) Which bridge will sag the most under the weight of the coins? (1)

b) If we are comparing them, why should we use the same thickness of card for both? (1)

c) Why do we need heavy books? (1)

d) Corrugated cardboard ∿∿∿∿∿ would be suitable for the flat bridge, but not the curved model. Why? (1)

11.5.1 Some forces in action

1 Your neighbour's car will not start and you are asked to go to give the car a push start.

a) Why would you NOT try to push it uphill? (1)

b) Why would you try to push it downhill? (2)

c) Why is it easier for three people to push a car than just one person? (1)

d) What is acceleration? (1)

e) When driving a car, what do you use to slow it down? (1)

2 The human flea is an amazing creature. It can jump 200 mm in height. This is 130 times the flea's own height. What is the height of the human flea (to two decimal places)? (1)

3 Alongside is a drawing which shows one possible way of making a simple hovercraft.

a) Explain how this hovercraft floats. (2)

b) Suggest a small change which would enable the hovercraft to move in one direction by jet power. (2)

4 Below are some ideas for model bridges.

a) If the same thickness of cardboard is used throughout, which bridge will sag the most as the coins are added, and why? (2)

b) Why do the books need to be heavy? (1)

c) Corrugated cardboard ∿∿∿∿∿ would be suitable for the flat bridge, but not the curved model. Why? (1)

11.5.3 Some forces in action

1 Below are 8 jumbled words.

ettrsch   aaeeccointrl   auqssh
ebdn   urnt   aainnblgc
alerotndceeei   eiidtcron

First, unjumble the letters and then place the word in the sentence best suited to it. All the sentences concern the effects forces have on the objects which they push or pull.

a) Forces cause objects to move and give them _____.

b) Forces stop objects which are moving and cause _____.

c) Forces distort or _____ objects.

d) Forces twist or _____ objects.

e) Forces _____ or compress objects.

f) Forces change the _____ in which an object moves.

g) Forces keep objects still by _____ other forces acting on the same object.

h) Forces _____ objects.   (8)

2

a) Where does the force come from to turn the spanner?

_____ (1)

b) What is the object on which the force acts?

_____
_____ (1)

c) What is the name given to this type of force?

_____ (1)

d) Does a spanner with a longer handle make the job easier or harder?

_____ (1)

e) Explain your answer to part (e)

_____
_____
_____
_____
_____
_____
_____ (4)

3 Below is a table of information about some flying insects.

Insect	Complete wingbeats per second	Flying speed (km/h)
Bumblebee	130	11
Cockchafer beetle	50	11
Dragonfly	40	55
Hawk moth	70	35
Honeybee	225	9
Housefly	200	7
Mosquito	600	1.5
Midge	1000	1.5

a) On the graph paper below, draw a bar chart showing the flying speed of each insect.   (4)

b) If you study the figures for the dragonfly and compare them with those for the midge, what conclusion can you draw?

_____
_____ (2)

Name: Class: Date:

11.6.2 Turning forces

1 The turning effect of a force is called the moment of the force. The point around which the force turns is called the pivot or fulcrum.

Size of force	×	perpendicular distance of force from fulcrum	=	moment of force around the fulcrum
Units: newton (N)		metre (m)		newton metre (N m)

The diagram alongside shows a door viewed from above. Two students take it in turns to close the door.

a) Which student, **A** or **B**, has to push the most? (1)

b) Student **A** pushes with a force of 10 N. What is the moment of this force about the hinge? (2)

c) The teacher asks student **B** to close the door with exactly the same moment as student **A**. With what force does student **B** need to push? (2)

2 When a see-saw is level it is said to be balanced. The person on the left has an anticlockwise moment. The one on the right has a clockwise moment. To balance they move until their moments are equal.

a) Student **A** weighs 200 N and sits at a distance of 4 metres from the fulcrum. What is student **A**'s moment about the fulcrum? (2)

b) Student **B** weighs 400 N and sits at a distance of 2 metres from the fulcrum. What is student **B**'s moment about the fulcrum? (2)

c) Will the see-saw be level? (1)

d) Explain your answer to part (c). (2)

e) Which student applies a clockwise moment about the fulcrum? (1)

f) Student **B** moves to 3 metres away from the fulcrum. What will happen? (2)

Anticlockwise moment = Clockwise moment

$A \times a = B \times b$

11.6.1 Turning forces

1 When you push a door closed, why does it take more force to push it nearer the hinge than at the edge of the door a long way from the hinge? (4)

2 A piece of wood is supported on a fulcrum at its centre **F**. Different weights are added at **A** and **B** at distances c and d centimetres from the fulcrum so that in all cases the wood is balanced.

a) A weight of 4 N is added at point **A**, 10 cm from **F**. What is the moment of this force around **F**? (1)

b) Is this movement a clockwise or anticlockwise moment? Explain your answer. (2)

c) If the weight is increased to 5 N, at what distance from **F** must it now be placed to produce the same moment? (1)

d) At what distance d cm must a weight of 2.5 N be placed at **B** to balance the wood? (1)

e) The following table of results was obtained by a student using the above apparatus. What are the values of **W**, **X**, **Y**, and **Z**? (4)

A(N)	c(cm)	B(N)	d(cm)
12	**W**	6	8
10	4	**X**	5
12	10	40	**Y**
Z	20	5	12

3 The piece of wood from question 2 is once again supported at its centre **F**. Now two separate clockwise moments are used to balance one anticlockwise moment. Below is a table of result, find the values of **S** and **T** needed for the piece of wood to balance. (2)

A(N)	c(cm)	B(N)	d(cm)	K(N)	l(cm)
12	**S**	5	4	10	10
10	8	8	5	**T**	20

11.6.3 Turning forces

1 Moments can be taken about any point. Look at the girl standing on a plank supported by two trestles. The girl's weight is 240 N.

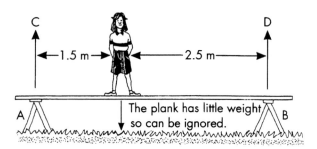

a) If you take moments about point **B**, what is the value of the upward force of the trestle **B**?

_____ (1)

b) Explain your answer to part (a).

_____
_____
_____
_____
_____
_____
_____ (3)

c) Calculate force **C** taking clockwise and anti-clockwise moments about **B**.

_____
_____ (2)

d) Now find the upward force **D** at trestle **B**.

_____
_____ (2)

2 There are three types of levers. For each type there are three components (parts), load (**L**), effort (**E**) and fulcrum (**F**). For each type of lever place the three components in their correct order.

First class lever _____
Second class lever _____
Third class lever _____ (3)

3 On the diagrams below, draw labelled arrows to mark the load, effort and fulcrum on each example, and then name the class of lever to which it belongs.

a) removing a tin lid with a screwdriver

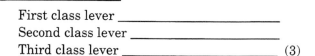

(3)

_____ class (1)

b) removing a nail

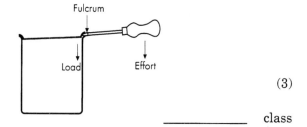

(3)

_____ class (1)

c) lifting a leaf on a table

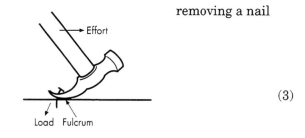

(3)

_____ class (1)

d) using tweezers to lift a small piece of paper

(3)

_____ class (1)

Name: Class: Date:

Science Companion © A Porter, M Wood, T Wood and Stanley Thornes (Publishers) Ltd 1995

11.7.1 Work and power

1 One joule of work is done when a force of one newton moves through one metre. Give the mathematical expression which scientists use to calculate work done. (1)

2 In scientific terms, why do you do no work when holding a heavy shopping bag at rest? (1)

3 In science, what special names are given to
   a) energy of movement (1)
   b) energy of height? (1)

4 On returning home from a shopping trip the man has to carry his shopping up a vertical height of 1.5 m and through a horizontal distance of 2 m. If he uses 100 N of force

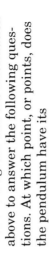

Total weight of person and shopping

   a) what work has he done? (1)
   b) Explain your answer to part a), in particular which figures you have used. (2)

5 Power is the rate of doing work. Give the mathematical expression which scientists use to calculate power. (1)

6 If you lift a heavy box from the ground on to a shelf 1 metre high, do you use more power if you do the job quickly, or slowly? Explain your answer. (3)

7 An ordinary domestic light bulb may have a power of 100 watts. A new energy-saving bulb which gives out as much light, has a power of only 20 watts.
   a) How much energy is saved every second by the new type of bulb? (1)
   b) Why are people better off buying the energy-saving bulbs despite the fact that they cost far more than the ordinary domestic light bulbs? (3)

11.7.2 Work and power

Kinetic energy is the energy of movement. Gravitational potential energy is the energy of height.

1 The diagram alongside shows a moving pendulum at three different times during its swing. Point **A** is as far left as it swings. Point **C** is as far right as it swings. Point **B** is its midway position.

Use the diagram and the information above to answer the following questions. At which point, or points, does the pendulum have its
   a) most kinetic energy   c) most gravitational potential energy
   b) least kinetic energy   d) least gravitational potential energy? (6)

2
$$\text{word done} \ = \ \text{force} \ \times \ \text{distance moved in the direction of the force}$$
units: joule (J)   newton (N)   metre (m)

Use the equation to help find the work done in each of the following.
   a) A cat of mass 2 kg (1 kg weights 10 N) jumps 2 m on to the branch of a tree. (1)
   b) You use a force of 20 N to lift a parcel from the floor to place it on a shelf 1.5 m above the floor. (1)
   c) A bird uses a force of 1 N to lift some bread 15 m to the chimney of a house (1)

3 Use the equation in question 2 to find the force used when
   a) 10 joules of work are used to move a brick a vertical distance of 25 metres. (1)
   b) 450 joules of work are used when a suitcase is moved 45 metres. (1)
   c) 90 joules are used by a bird to lift a twig 100 metres to its nest. (1)

4 Use the equation alongside to help you find the following.

$$\text{power} = \frac{\text{work done}}{\text{time taken}}$$

   a) How many watts does Amy use to lift a shopping bag for 2 seconds using 100 joules? (1)
   b) Alice now lifts the same bag for 4 seconds using 100 joules. How many watts does she use? (1)
   c) From your answers to (a) and (b), is less energy used in doing a job quickly or slowly? (1)

11.7.3 Work and power

1 Alan is pulling a sledge through the snow. He uses a force of 60 N to pull the sledge the first 8 m on the flat.

a) How much work has he done?

_____ (1)

b) He reaches the slope, and pulls the sledge up it for 6 m in length and only 2 m in height. He uses 90 N of force to get up the slope. What work does he do against gravity to get up the slope?

_____ (1)

c) Alan's friend Asif asks Alan to pull the sledge again and now he will time Alan. Alan takes 4 seconds on the flat and 6 seconds up the slope, using the same forces as before. What is his total power?

_____ (1)

_____ (1)

_____ (1)

2 Lynne runs up 3 flights of stairs at school. It takes her 12 seconds. There are 60 steps, and each step is 21 cm high. Lynne's mass is 50 kg (1 kg weighs 10 N).

a) What was the vertical distance Lynne climbed?

_____ (1)

b) Find Lynne's weight in newtons.

_____ (1)

c) Find Lynne's work done in joules.

_____ (1)

d) Find Lynne's power.

_____ (1)

3 Given the work done in each of the following examples, find the distance moved by the force in the direction of the force.

a) 150 J of work were done in moving a force of 50 N.

_____ (1)

b) 1 kJ of work was done in moving a force of 100 N.

_____ (1)

c) A removal man did 30 kJ of work in lifting a chair of mass 20 kg to his van

_____
_____ (2)

d) A climber does 60 kJ of work lifting a rucksack of 15 kg up a rock face.

_____
_____ (2)

e) A toy train does 45 J of work in hauling 500 g of train around the track.

_____
_____ (2)

4 Use the information in each of the following questions to find the time taken for each activity.

a) A model boat has an engine of power 100 watts. Its total weight is 1000 N and it moves a distance of 1.5 m.

_____ (1)

b) A crane lifts a weight of 8000 N over a height of 10 m using a motor of power 800 watts.

_____ (1)

c) A shopkeeper picks up a box of apples weighing 500 N and carries it 20 m using 1000 watts.

_____ (1)

Name: Class: Date:

Science Companion © A Porter, M Wood, T Wood and Stanley Thornes (Publishers) Ltd 1995

11.8.1 Under pressure

Pressure = force ÷ area

1 The frog in the picture alongside has a mass of 200 grams. The total area of its feet is 8 cm^2. The area of the lily pad is 100 cm^2. The mass of the lily pad can be ignored.

a) What is the weight of the frog in newtons? (1)
b) What pressure (in newtons per square centimetre) would its feet exert if it tried to stand on the water? (2)
c) When it stands on the lily pad, what pressure does it exert on the water? (2)
d) Why does the frog not sink into the water when it sits on the lily pad? (1)

One snow shoe area = 500 cm^2

One ordinary shoe area = 125 cm^2

2 Explain in your own words why people wear snow shoes to walk over deep snow. (3)

3 The volume of gas inside a syringe was measured at different temperatures. Here are the results.

Temperature in kelvins	200	300	400	500
Volume in cm^3	70	105	140	175

a) Use some graph paper to plot the results of the experiment. Start each axis at zero and put temperature along the bottom. (7)
b) In this case what happened to the pressure of the gas inside the syringe? (1)
c) What happened to the movement of the gas particles as the temperature increased? (1)
d) If the syringe had been sealed tight so the plunger could not move, what would happen to the gas as the temperature increased? (1)
e) What happens to this gas when it reaches zero kelvins? (1)

Science Companion © A Porter, M Wood, T Wood and Stanley Thornes (Publishers) Ltd 1995

11.8.2 Under pressure

Pressure = force ÷ area

1 The frog in the picture alongside has a mass of 200 grams. The total area of its feet is 8 cm^2. The area of the lily pad is 100 cm^2. The mass of the lily pad can be ignored.

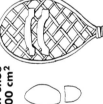

a) Why does the frog have webbed feet? (1)
b) What pressure (in newtons per square centimetre) would its feet exert if it tried to stand on the water? (2)
c) When it stands on the lily pad, what pressure does it exert on the water? (2)
d) Why does the frog not sink into the water when it sits on the lily pad? (1)

2 The picture alongside shows an ordinary shoe and a snow shoe.

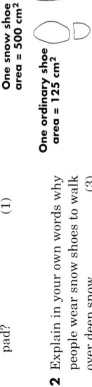

One snow shoe area = 500 cm^2

One ordinary shoe area = 125 cm^2

a) What area would a girl cover if she stood in a pair of ordinary shoes? (1)
b) What area would she cover if she stood in a pair of snow shoes? (1)
c) How does the area of the snow shoes compare with the area of the ordinary shoes? (2)
d) Why are snow shoes useful for walking on snow? (1)

3 The volume of gas inside a syringe was measured at different temperatures. Here are the results.

Temperature in kelvins	200	300	400	500
Volume in cm^3	70	105	140	175

a) What happens to the volume as the temperature increases? (1)
b) What happens to the pressure as the temperature increases? (1)
c) Use some graph paper to plot the results of the experiment. Start each axis at zero and put temperature along the bottom. (7)

Science Companion © A Porter, M Wood, T Wood and Stanley Thornes (Publishers) Ltd 1995

11.8.3 Under pressure

A pupil was investigating the effect pressure has on some plasticine. She set up two experiments using a heavy iron mass and small wooden cubes. Each of the cubes had a face of 1 cm².

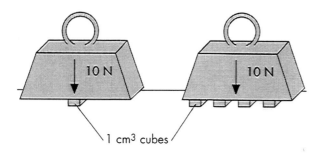

1 cm³ cubes

a) What force does the iron mass exert?

_____ newtons. (1)

b) What is the pressure on the plasticine from the cube in the first experiment?

_____ newtons per square centimetre. (1)

c) What is the pressure on the plasticine from the cubes in the second experiment?

_____ newtons per square centimetre. (1)

d) How does the first pressure compare with the second experiment?

_____ (2)

e) What results would you expect to see in the plasticine?

_____
_____
_____ (1)

2 Use the following words to complete the paragraph below. Words can be used once, more than once or not at all.

rises	falls	expands
contracts	more	less
increases	decreases	

When the air inside a hot air balloon is heated, the volume of the air _____. We say that the air _____. This makes the air _____ dense than before and the balloon _____. If the air cools down inside the balloon, the speed of the air particles _____. The air inside the balloon takes up _____ volume and the balloon _____. (7)

3 Work out the missing words in the passage below and enter them on the grid. 6 across is two words – 3 and 5 letters. (9)

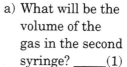

The French scientist, Blaise (*1 across*) carried out research into (*1 down*). Weight is the mass of an object being pulled by the (*4 down*) of gravity. The units are named after (*6 across*) Newton.
100 g = (*8 across*) newton.
The pressure in a liquid is the same, or (*5 across*) in (*3 down*) directions. Pressure is measured as a force applied over an (*7 down*) – usually measured in (*2 down*) metres.

4 Look at the diagram of the three sealed syringes. The first syringe has a volume of V cm³.

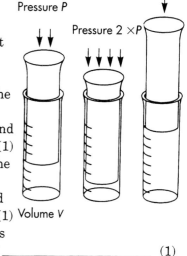

a) What will be the volume of the gas in the second syringe? _____ (1)

b) What will be the volume of the gas in the third syringe? _____ (1)

c) Whose law does this illustrate? _____ (1)

Name: Class: Date:

Science Companion © A Porter, M Wood, T Wood and Stanley Thornes (Publishers) Ltd 1995

12.1.2 All done with mirrors

1 a) The image we see in a mirror is called a virtual image. What does this mean? (1)

b) If you stand one metre away from a mirror, how far away does your image appear to be? (1)

2 The diagram shows how a ray of light is reflected off a mirror.

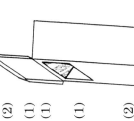

a) What is the angle of incidence? (1)
b) What is the angle of reflection? (1)
c) What can you say about these two angles? (1)

3 a) If you look at the reflection on the back of a spoon, what does it look like? (1)
b) What is different when you look at the reflection on the inside of the spoon? (2)
c) Which is a concave mirror, the inside or the back of a spoon? (1)
d) What word do we give to the opposite of a concave mirror? (1)

4 Some power stations use mirrors to generate electricity.
a) What is the source of heat in the power station? (1)
b) What are the mirrors used for? (1)
c) How does the heat generate electricity? (1)

5 a) How many mirrors are needed to make a periscope? (1)
b) Why does a submarine need a periscope? (1)
c) When might you need to use a periscope? (1)

6 What are mirrors used for in the following devices?
a) kaleidoscope (2)
b) torch (2)

12.1.1 All done with mirrors

1 a) The image we see in a mirror is not a real image. What name do we give to this type of image? (1)

b) If you stand one metre away from a mirror, how far away does your image appear to be? (1)

c) Write a few sentences explaining the diagram below. (4)

2 Curved mirrors distort the reflections. How does the image seen in a concave mirror compare with the image in a convex mirror? (3)

3 Mirrors have been used to generate electricity in power stations. In your own words explain how this is possible. (4)

4 Optical fibres can carry light over long distances and round corners. Light travels in straight lines. How do optical fibres manage to make the light turn corners? (2)

5 a) How many mirrors are needed to make a periscope? (1)
b) Why does a submarine need a periscope? (1)
c) When might you need to use a periscope? (1)
d) Explain how you could make a periscope using a cardboard box. (2)

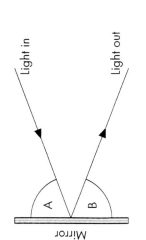

12.1.3 All done with mirrors

1 Say whether the statements below are true or false. Tick the correct box beside each one. (10)

	true?	false?
The reflection in a mirror is a virtual image.		
The angle of incidence on flat mirrors is greater than the angle of reflection.		
The image on the back of a spoon is upside down.		
Mirrors can be used in electricity power stations.		
Optical fibres can curve the rays of light inside them.		
The speed of light is 300 000 metres per second.		
Light can travel to the Moon and back in 2½ seconds.		
Laser beams are used to measure the distance between the Earth and Moon.		
A periscope produces colourful patterns with mirrors.		
The speed of light is slower in air than in a vacuum.		

2 A mirror was used to look at letters of the alphabet.

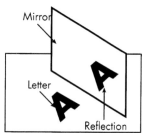

Say whether the reflection shown in the table below is a true or false picture of the letter. (4)

LETTER	REFLECTION	TRUE OR FALSE
B	ᗷ	
N	И	
Q	Ϙ	
S	ꙅ	

3 The picture below shows a boy looking into a mirror. On the right hand side draw his reflection. (2)

View from top

mirror   Virtual image

4 The Hubble telescope was launched into orbit around the Earth in 1990. It was designed to beam clear pictures of the stars and planets back to Earth. Part of the telescope used mirrors to magnify and focus light from stars and planets. The telescope is pointed towards an area of the sky. The light from the sky hits the large mirror. The light is reflected on to the second mirror which in turn reflects the light to the receptor. The receptor sends the signals down to Earth.

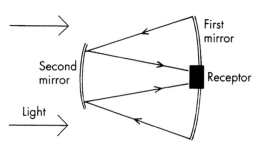

a) What is the advantage in putting a telescope into space rather than having it on the Earth?

_____
_____
_____ (1)

b) What does 'magnify' mean?

_____ (1)

c) What does 'focus' mean?

_____ (1)

d) Why are the mirrors curved?

_____ (1)

Name: Class: Date:

12.2.1 Light and shadow

1 Explain what the following words mean.
  a) opaque (1)
  b) transparent (1)
  c) translucent (1)

2 Describe what your shadow will look like if you stand
  a) close to a wall with a light behind you (2)
  b) close to a light with your shadow cast on to a wall. (2)
  c) Why do the edges of shadows appear blurred? (1)

3 What is a silhouette and how is it produced? (2)

4 a) In your own words describe how a pinhole camera is made and how it works. (4)

Screen made of greaseproof paper

Back

  b) What happens to the image if the pinhole is made larger? (2)
  c) How can you make the image appear larger? (1)
  d) What happens if there are two pinholes in the camera? (1)
  e) In what two ways do real cameras have to control the amount of light hitting the film in them? (2)

12.2.2 Light and shadow

1 Match the words in the first column to their correct meanings in the second column. (4)

  opaque   spread out
  absorb   see-through
  diffused   soak up
  transparent   not see-through

2 Use the words below to help you answer the question.
  larger   smaller   sharp   blurred
  Describe what your shadow will look like if you stand
  a) close to a wall with a light behind you (2)
  b) close to a light with your shadow cast on to a wall (2)

3 When the Sun is behind an object, it appears in silhouette. What do we mean by silhouette? (1)

4 A pinhole camera can be used to see images of bright objects.

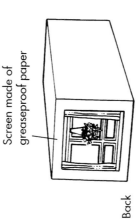
Screen made of greaseproof paper

Back

  a) Where is the pinhole in a pinhole camera? (1)
  b) Why does the hole have to be small? (1)
  c) Where is the image seen? (1)
  d) How is the image different from the object you look at? (2)
  e) What could you do to the camera to make the image brighter? (1)
  f) What is the disadvantage of doing this? (1)
  g) What happens to the image if you move the camera closer to the object? (1)
  h) How could you get two images on the camera's screen? (1)
  i) What do commercial cameras use instead of a pinhole? (1)
  j) How can cameras control how much light hits the film? (1)

12.2.3 Light and shadow

1 The diagram below shows the rays of light which pass through a pinhole camera when it is pointed towards a glowing bulb. An image can be seen on the screen at the back of the camera.

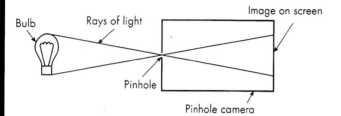

a) Draw a picture of the image you would see on the screen. (1)

b) The camera was moved further away from the bulb. Draw a picture showing how the image would be different from that in part a). (1)

c) Two holes were put in the front of the camera. Draw a picture of what would now be seen on the screen. (1)

d) What happens to the image if the pinhole is made larger? (*Delete the incorrect word*)

The brightness of the image **increases/decreases**

The image becomes more **sharp/blurred** (2)

e) How will a lens in front of the pinhole help if the pinhole is too large?

_____

_____ (1)

f) How will a lens in front of many pinholes in the camera improve the image?

_____

_____ (1)

2 Enter the answers to the clues into the crossword grid. (15)

ACROSS
1. Transparent (two words 3 and 7)
5. This means light will not pass through (6)
6. Another word for beams of light (4)
7. Glass is more _____ than water (5)
10. A lighthouse sends out _____ of light. (6)
13. Light will not turn a _____ without a mirror (6)
14. A room will light up when you _____ the lamp switch (5)

DOWN
2. This means light will pass through but it will be diffuse (11)
3. The surface of a lens is (5)
4. A transparent substance (5)
6. What light will do off a mirror (7)
8. The part of our body we use to see (3)
9. When our _____ close we cannot see things (7)
11. A word which means in focus (5)
12. This means not in direct light (5)

3 Delete the incorrect words in the passage below.

Light does not pass through our skin because our skin is **opaque/transparent**. One form of light which will pass through skin is **UV light/X-rays**. This will not pass through bones which form a **virtual image/shadow**. (3)

Name: Class: Date:

12.3.1 Light

1 The evening primrose flower looks different when photographed using ultraviolet light. Explain why this is so. (3)

2 a) What are the seven colours which can be seen in the visible spectrum? (7)
  b) Why does a prism produce a spectrum of colours from white light? (2)
  c) Which of the seven colours has the shortest wavelength? (1)
  d) Which of the seven colours has the highest frequency? (1)
  e) Which property do all the seven colours have in common? (1)

3 To what use are the following electromagnetic rays put?
  a) gamma radiation (1)
  b) infra-red radiation (1)
  c) microwave radiation (1)
  d) radio waves (1)
  e) ultraviolet radiation (1)
  f) X-rays (1)
  g) List the 6 types of radiation above in order from the shortest wavelength to the longest. (1)

4 The photograph shows pencils standing in a glass which is half filled with water. In your own words explain why the pencils appear bent. (3)

12.3.2 Light

1 Bees can see light which humans cannot.
  a) What kind of light can bees see which humans cannot? (1)
  b) How do flowers like the evening primrose make use of this fact? (1)
  c) Why do flowers need to attract bees? (1)

2 The diagram shows how white light can be split into coloured light.
  a) What name is given to the glass equipment used in the experiment? (1)
  b) In which direction is the light travelling in the diagram? (1)
  c) Name the seven colours which white light is made from. (7)
  d) What is the scientific name for the rainbow of colours? (1)

3 a) Match the electromagnetic rays in the first column to their uses in the second column. (6)

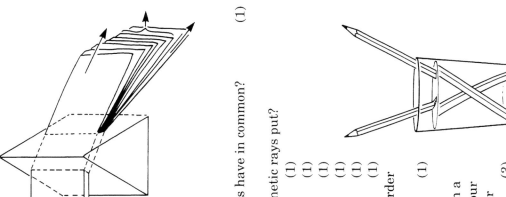

gamma radiation	TV remote control
infra-red radiation	photographing bones
microwave radiation	sterilising surgical instruments
radio waves	sun beds
ultraviolet radiation	heating food
X-rays	radio signals

  b) Which of the rays above has the shortest wavelength and which has the longest? (2)
  c) Why is ultraviolet radiation considered dangerous to humans? (1)

4 Rescue workers use cameras sensitive to infra-red radiation when looking for people trapped under collapsed buildings.
  a) How can the camera detect people? (1)
  b) Why will the camera not pick up the stones of the building? (1)
  c) Give another use for this camera. (1)

12.3.3 Light

1. Colour in or name the seven colours in the spectrum made when white light passes through a prism. (7)

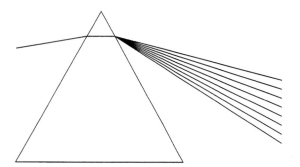

2. Many radio stations are listed in newspapers with their wavelength (in metres) or their frequency (in kilohertz). You can convert between the two as long as you know what the speed of the radio waves are.

speed (km/s) = wavelength (m) × frequency (kHz)

The speed of the radio waves is 300 000 kilometres per second. Use this to work out the missing numbers in the following table. (6)

Wavelength (m)	Frequency (kHz)
250	
330	
433	
	1215
	198
	648

3. The answers to the clues (a) to (h) will fit into the grid. Use the numbers above the grid and the letters by the side of the grid to complete the phrase under the grid. (9)

a) Laser beams read bar codes at supermarket _____-outs
b) The _____ is the unit of frequency
c) Sun beds use _____ violet light
d) _____ light lies between yellow and blue
e) All colours combine to make _____ light
f) Food can be heated in _____ wave ovens
g) TV signals travel on _____ waves
h) A glass _____ will split light into colours

	1	2	3	4	5
a					
b					
c					
d					
e					
f					
g					
h					

d3	c2	d4	f3	c3	h2	g5	f1	g2	d1	d5	e5	b4	g4	a1
h4	h1	b2	a4	e4	f4	c1	h5							

4. The table shows the speed of light through different materials.

Material	Speed of light
vacuum	300 000 km per second
glass	200 000 km per second
water	225 000 km per second

The refractive index of a material is the speed of light in a vacuum divided by the speed of light through the material.

Refractive index = $\dfrac{\text{speed of light in a vacuum}}{\text{speed of light in the material}}$

a) Work out the refractive index for glass.
_____ (1)

b) Work out the refractive index for water.
_____ (1)

c) The refractive index for ice is 1.31. What is the speed of light through ice?
_____ (1)

Name: Class: Date:

12.4.1 Recording pictures

1 Read the article below and then answer the questions which follow.

> The world's oldest known photograph was taken by a French army officer Nicéphore Niepce (1765–1833). It showed buildings next to his home at Chalon-sur-Saône near Beaune. The image took eight hours to expose to the subject and was recorded on a pewter plate.
>
> Clearly eight hours was too long to ask people to stand to have their photos taken. It was left to Louis Daguerre in 1833 to find a method of making a film sensitive enough to record a person who had to remain motionless for a minute or two. The film became known as daguerrotype. The picture was recorded on to copper plates coated in silver. After exposure, the plates were developed by treatment with highly poisonous mercury vapour and fixed with salt.

a) Who took the oldest known photograph? (1)
b) What was the subject of the photograph? (1)
c) How long was the film left to receive light from the subject? (1)
d) Who took the first known photograph of a person? (1)
e) What nationality were both the early photographers? (1)
f) Of what were daguerrotypes made? (2)
g) What poisonous substance was used to develop daguerrotypes? (1)
h) How were daguerrotype pictures made permanent? (1)

2 In 1947 the first Polaroid film and camera were introduced. What was special about them? (2)

3 Pictures need not always be stored on film. The video home system of recording uses magnetic tape.
a) By what three initials is this type of recording best known? (1)
b) On which part of the tape does the picture go? (1)
c) Where is the sound placed on the tape? (1)

4 A single-lens reflex camera allows the photographer to see through the lens before taking the picture.
a) What part of this camera flicks up when a picture is taken? (1)
b) What part allows light to hit the film for a given time? (2)
c) Given the same film, would the exposure be longer at night or in bright sunlight? (1)
d) Explain your answer to part (c). (2)

Science Companion © A Porter, M Wood, T Wood and Stanley Thornes (Publishers) Ltd 1995

12.4.2 Recording pictures

1 Read the article below and then answer the questions which follow.

> Two early photographers were French. They were called Niepce and Daguerre. Niepce probably took the oldest photograph known today. It was of buildings next to his home. The film he used was very slow. This meant light had to hit it for a long time before the chemicals changed to record the subject. He let light fall on the film for eight hours.
>
> Daguerre used a faster film. He was able to take the first known picture of a person in 1833. He needed only a minute or two of light to fall on the film. His film became known as daguerrotype. It was not film as we know it today. The film was individual plates made from copper. The copper was coated with silver. Highly poisonous mercury vapour was used to develop the film which was later fixed with salt.

a) What nationality were the two early photographers? (1)
b) Who took the earliest known photograph? (1)
c) What was the subject of this first photograph? (1)
d) The shutter is used to let light into the camera on to the film. For how long was the shutter open for this first photograph? (1)
e) Why is Daguerre's film said to be faster than that used by Niepce? (1)
f) How was daguerrotype 'film' made? (1)
g) Which poisonous chemical was used to develop the film? (1)
h) Which common chemical fixed the film? (1)
i) What do we mean by 'fixing a film'? (1)

2 In 1947 the first Polaroid film and camera were introduced. What was special about them? (2)

3 Pictures need not always be stored on film. The video home system of recording uses magnetic tape.
a) By what three initials is this type of recording best known? (1)
b) On which part of the tape does the picture go? (1)
c) Where is the sound placed on the tape? (1)

4 A single-lens reflex camera allows the photographer to see through the lens before taking the picture.
a) What part of this camera flicks up when a picture is taken? (1)
b) What part allows light to hit the film for a given time? (2)
c) Given the same film, would the exposure be longer at night or in bright sunlight? (1)
d) Explain your answer to part (c). (2)

Science Companion © A Porter, M Wood, T Wood and Stanley Thornes (Publishers) Ltd 1995

12.4.3 Recording pictures

1 Use clues (a) to (l) below to complete the horizontal words in the spaces which follow. Then find the word in column 4 which reads from top to bottom. (12)

CLUES

a) We can buy or hire our favourite films on this tape.
b) Dr. Edwin Land invented this first instant picture camera.
c) When this gets on a film in a camera, it makes a picture.
d) A camera needs to be in _____ to get a sharp picture.
e) A fast film needs less _____ to make a picture.
f) Film can be developed in a dark _____.
g) This lifts up in an SLR camera when a picture is taken.
h) You need a short _____ time to take a picture of a person.
i) This recording material is found in a cassette.
j) Once a recording is made, we want to use the _____ button on the machine to see what was recorded.
k) Short for photograph.
l) If you buy blank video tape you first _____ something and then watch it.

	1	2	3	4	5	6	7	8
a)								
b)								
c)								
d)								
e)								
f)								
g)								
h)								
i)								
j)								
k)								
l)								

Column 4 is the word _____ an early type of photograph. (1)

2 Use the following words to complete the passage about recording on video (VHS). Each word may be used once, or more than once.

**capstans   cassette   forwards
heads   magnetic   picture
playback   remove   roller
synchronises**

On loading, the tape is pulled out of the _____ into contact with the recording and _____ mechanism. As the tape moves round it passes the erase head which uses a _____ field to _____ any previous recording. Two recording _____ put the picture on the tape by _____ signal. On playback these recording _____ turn the signal on the tape back into a _____. The sound recording is put on by a separate head which _____ the sound to the picture. The pinch _____ pushes the tape against the _____ which move the tape _____. (12)

3 From the list of words below match **one** to each of the sentences.

**developed   exposure   lens
shutter   viewfinder**

a) The length of time the light shines on a film to make a picture.

b) The part of the camera which moves to let a known amount of light to get to the film.

c) This part of the camera is glass or plastic and focuses the light on to the film.

d) You look through this when taking a picture.

e) Films have to be _____ using chemicals to produce a picture.

(5)

Name: Class: Date:

Science Companion © A Porter, M Wood, T Wood and Stanley Thornes (Publishers) Ltd 1995

12.5.1 Seeing and hearing

1. Use the diagram alongside to help you answer the following questions. Illustrate your answers with diagrams whenever they help.

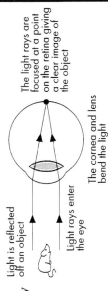

2. a) Explain why a short-sighted person can only focus on objects close to them. (2)
   b) What happens to rays of light in the eye when a short-sighted person tries to focus on a distant object? (2)
   c) What type of lens is fitted in spectacles to help a short-sighted person see distant objects? (1)
   d) What does this type of lens do to parallel light which enters it? (1)

3. a) A long-sighted person can see distant objects clearly. Where does the light from distant objects focus in their eyes? (1)
   b) A long-sighted person cannot focus on objects which are close. Where does the light focus without the aid of spectacles? (1)
   c) What type of lens is fitted in spectacles to correct the vision of a long-sighted person? (1)
   d) In what way does this lens change light as it passes through the lens? (1)

4. The ear may be divided into three main sections, the outer, the middle and inner ear. For each part of the ear given below say which main section it is in. Present your answer in the form of a table.

   auditory nerve   cochlea   eardrum   pinna   semicircular canals   oval window   ossicles   eustachian tube (8)

5. Give the function of each of the following parts of the ear.
   a) ossicles   b) cochlea   c) auditory nerve   d) eustachian tube (5)

6. Name **two** parts of the middle or inner ear which if damaged may lead to permanent deafness. (2)

12.5.2 Seeing and hearing

1. The diagram alongside shows how a person with normal eyesight sees an object. Copy the diagram, without the writing and then label the following things on your diagram.
   a) lens, b) cornea, c) retina d) point of focus. (8)

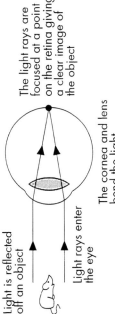

2. The diagram alongside shows what happens when a short-sighted person tries to focus on a distant object. Is the light focused:
   a) in front of the retina
   b) on the retina
   c) behind the retina? (2)

3. A concave lens bends light outwards.   A convex lens bends light inwards.

   For a short-sighted person, which type of lens is used in spectacles to correct their sight? (2)

4. Below are five main parts of the ear.

   auditory nerve   cochlea   eardrum   pinna   semicircular canals

   For each of the functions below name the part of the ear which does that job.
   a) collects sound waves (2)
   b) vibrates when hit by sound waves (2)
   c) converts the vibrations to nerve messages (2)
   d) allow us to balance (2)
   e) sends messages to the brain (2)

5. Name **three** parts of the middle or inner ear which if damaged may lead to permanent deafness. (3)

12.5.3 Seeing and hearing

1 Below is a diagram of the eye with eight important parts of the eye labelled. Following the diagram are eight functions of parts of the eye. In the space provided, match the part of the eye to its function.

Diagram of the eye labelled: Cornea, Iris, Pupil, Lens, Aqueous humour, Vitreous humour, Retina, Optic nerve

Function of the part	Name of the part
allows light into the eye	
focuses the light to form a sharp image on the retina	
supports the shape of the eye	
bends light before it enters the eye	
contains light-sensitive cells	
feeds the cornea and the lens	
adjusts the size of the pupil to allow different amounts of light to enter	
carries messages from the retina to the brain	

(8)

2 Complete the following passage about how we see selecting words from the list below. Each word may be used once only.

**brain   cells   cornea   focus   image
lens   rays   reflected   refracted
retina   Sun**

We see something when light from the _____ or a lamp is _____ off the surface of the object and reaches our eyes. The light enters the eyes through the _____ and _____. They bend the light inwards to _____ it on the _____. When light _____ are bent it is called _____. There is now an _____ of the object on the retina and light-sensitive _____ send a message to the _____. (11)

3 Name the three main parts into which the ear may be divided.
_____ (3)

4 Describe what each of the following parts of the ear does.

a) *PINNA* _____
_____ (1)

b) *EARDRUM* _____
_____
_____ (1)

c) *COCHLEA* _____
_____
_____ (1)

d) *SEMICIRCULAR CANALS* _____
_____ (1)

5 What device do some people use to enable them to hear more clearly when they suffer from some types of deafness?
_____ (1)

Name: **Class:**   **Date:**

Science Companion © A Porter, M Wood, T Wood and Stanley Thornes (Publishers) Ltd 1995

12.6.1 How sound is made

1. Vibrations cause sound. What are vibrations? (1)

2. A bell placed inside a glass container can be seen and heard ringing. The air is now removed from the glass container creating a vacuum. Why can the bell no longer be heard to ring? (4)

3. Why are science fiction films wrong when we hear a large spaceship's engines as it approaches through outer space? (2)

4. What type of wave is produced by sound? (1)

5. In the air, particles are quite a distance apart. What happens to air particles in the part of a sound wave where there is
   a) compression (1)
   b) expansion? (1)

6. Look at the diagram alongside. Explain why this young person's voice sounds louder when a megaphone is used. (2)

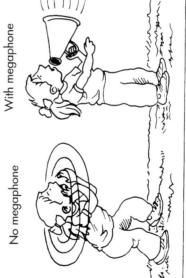

No megaphone   With megaphone

7. Why can fish detect sound underwater? (1)

8. a) Would you expect sound to travel faster through air or glass? (1)
   b) Give a reason for your answer to part (a). (1)

12.6.2 How sound is made

1. Look at the diagrams alongside to help you answer the questions which follow.

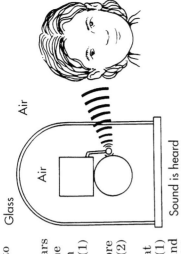

Glass — Air — Air — Sound is heard

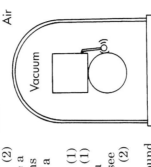

Glass — Vacuum — Air — No sound heard

   a) The bell rings and the girl hears sound. After the bell, name the next substance through which the sound travels. (1)
   b) Before it reaches her ear, the sound travels through two more substances. What are they? (2)
   c) Now look at the diagram where no sound is heard. What is a vacuum? (1)
   d) Why does the girl hear no sound through the vacuum? (2)
   e) Sound passes through gas and solid. Will it pass through a liquid, do you think? Explain your answer. (2)
   f) In the air, particles are quite a distance apart. What happens to air particles in the part of a sound wave where there is
      i) compression (1)
      ii) expansion? (1)

2. If you have a still pond, and you drop a stone in it, what do you see happening? (2)

3. Look at the diagram alongside.

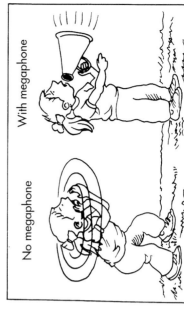

No megaphone   With megaphone

   a) In what direction does the sound travel when the person shouts without a megaphone? (1)
   b) Why does the shouting seem louder to people in front when the megaphone is used? (1)
   c) If you stand behind the shouter, what will you notice when the megaphone is used? (1)

12.6.3 How sound is made

1 Below is a word search. Find the NINE words connected with 'How sound is made'. There is one word which is in twice. Write out all NINE words on the lines below and say which ONE is used twice by writing (×2) after it.

S	C	T	U	J	B	C	S	R	E	V	K
L	O	N	G	I	T	U	D	I	N	A	L
F	M	E	Y	H	N	M	J	U	I	C	K
G	P	A	R	T	I	C	L	E	S	U	Y
H	R	R	S	W	X	Z	A	Q	W	U	F
J	E	X	P	A	N	S	I	O	N	M	T
K	S	R	F	V	X	C	V	B	U	I	E
L	S	V	H	E	A	R	E	R	E	Y	U
V	I	B	R	A	T	I	O	N	A	I	O
O	O	S	E	R	F	V	E	W	R	M	K
P	N	E	W	E	D	G	F	A	J	E	Y

_____
_____
_____
_____ (10)

2 Solve the clues to find the missing phrase. The first letter of each word fits into the phrase. Write this first letter above the number of the clue to find the phrase.

PHRASE $\overline{1}$ $\overline{2}$ $\overline{3}$ $\overline{4}$ $\overline{5}$ $\overline{6}$ $\overline{7}$ $\overline{8}$ $\overline{9}$ (1)

CLUES

1 There are three states of matter. They are _ _ L _ _, liquid and gas.

2 When a stone is dropped into a pond, waves spread _ U _ in all directions.

3 When a noise is not heard, we say it is _ N _ _ _ _ D.

4 A sound is louder when you are _ _ _ R _ _ than further away.

5 A megaphone can be used to point sound in one _ _ _ E _ _ _ _ _ .

6 A _ _ IS _ _ _ is a quiet sound usually made between two people.

7 Sound, rain, and wind all pass through the _ _ R.

8 Sound cannot travel through a _ _ _ U _ _.

9 A longitudinal wave contains alternate places of _ _ _ _ _ _ O _ and compression. (9)

3 Unscramble the words in each line below and place the correct words in the matching blank. You can use the word clues to help you. Then find the word in the lightly shaded column.

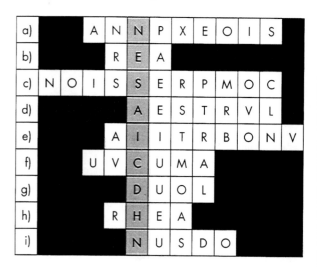

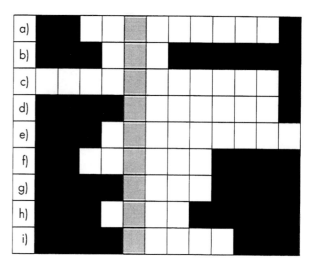

CLUES

a) To make larger.
b) To hear with.
c) Squeezed together.
d) Sound does this through the air.
e) To move backwards and forwards very quickly.
f) No particles, usually no air.
g) The opposite of quiet.
h) Listen.
i) Noise.

The word in the lightly shaded column is
_____ (1)

Name:　　　　　　　　　　　　**Class:**　　**Date:**

12.7.2 Frequency, pitch and volume

1 a) Sound travels in longitudinal waves. When these reach a microphone what are they turned into? (1)
  b) If this microphone is connected to an oscilloscope the sound wave can be shown on a screen. What sort of wave is it now? (1)

2 Use the words below to complete the following sentences. On your answer sheet you can just write the letter of the question followed by the word which should go in the sentence.

  frequency   hertz   pitch   wavelength

  a) The distance between the top of adjacent waves is called the _____ (1)
  b) The number of wavelengths that go by per second is the _____ (1)
  c) The higher the frequency, the higher the _____ of a musical note. (1)
  d) One complete wave in one second gives a frequency of one _____ (1)

3 The diagram alongside shows how a piano string behaves when the hammer hits the string. To do this the pianist plays a note on the piano.
  a) Which diagram, left or right, shows a loud note, and which a quiet note? (2)
  b) How do you think a pianist makes the note louder or quieter? (4)

4 Volume means the intensity of the sound. This depends on the amount of energy in the sound.
Two sounds are played through a microphone and shown on an oscilloscope. They are drawn alongside. a is the amplitude of the note.
  a) Which drawing shows the loud sound and which the quiet one? (2)
  b) How does amplitude link with volume? (1)

12.7.1 Frequency, pitch and volume

1 What are sound waves turned into when they go into a microphone? (1)

2 a) Name the electrical device which can show sound waves on a screen. (1)
  b) The waves shown on this device are not longitudinal. What are they? (1)

3 Use the diagram below to help answer the following questions.

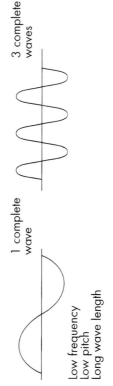

low frequency
low pitch
long wave length

  a) Define
    i) wavelength
    ii) frequency
    iii) pitch
    iv) hertz. (4)
  b) In the diagram above, the two drawings are to the same scale. Given statements about the left-hand one, write three similar statements about the right-hand drawing. (3)

4 Hearing varies a lot from person to person and between people and animals.
  a) What happens to hearing as people get older? (1)
  b) How does the range of frequencies which can be heard differ between people and dogs? (1)

5 What do you think the difference would be in the vibrations of a drum skin for a quiet note compared with a loud note? (1)

6 On the wave as shown above for questions 3, how is volume depicted? In your answer draw one complete wave for a quiet note compared with one for a loud note. Label the part of each wave which illustrates the difference in volume. (2)

12.7.3 Frequency, pitch and volume

1 For a string on a guitar, the musical note it can make gets higher in pitch as the string gets shorter.

You are given eight straws each of which has been cut to a different length. In the space below draw eight straws of different lengths arranged so that on blowing over them you can produce notes which get higher in pitch as you go from left to right.

```
1   2   3   4   5   6   7   8
    ── higher in pitch ──→
```
(3)

2 Select words from the list below to complete the following sentences. Use each word once, more than once, or not at all.

**decreasing   higher   increasing
lower   lowers   raises**

a) On stringed musical instruments, the longer the string, the _____ the musical note.

b) On a violin, the thinner the string, the _____ the musical note.

c) On a stringed musical instrument raising the tension in a string _____ the note.

d) On a trombone you can lower the musical note by _____ the length of the tube.

e) On a church pipe organ, the shorter pipes produce notes of _____ pitch. (5)

3 When big speakers are producing loud low notes at pop concerts, why can the speaker cones be seen moving backwards and forwards?

_____
_____
_____
_____
_____
_____ (6)

4 Find the ten horizontal words in the grid below by using the clues which follow. Then identify the word in column 3.

	1	2	3	4	5	6	7	8	9	10	11
a)											
b)											
c)											
d)											
e)											
f)											
g)											
h)											
i)											
j)											

(10)

CLUES

a) Sound travels as longitudinal _____.

b) In a transverse wave, if the wave has more energy it has a bigger _____.

c) The _____ of a tuning fork produces a single musical note.

d) The number of wavelengths that go by per second.

e) Opposite of quietness.

f) The intensity of sound depends on the amount of _____ in the sound.

g) A lot of sound all mixed up is often described as _____.

h) A musical instrument usually with six strings used by many 'pop' musicians.

i) The higher the frequency, the higher the _____ is of a musical note.

j) Short wavelengths have _____ frequencies.

The word in column 3 is _____ (1)

Name: **Class:**   **Date:**

12.8.1 The speed of sound

1 The speed at which sound travels through a material depends on the density and elasticity of the material. Explain why sound usually travels faster through liquids than through gases, and faster still through solids. (4)

2 In a thunderstorm the time between the lightning and the thunder increases by 3 seconds for every 1 km from the centre of the storm. Below are times between the lightning and thunder reaching you in a thunderstorm. How far are you from the centre of the storm in each case?
   a) 9 seconds
   b) 5 seconds
   c) at the same time. (3)

3 Use the drawing alongside to help answer this question.
   a) How much faster does sound travel along the grain of a piece of wood compared with across the grain? (1)
   b) How may this relate to the shape of a violin? (1)

4 What word is used to describe a plane flying
   a) slower than the speed of sound (1)
   b) faster than the speed of sound? (1)

5 a) What name was given to the speed at which a plane had to fly to go as fast as the speed of sound? (1)
   b) In the 1940s as planes tried to go faster than the speed of sound, why did it seem impossible before 1947? (1)
   c) As scientists studied plane design and the way a plane behaved at high speed, what did they find these early designs of planes were doing to the air, which caused the problem? (1)
   d) How did plane designers overcome the problem to make a plane fly faster than the speed of sound? (3)
   e) Which plane first flew faster than the speed of sound? (1)
   f) Who flew this record-breaking plane? (1)
   g) When did this event happen? (1)

12.8.2 The speed of sound

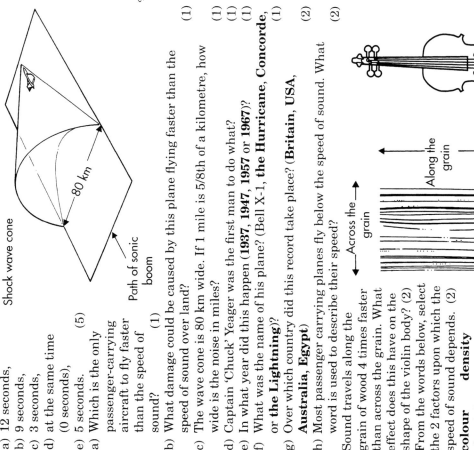

1 Sound travels much slower than light. In a thunderstorm the time between the lightning and the thunder increases by 3 seconds for every 1 km from the centre of the storm. In each case below you are given the time between the thunder and lightning reaching you. How far are you from the storm's centre?
   a) 12 seconds,
   b) 9 seconds,
   c) 3 seconds,
   d) at the same time (0 seconds),
   e) 5 seconds. (5)

2 a) Which is the only passenger-carrying aircraft to fly faster than the speed of sound? (1)
   b) What damage could be caused by this plane flying faster than the speed of sound over land? (1)
   c) The wave cone is 80 km wide. If 1 mile is 5/8th of a kilometre, how wide is the noise in miles? (1)
   d) Captain 'Chuck' Yeager was the first man to do what? (1)
   e) In what year did this happen (**1937**, **1947**, **1957** or **1967**)? (1)
   f) What was the name of his plane? (Bell X-1, **the Hurricane, Concorde,** or **the Lightning**)? (1)
   g) Over which country did this record take place? (**Britain, USA, Australia, Egypt**) (2)
   h) Most passenger carrying planes fly below the speed of sound. What word is used to describe their speed? (2)

3 Sound travels along the grain of wood 4 times faster than across the grain. What effect does this have on the shape of the violin body? (2)

4 From the words below, select the 2 factors upon which the speed of sound depends. (2)
   colour   density
   elasticity   pressure
   solubility   volume

12.8.3 — The speed of sound

1 Read the following newspaper report very carefully and then answer the questions which follow.

PENCIL SHAPE FOR NEW 'CONCORDE' TO SILENCE SONIC BOOMS

DESIGNERS at America's National Aeronautics and Space Administration (NASA), claim to have beaten the major environmental objections to a successor for Concorde, by reducing the sonic boom to a rumble.

The boom is caused when a plane breaks the sound barrier. The designers have given their plane a flatter nose, more like that of the high speed train, rather than the famous droop of Concorde. This nose will be built from a new generation of plastics.

Their first design was put forward two years ago, but is now being modified to a pencil shape. This will produce a plane which is one third longer than the present giant passenger plane, the Boeing 747-400.

They claim to have reduced the sonic boom to a rumble. If true, this will enable 'the pencil' to fly over land at supersonic speeds. Concorde is restricted to supersonic speed over the water for the sonic boom is sudden and has been known to cause damage to buildings.

'The pencil' could be flying by 2005 at a cost of £10 billion if built by an international consortium, pooling talent from the United States, Britain, Germany, Japan and the former Soviet Union. If this international consortium holds together then they hope to build around 2000 planes.

World News 19/1/92

a) Who has designed this new supersonic plane?

_____ (1)

b) In what way have the designers changed the shape, from that of Concorde, to lower the noise from the sonic boom?

_____
_____ (2)

c) How much larger will this plane be than the Boeing 747-400?

_____ (1)

d) From what will the nose of the plane be made?

_____ (1)

e) What will this plane be able to do which Concorde cannot, and why?

_____
_____ (2)

f) Which countries are to build it?

_____
_____ (2)

2 Use a calculator, and the equation
speed = frequency × wavelength
to find the values to fit in the lightly shaded boxes and write them in.

Speed (m/s)	Frequency (Hz)	Wavelength (m)
	9	14
	5	23
344		34.4
344		1.3
200	50	
5000	650	
	440	0.1
	440	0.3
800		125
1500	440	

(10)

Name: Class: Date:

12.9.1 Noise in the environment

1. Explain what is meant by
   a) the intensity of sound. (1)
   b) the loudness of sound. (1)

2. a) What is the decibel (dB)? (1)
   b) What is meant by 'the threshold of audibility'? (1)
   c) How many decibels is 'the threshold of audibility'?
      0 dB   10 dB   50 dB   100 dB   150 dB (1)
   d) What is 'the threshold of feeling'? (1)
   e) How many decibels is 'the threshold of feeling'?
      0 dB   20 dB   60 dB   120 dB   160 dB (1)

3. When measuring the intensity of sound produced by a musical instrument, why is the measurement always made at the same distance from the instrument (3 metres), and in the open air? (2)

Trombone 107 dB

4. Why are all petrol and diesel cars fitted with silencers on their exhausts? (2)

5. Suggest two methods by which motorways planners try to reduce the noise from motorways for houses which are built at the side of the road in built-up areas. (2)

6. What can be done with house windows to help reduce noise from outside for the people inside? (1)

7. What effect does loud noise have on a person's pulse and breathing rates? (1)

12.9.2 Noise in the environment

1. The decibel (dB) is the unit used to measure the intensity of sound.
   a) When measuring the sound from a musical instrument why is it always done at the same distance and in the open air? (2)
   b) Place the following four instruments in increasing order of sound intensity.
      Bass drum   Clarinet   Trombone   Trumpet
      113 dB   85 dB   107 dB   94 dB (1)
   c) A sound of 0 decibels is at the 'threshold of audibility' (hearing). What do you think this means? (1)

Trombone 107 dB

2. Motorways which go through towns are often sunk into cuttings in the ground or have fences at the side. Why do you think they are built like this? (2)

3. Why do car exhausts have silencers fitted? (1)

4. Here are five words:

   hearing   intensity   loudness   noise   soundproofing

   Which of the above words is defined by:
   a) The amount of energy in a sound wave. (1)
   b) The apparent strength of sound as it reaches the eardrum and is sent to the brain. (1)
   c) Material with the ability to absorb sound. (1)

5. A new airport is being built near your home. Suggest two possible ways to cut down the noise you would hear from the aircraft when you are sitting inside the house. (2)

6. Suggest three sources of sound which help to make the inside of a car noisy at high speed (70 mph) on a motorway. (3)

12.9.3 — Noise in the environment

1. Read the following report and then answer the questions which follow using the information in the report wherever possible.

> **Noise reduction brings peace and quiet**
>
> Reducing noise at source is often easier than trying to reduce it on its way to the human ear.
>
> Moving or rotating parts cause vibrations and these in turn make sound. Floors, buildings or vehicles can all magnify this sound because they resonate in time with the vibrations.
>
> A rotating shaft on an engine causes a lot of vibration in the engine, and so a lot of noise. This in turn produces excess wear on the bearings, and causes a loss of efficiency. If the shaft is removed, balanced, and refitted, it will result in a quieter, smoother and more efficient engine.
>
> If the engine cannot be made quieter at source, action is taken on the surroundings. An engine can be supported on rubber mountings which absorb the vibrations instead of transmitting them. The engine can be placed in an airtight enclosure if practical. As a last resort, the sound can be allowed 'out' and the operator and nearby people forced to wear ear defenders to prevent permanent damage to their ears.
>
> At home we can do many simple things to cut down on outside noise spoiling our quietness. Double glazing with a wide air gap of several centimetres helps, and the inside becomes even quieter when heavy curtains are closed. Thick carpets help too. Wall coverings of cork or polystyrene absorb noise and prevent it reflecting around a room.
>
> Next time you clear a room totally for redecoration, notice how noisy it becomes. Speedy redecoration will soon return the peace and quiet we all like.
>
> A. Noy (World Sound News 6/1/95)

a) How can noise be made in some machines?

_____ (1)

b) Name **three** things which can magnify unwanted sound.

_____ (3)

c) Name **three** things produced as a result of a badly balanced rotating shaft.

_____ (3)

d) How do rubber engine mountings help reduce noise in a motorcar?

_____ (2)

e) What must people wear if working in a noisy environment?

_____ (1)

2. Why do you think sound recording studios reduce their glass areas to a minimum and have cork or polystyrene ceiling and wall coverings and good carpets?

_____ (4)

3. Firing a gun can produce a noise of 160 dB in the ears of the person using the gun.

a) What do you think would be the result if this noise was allowed into someone's ears?

_____ (1)

b) What precautions must people take against noise on shooting ranges?

_____ (1)

Name: **Class:**   **Date:**

12.10.1 A brief history of sound recording

1 Study the following passage and then answer the questions which follow.

> In 1877 American Thomas Edison became the first person to record sound. He used a needle to mark the sound as grooves in tin foil on a cylinder. When the needle moved back through the grooves the original sound was heard. In 1886 he introduced a wax cylinder. When used with records, the needle is usually called a stylus.
>
> The first flat disc recording was made in 1880 by Emile Berliner, a German, working in America. He called the machine a gramophone.
>
> 1899 saw the introduction of a 'tape recorder'. The inventor was Danish born Valdemar Poulsen. His machine was called a ribbon telegraphone. The tape was thin piano wire normally used to make the strings on pianos. This machine was first demonstrated publicly at the Paris Exhibition of 1900.
>
> In 1929 a German, Dr. Fritz Pfleumer developed the first plastic tape with a magnetic coating. This made tapes much lighter.
>
> The idea of tape recordings has remained almost unchanged since 1929.
>
> Scientists have made recording and playback equipment smaller and lighter over the years. The Dutch company Philips introduced the cassette tape in 1963. The Sony company of Japan introduced the Walkman in 1980 which along with similar equipment is really the most popular lightweight tape system available today.
>
> On the disc side, flat discs were made from plastic in 1948, and CDs have followed since their launch in 1983. A CD is plastic, coated with a thin layer of aluminium, followed by a thin layer of lacquer to protect the surface.

a) Who first recorded sound, and what was his nationality? (2)
b) On what was the first sound recordings made? (1)
c) What was the first improvement made in early recordings? (1)
d) What name was given to the first machine to play flat discs? (1)
e) What type of recording 'tape' was used on the first 'tape recorder'? (1)
f) Where and when was the first public demonstration of this 'tape recorder'? (2)
g) How were recording tapes made lighter? (1)
h) When was the Walkman introduced? (1)
i) What three materials are used in the construction of a CD? (3)

2 In 1967 R M Dolby in America introduced the Dolby system for tape recording. How has this system improved the sound quality of cassette tape recordings? (2)

12.10.2 A brief history of sound recording

1 Read the following about records and then answer the questions which follow.

> In 1877 American Thomas Edison became the first person to record sound. He used a needle to mark the sound as grooves in tin foil on a cylinder. When the needle moved back through the grooves the original sound was heard. In 1886 he introduced a wax cylinder. The first flat disc recording was made in 1880 by Emile Berliner, a German, working in America. He called the machine a gramophone.
>
> Flat discs were made of plastic for the first time in 1948, and this soon became the normal material, and CDs have followed since their launch in 1983. A CD is plastic, coated with a thin layer of aluminium which reflects the laser light, followed by a thin layer of lacquer to protect the surface.

a) In what year was sound first recorded? (1)
b) Who first recorded sound, and what was his nationality? (2)
c) What material was first used to record sound on? (1)
d) What improvement was made in 1886? (1)
e) Who invented the flat disc recorder, and when? (2)
f) What name was given to the first flat disc machine? (1)
g) CDs are made from plastic. Which metal is put on top of the plastic to reflect light? (1)

2 Read the following about tape recording and then answer the questions which follow.

> 1899 saw the introduction of a 'tape recorder'. The inventor was Danish born Valdemar Poulsen. The tape was thin piano wire normally used to make the strings on pianos. This machine was first demonstrated publicly at the Paris Exhibition of 1900.
>
> In 1929 a German, Dr. Fritz Pfleumer developed the first plastic tape with a magnetic coating. This made tapes much lighter.
>
> Scientists have made recording and playback equipment smaller and lighter over the years. The Dutch company Philips introduced the cassette tape in 1963. The Sony company of Japan introduced the Walkman in 1980.

a) In 1899 the first 'tape recorder' was invented. From what was the first tape made? (1)
b) Where and when was the first 'tape recorder' demonstrated in public? (2)
c) When was the first plastic recording tape invented? (1)
d) What advantage did the plastic tape have over the earlier type? (1)
e) Who introduced the Walkman? (1)

12.10.3 A brief history of sound recording

1 Study the information below and then answer the questions which follow.

> Until the 1950s sound recordings were made with just one microphone. Records were described as mono, which stood for monophonic. If the sound was sent out through more than one speaker, all speakers sent out the same sound.
>
> The introduction of a worldwide stereophonic recording system in 1958 meant that two microphones could be used. The sound entering each is different. It is stored on tape or disc and then reproduced through one of two speakers. This is known as stereophonic sound. It is possible to have a singer or one particular instrument just from one of the two speakers, matching their original positions in the recording session.
>
> Today the Dolby system uses many different channels. In cinemas speakers almost surround the audience and can make an actor's voice move across the screen exactly as the actor moves.

a) What is the full name given to recordings made with just one microphone? _____ (1)

b) What shortened version of your answer to part (a) is commonly used to describe this type of recording? _____ (1)

c) When did a worldwide standard for stereo records come into use? _____ (1)

d) How many speakers are needed to demonstrate a stereo sound system? _____ (1)

e) Which system is used today in many cinemas to place a sound at a particular point using many sound channels? _____ (1)

2 Read the following facts and use them as well as your own knowledge to answer the questions which follow.

Facts:
CDs — virtually lifelong if used normally.
— produce near perfect sound recordings.
Tapes — need careful handling to last a long time
— wear out as the magnetic particles rub off the plastic tape.
LPs — can be scratched easily.
— wear out because a needle runs over the surface for each playing.

a) Suggest why CDs have taken over from LPs. _____ (4)

b) Why are cassette tapes the cheapest form of music recordings available today? _____ (2)

c) Why do CDs last longer than LPs? _____ (4)

d) Why does it take longer to find a particular song on a tape than on a CD? _____ (2)

Name: **Class:** **Date:**

Science Companion © A Porter, M Wood, T Wood and Stanley Thornes (Publishers) Ltd 1995

13.1.1 Voyage to the Moon

1. This is a research investigation. You will need to visit a library to find out the answers to many of these questions. Look out for a book on space, astronomy, astronauts or the Moon. Use the index at the back of any book to see if the clue words (in bold) are there. You may think of your own clue words.

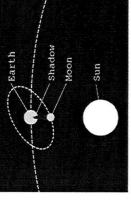
Y. Gagarin
first man in orbit

a) When did the first human being orbit the Earth in space? (clue word: **Gagarin, Y.**) (1)

b) How far away is the Moon from the Earth? (clue word: **Moon**) (1)

c) What was the date when the first human being set foot on the Moon? (clue words: **Moon landings**) (1)

d) Neil Armstrong was the first human to set foot on the Moon. Who were his crew mates and what were his first words as he jumped onto the surface of the Moon for the first time? (clue words: **Armstrong, Neil**) (3)

e) What happened to the third mission to the Moon and how did the astronauts get back to Earth? (clue words **Apollo 13**) (2)

2. This question is about eclipses.

a) How long does it take the Moon to travel around the Earth? (1)

b) In your own words explain how an eclipse of the Sun happens. (2)

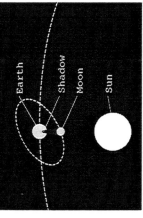

c) What happens in the following type of eclipses?
  i) total eclipse (1)
  ii) partial eclipse (1)

d) What is the difference between a lunar eclipse and a solar eclipse? (2)

Science Companion © A Porter, M Wood, T Wood and Stanley Thornes (Publishers) Ltd 1995

13.1.2 Voyage to the Moon

 V. Tereshkova — first woman in orbit

 Laika — first animal in space

 Colombia — first shuttle in space

 J. Glenn — first American in orbit

 Voyager — first probe to Neptune

 Y. Gagarin — first man in orbit

 Apollo 11 — first men on the Moon

 Sputnik — first satellite in orbit

1. Use your judgement to match the eight space 'firsts' shown above to the dates when they happened. (8)

   October 1957, March 1961, April 1961, February 1962, June 1963, July 1969, April 1981, September 1989.

2. A total eclipse of the Sun is due in 1999. It will be seen from parts of Devon and Cornwall.

   a) What will people on the ground see during the eclipse? (2)

   b) Put the Earth, Moon and Sun in order of size starting with the largest. (3)

   c) Use this picture to help you to explain how an eclipse works. (2)

Science Companion © A Porter, M Wood, T Wood and Stanley Thornes (Publishers) Ltd 1995

13.1.3 Voyage to the Moon

FAMOUS ASTRONAUTS

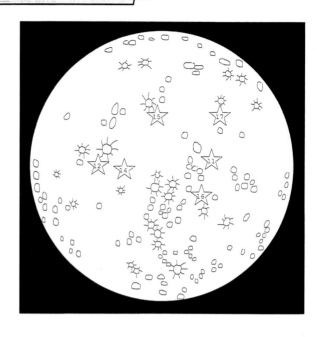

The map above shows the landing sites for all the Apollo missions which made it to the Moon. Complete the missing words in the passage below.

The words in bold can be found on the word-search grid. Circle each word as you find it. (16)

Yuri **Gagarin** was the first human in space. His mission lasted only 1 hour 48 minutes. About four months later Gherman **Titov** spent more than a day in space. The American, Gus **Grissom** entered space in 1961 but did not orbit the Earth. In August 1962 Pavel **Popovich** made 48 orbits of the Earth. The Americans were still far behind the Soviet astronauts. Two months after Popovich, Walter **Schirra** made only 6 orbits but in 1963 Leroy **Cooper** made 22 orbits in America's longest mission so far. The first woman in space was Valentina **Tereshkova** in June 1963.

The first two men on the Moon were part of the Apollo 11 mission. They were Neil **Armstrong** and Edwin **Aldrin** and spent about two and a half hours exploring the Moon. Apollo 12 followed with Charles **Conrad** and Alan **Bean**. Alan **Shepherd** was aboard Apollo 14. The two astronauts in Apollo 15 were David **Scott** and James **Irwin**. Apollo 16 astronauts spent over 20 hours walking and driving about the Moon on a rover vehicle. They were John **Young** and Charles **Duke**.

The first Apollo mission to land on the Moon was Apollo _____ which landed in July 1969. There were _____ landings on the Moon. The one mission which did not make it to the Moon was Apollo _____. Each mission carried two astronauts to the Moon. In all _____ people have walked on the Moon. The last mission was Apollo _____ which landed in December 1972. On the Moon the astronauts needed to wear _____ _____ because there is no _____ there. The landing sites had to be chosen carefully. They had to avoid any large _____ on the surface which could have damaged the lander. Since 1972 there have been _____ more Moon landings.

Name: Class: Date:

13.2.2 The planets

Planet	Distance from the Sun (millions km)	Time to go round the Sun	Time to rotate round its axis once	Diameter (thousands km)
Earth	150	365 days	24 hours	12.7
Jupiter	778	12 years	10 hours	144
Mars	228	780 days	24 hours	6.8
Mercury	58	88 days	59 days	4.9
Neptune	4496	165 years	16 hours	50
Pluto	5900	248 years	6 days	2.4
Saturn	1427	165 years	11 hours	121
Uranus	2870	84 years	17 hours	51
Venus	108	225 days	243 days	12.1

1 Which planet is the closest to the Sun? (1)
2 Which planet is furthest away? (1)
3 Which planet lies about half way between the Sun and Pluto? (1)
4 Which planet is nearly twice as far from the Sun as Jupiter? (1)
5 A year on Earth lasts 365 days. Which planets have a shorter year than Earth? (2)
6 A day on Earth lasts 24 hours. Which planets have a shorter day than Earth? (4)
7 Which planet has the longest day? (1)
8 Which planet's day is nearly ten times longer than a day on Pluto? (1)
9 Which planet has a day which is longer than its year? (1)
10 Which is the largest planet? (1)
11 How many planets are smaller than Earth? (1)
12 Make a list of the nine planets in order of size from the largest to the smallest. (1)
13 Pluto belongs to the five outer planets. Using the data in the table say what is unusual about Pluto compared with the other outer planets. (1)
14 Get a pair of compasses and set the pencil 72 mm from the point. Draw a circle using the compasses and label this Jupiter. (1)
15 Now draw a second circle with the compasses only 6 mm apart. Label this Earth. These pictures are to scale. (1)
16 Which planet would you be drawing if the compasses were set at 25 mm apart? (1)

13.2.1 The planets

Planet	Distance from the Sun (millions km)	Time to go round the Sun	Time to rotate round its axis once	Mass compared to Earth
Earth	150	365 days	24 hours	1
Jupiter	778	12 years	10 hours	318
Mars	228	780 days	24 hours	0.1
Mercury	58	88 days	59 days	0.06
Neptune	4496	165 years	16 hours	17
Pluto	5900	248 years	6 days	0.003
Saturn	1427	165 years	11 hours	95
Uranus	2870	84 years	17 hours	15
Venus	108	225 days	243 days	0.8

1 Make a list of the nine planets in order of:
  a) their average distance from the Sun. (1)
  b) the time to rotate once (1)
  c) their mass (1)
2 What is the relationship between the distance from the Sun and the time it takes to go round the Sun? (2)
3 What is the total mass of the eight planets Earth, Mars, Mercury, Neptune, Pluto, Saturn, Uranus and Venus and how does this compare to the mass of Jupiter? (2)
4 The average density of Earth is 5.5 g per cm³ and Jupiter is 1.3 g per cm³. Explain why this is so given that Jupiter is 318 times as massive as Earth. (2)
5 Which planets would you expect to be the hottest and the coldest? Explain why you have chosen these planets. (3)
6 The hottest planet is Venus. It has an atmosphere which is mainly carbon dioxide. Why would this make it the hottest? (2)
7 On which planet would a day seem to be the same as on Earth? (1)
8 There are four planets called gas giants. Which planets are these? (4)
9 The distance of Mercury from the Sun is an average figure. Why does the distance vary? (1)

13.2.3 The planets

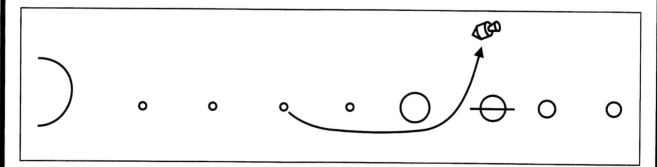

1 The drawing above was fixed on to the side of the Voyager 1 spacecraft. It is meant to show the path that the spacecraft took past some of the planets in the solar system. Put the following labels on the drawing.

**Earth   Jupiter   Mars   Mercury
Neptune   Saturn   Sun   Uranus
Venus   Voyager**   (10)

2 The picture below shows the nine planets in order from the Sun and roughly to scale. Use your knowledge or a text book to colour in the planets as accurately as you can. The best pictures to use are those taken by space probes which have visited the planets.   (9)

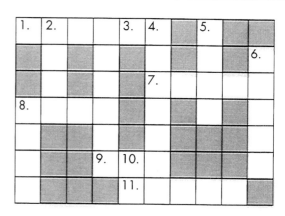

3 CROSSWORD – *Use the clues below to fill in the crossword above.*   (11)

1. The planet with the largest rings (6)
2. Comes before dynamic, plane and space (4)
3. The seventh planet from the Sun (6)
4. Planet named after the god of the sea (7)
5. The number of inner planets (4)
6. An icy lump orbiting the Sun (5)
7. The coldest planet (5)
8. *across* The Earth's partner (4)
8. *down* The red planet (4)
9. The star at the centre of the solar system (3)
10. Harmful light from this star (2)
11. The planet roughly the same size as the Earth (5)

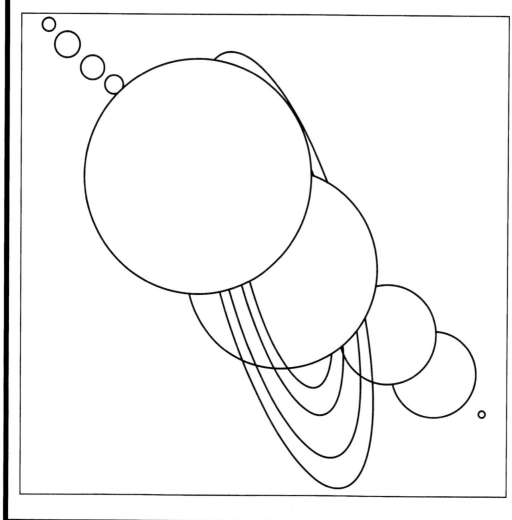

Name: **Class:**   **Date:**

Science Companion © A Porter, M Wood, T Wood and Stanley Thornes (Publishers) Ltd 1995

13.3.1 The Sun as a star

1. a) What is the main fuel which the Sun uses? (1)
   b) How is energy produced from this fuel? (2)
   c) Why is it difficult to see other stars during the day? (1)

2. a) What are sunspots? (1)
   b) How can sunspots tell us about the speed of rotation of the Sun? (1)

3. The Sun is 150 million km away from the Earth. Light from the Sun travels at 300 000 km per second.
   a) How many seconds does it take the light from the Sun to reach the Earth? (1)
   b) To the nearest minute, how long does the light take to travel this distance? (1)

4. The Sun and planets formed from clouds of dust and gas.

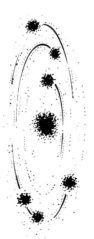

   a) What force caused the dust and gas to form the Sun? (1)
   b) When particles are squeezed together, what can happen to the temperature? (1)
   c) How did the planets which orbit the Sun form? (1)
   d) What do we call the collection of the planets and the Sun? (1)

5. a) Why will the Sun not go on shining forever? (1)
   b) The Sun will collapse into a very small object. What do we call these objects? (1)
   c) What happens to much larger stars at the end of their lives? (1)

13.3.2 The Sun as a star

1. The Sun uses hydrogen as a fuel but does not burn it.
   a) What happens to the atoms of hydrogen in the Sun? (10)
   b) What kind of reaction do we call this? (1)

   combustion   fission   fusion   decay

   c) The star Rigel is 50 000 times brighter than the Sun. Why can we not see Rigel in the daytime? (1)
   d) The planet Jupiter is mainly made from hydrogen. Why does Jupiter not shine like the Sun? (1)

2. What do we call cooler areas of the surface of the Sun? (1)

3. Why must you never look directly at the Sun? (1)

4. Say which is the odd one out in each of the following lists. Explain why you think it is the odd one out.

Mercury	Mars	Sun	Venus	(2)
Planets	Comets	Asteroids	Galaxies	(2)
Jupiter	Moon	Saturn	Venus	(2)

5. The Sun was formed from a swirling cloud of gas and dust.

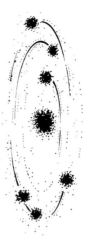

   a) What effect did gravity have on the cloud? (1)
   b) What happened to the temperature as this happened? (1)
   c) What do we call the collection of the planets and the Sun? (1)
   d) What will the Sun not go on shining forever? (1)

13.3.3 The Sun as a star

Read the following account of the life of stars and then use the words in the box to complete the flow diagram opposite. (11)

In the space between stars and galaxies there are huge clouds of dust and gas which are mostly remains of previous stars which have exploded. These gas clouds begin to contract as, over a long period of time, gravity pulls the pieces together. As this force makes the particles move faster, the cloud begins to increase in temperature. The centre of what will become the star can be seen and as it spins a disc of material spreads out from its equator. The next stage depends on how much dust and gas were present in the original cloud.

A small amount of material (less than 8% of the Sun's mass) means that the temperature will never rise high enough to start the nuclear reactions which make the star shine. It will remain a brown dwarf.

More than 8% of the Sun's mass means the star will begin to shine. The material close to the star will be blown off, leaving only small rocky planets. Away from the star planets may keep their large gassy atmospheres. The Sun has enough material to shine for 10 000 million years and it is about half way through its life cycle. When it runs out of material to burn it will expand to a size beyond the orbit of the Earth and then collapse to a planet sized white dwarf which will eventually cool down to a black dwarf.

If the star is very large (over 60 times the size of the Sun) then its life is much shorter. Higher temperatures mean that it gets through its fuel supply much faster and when it runs out of fuel (after only perhaps 10 million years) it explodes in a violent way (called a supernova). The material that is left collapses so dramatically that the atoms it is made of collapse in on themselves and all that remains are neutrons. This neutron star is so dense that one cubic centimetre would be about 300 million tonnes.

WORDS TO USE	dust cloud
neutron star	supernova
below 8% Sun mass	contracting cloud
steady use of fuel	brown dwarf
white dwarf	black dwarf
over 60 times Sun mass	rapid use of fuel

START HERE

[Flow diagram with boxes branching into SMALL MASS, LARGE MASS, and MEDIUM MASS pathways]

1 How are planets formed from a star?
   _____
   _____ (2)

2 What is a brown dwarf?
   _____
   _____ (1)

3 What is a white dwarf?
   _____
   _____ (1)

4 What is a black dwarf?
   _____
   _____ (1)

5 How large does a star have to be (compared with the Sun) in order to explode as a supernova?
   _____ (1)

Name: Class: Date:

Science Companion © A Porter, M Wood, T Wood and Stanley Thornes (Publishers) Ltd 1995

13.4.2 Our views of the universe

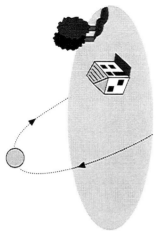

1 a) In which general direction does the Sun appear to rise in the morning? (1)
   b) In which direction does the Sun set? (1)
   c) At midday, towards which general direction are shadows cast? (1)
   d) Are shadows longer or shorter in winter than in summer? (1)
   e) Explain your answer to part (d) (1)

2 Unscramble the names of these scientists who helped to explain the way our universe works. (4)

   peculiar concussion
   plane has jerk on
   I age illegal oil
   nice as a town

3 a) What invention allowed Galileo to take a closer look at Jupiter? (1)
   b) What did he see orbiting Jupiter? (1)
   c) Why did his observations get him into trouble with the Church? (2)

4 Stars are recognised by the patterns they make in the night sky.
   a) What name do we give to these patterns? (1)
   b) What is this pattern of stars called? (1)

   the plough   orion
   gemini   libra

13.4.1 Our views of the universe

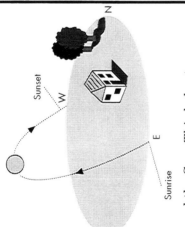

1 Why are shadows longer in the morning and afternoon, but shorter at midday? (2)

2 a) Why are days longer in the summer than in the winter? (2)
   b) Why is the Sun warmer when it is higher in the sky? (1)

3 Today we believe that the Earth goes round the Sun. This helps to explain many of our observations.
   a) What is unusual about the movement of planets across the sky? (1)
   b) Who was the first man to explain the solar system by putting the Sun at the centre and making the Earth orbit? (1)
   c) Who worked out the detailed movements of the planets in the solar system? (1)
   d) The moons of which planet did Galileo discover? (1)
   e) Why did this destroy the idea that the Earth was the centre of the universe? (1)
   f) What invention allowed Galileo to make his observations? (1)
   g) Which scientist worked out the laws of gravity which keep the planets in orbit round the Sun? (1)

4 The photograph was taken by leaving the camera lens open for several minutes on a clear night.
   a) What are the streaks of light caused by? (1)
   b) Why are the streaks in a circular pattern? (1)
   c) Which star is at the centre of the picture? (1)

13.4.3 Our views of the universe

1. The diagram alongside shows the shadow cast by a flagpole at noon on 30th June. Read the three different times below and draw your own shadow on the three diagrams.
   a) 8 am 30th June (2)
   b) 4 pm 30th June (2)
   c) noon 30th December (2)

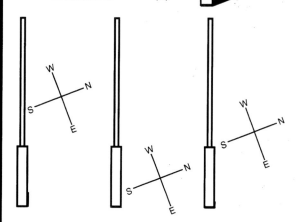

2. Use the words below to complete the passage.

   longer   shorter   higher   lower

   In winter the Sun is _____ in the sky at noon than in summer. This means that shadows appear _____ in winter. The Sun takes a _____ time to cross the sky which is why the nights are _____ in winter than in summer. At noon the Sun is _____ in the sky than it is in the morning. (5)

3. Put the answers to the clues into the crossword grid. (19)

Across
1. Nicolaus _____ was the first man to say the Earth goes round the Sun.
6. Sir Isaac _____ explained how gravity affected the planets.
7. In 24 hours the Earth turns on its axis _____.
8. The pull of the Moon's gravity has an _____ on the tides on Earth.
10. Shadows are _____ in the evening.
11. The place to see stars is in the _____ .
13. The sundial, candle and egg _____ are early types of clock.
16. A spacecraft needs less energy to _____ from the Moon's gravity than the Earth's.
17. The time to observe the stars is at _____.
18. The time for the Earth to go round the Sun once.

Down
1. The patterns the stars make in the sky.
2. Modern telescopes are more _____ than the one Galileo used.
3. There are _____ major planets in the solar system.
4. We keep track of the months and years by using _____.
5. The Sun takes longer to cross the sky in _____ than in winter.
9. The Sun is at the _____ of the solar system.
12. Johannes _____ worked out the orbits of the planets round the Sun.
14. When galaxies collide they _____ into one large galaxy.
15. Astronomers use a tele _____ .

Name: Class: Date:

Science Companion © A Porter, M Wood, T Wood and Stanley Thornes (Publishers) Ltd 1995

Answers

❑ 1.1.1 What do living things do? ❑

1 Activities are in this order:
Moving
Excretion
Reproduction
Sensitivity
Respiration
Feeding
Growing (7)

2 **Plants** **Animals**
have chlorophyll unable to make food
do not move around have no chlorophyll
have roots have nerves and
have leaves   muscles
have no nerves or move around
   muscles have no roots
make food have no leaves (6)

3 A snake needs to move around to find food. It has muscles which enable it to move along the ground. A plant can make food, so it does not need to move around for food. It can move its flowers and leaves. (4)

4 a) A change in the environment which initiates a response in an organism. (2)
b) Temperature, pain, sound and touch. (2)
c) Light, temperature, gravity. (2)

5 The removal of waste substances from the body. (2)
a) water, carbon dioxide, urea. (2)
b) respiration. (2)

6 Sight, touch, taste, smell, hearing. (5)

7 Diagrams. (4)

8 A leaf cell would contain a cellulose cell wall, chloroplasts and a sap-filled vacuole in addition to the other parts, i.e. nucleus, cell membrane and cytoplasm. The muscle cell would not. (2)

❑ 1.1.2 What do living things do? ❑

1 As 1.1.1 Q1. (7)

2 Explain that animals take in food and break it down into simple substances. Plants do not. (4)

3 Make suitable notes, e.g. snake has muscle to enable it to move. Plants do not move around but leaves and flowers respond to stimuli. (14)

4 Because they make their own food. (2)

5 Cell wall*
cell membrane
vacuole*
nucleus
chloroplasts*
*only in plants (12)

❑ 1.2.1 Organs and organ systems ❑

1 A cell is one unit of a living thing. A tissue is a collection of cells which make a layer. They carry out jobs like lining organs. An organ is a part of the body which carries out a particular job (3)

2 a) kidneys (1)
b) lungs (1)

3 Description as in 1.2.2 Q5. (2)

4 An organ system carries out a larger job in the body, e.g. digestion. (2)

5 a) mitochondria
b) chromosomes
c) chloroplasts
d) starch grains
e) vacuoles (5)

6 Any, e.g.   leaf   stem   flower (2)

7 a) E liver   F stomach   G pancreas
H small intestine   J large intestine (5)
b) Food is broken down into simpler substances which can be absorbed by the blood. (2)
c) Large intestine (1)
d) Small intestine (1)
e) Liver (1)
f) Pancreas (1)

Science Companion © A Porter, M Wood, T Wood and Stanley Thornes (Publishers) Ltd 1995

Answers

8 a) X is kidney   Y is bladder (1)
b) Kidneys filter the blood and the bladder stores urine. (1)
c) Bring blood to and from the kidneys. They are the main artery and vein. (1)

❑ 1.2.2 Organs and organ systems ❑

1 Cell (2)

2 Part of an organism which carries out a particular job. (2)

3 Any 3 organs (3)

4 Brain, heart, lungs, spine. (3)

5 a) kidney
b) lungs
c) heart
d) liver
e) brain (5)

6 a) As 1.2.1 Q7. (5)
b) 6 metres (2)
c) large intestine (2)
d) small intestine (2)
e) liver (2)

7 As 1.2.1 Q8. (2)

❑ 1.3.1 Human reproduction ❑

1 Sperm in the testes and eggs in the ovaries. (4)

2 Egg   A membrane
        B nucleus
        C cytoplasm
Eggs are larger than sperms. They contain food for the embryo.
Sperms   A head
         B middle
         C tail
         D nucleus

Sperms are smaller than eggs. They have a tail to swim in the watery lining of the uterus to reach the egg. (8)

3 Nuclei (1)

4 Eggs contain food for the embryo as it begins to develop. (2)

5 Puberty (1)

6 11–16 in boys and 8–15 in girls (2)

7 Boys grow body hair, their voice deepens and they produce sperms.
Girls start to have periods, their breasts and hips grow. (4)

8 One every 28 days. (1)

9 When a sperm and an egg meet and join. (2)

10 a) D
b) B
c) A
d) C (4)

11 One (1)

❑ 1.3.2 Human reproduction ❑

1 Eggs (1)

2 Sperms (1)

3 As 1.3.1 Q2. (7)

4 a) sperm (1)
b) egg (1)
c) egg (1)

5 As 1.3.1 Q7. (3)

6 a) once in 28 days
b) fertilisation
c) fallopian tubes
d) uterus
e) testes
f) one
g) embryo (14)

7 9 months (1)

Science Companion © A Porter, M Wood, T Wood and Stanley Thornes (Publishers) Ltd 1995

Answers

❏ 1.4.1 Blood and circulation ❏

1 A liquid called plasma with blood cells floating in it. (1)

2 Carry substances and food. Carry oxygen and carbon dioxide and fight diseases. (3)

3 In tubes called vessels. (1)

4 a) A platelet   B white cell   C red cell (3)
   b) i) C ii) B iii) C iv) A v) B vi) B (6)

5 It means that you cannot get a disease because cells in your blood prevent it from developing. (2)

6 An injection of a mild form of a disease which activates the white cells to produce antibodies for that disease. This stops you getting the disease. (2)

7 a) heart (1)
   b) cardiac (1)

8 **Veins**   **Capillaries**
   Thin walls and valves   Walls one cell thick
   Carry blood to the   Carry blood to
   heart   every cell (6)

9 a) Copy diagram (1)
   b) arteries right (1)
   c) veins left (1)
   d) arteries away from the heart and towards veins (1)

❏ 1.4.2 Blood and circulation ❏

1 a) red cells
   b) dissolved
   c) oxygen
   d) disease (4)

2 As 1.4.1 Q4. (3) (6)

3 a) immunity
   b) antibodies
   c) antigens
   d) vaccination
   e) heart (10)

4 a) oxygenated (1)
   b) to withstand the pressure of oxygenated blood as it is forced out of the heart (2)
   c) away from (1)
   d) they have valves and thin walls (2)
   e) capillary (1)

❏ 1.5.1 Respiration ❏

1 One is the release of oxygen from food and the other the exchange of gases. (2)

2 Mitochondria (1)

3 Aerobic and anaerobic (2)

4 Anaerobic (1)

5 a) i) oxygen
       ii) glucose
       iii) water, energy and carbon dioxide (in any order) (5)
   b) carbon dioxide and water (2)
   c) moving our muscles (1)
   d) glucose + oxygen → carbon dioxide + water + energy (2)

6 a) Adenosine triphosphate (2)
   b) The A.T.P. molecules store chemical energy in mitochondria until it is needed. (2)

7 a) A is combustion and B is respiration (2)
   b) A could be a fuel burning and B could be occurring in muscle cells. (2)

8 a) fungus (2)
   b) By converting sugar into alcohol and carbon dioxide. It is called alcoholic fermentation. (2)
   c) Carbon dioxide makes the bread rise and makes the bubbles in the beer. Alcohol gives beer its taste. (2)

❏ 1.5.2 Respiration ❏

1 Anaerobic and aerobic. (2)

2 Cells which produce energy. (2)

Answers

3 a) i) oxygen
  ii) glucose
  iii) water, carbon dioxide and energy. (5)
  b) lungs (2)
  c) food (2)
  d) working the body (2)

4 oxygen, carbon dioxide and water. (3)

5 a) combustion
  b) internal respiration
  c) breathing
  d) alcoholic fermentation (4)

6 a) fungus (2)
  b) alcoholic fermentation (2)
  c) i) carbon dioxide and alcohol (2)
  ii) carbon dioxide (1)
  iii) alcohol (1)

❏ 1.6.1 Breathing ❏

1 Lungs, respiratory. (2)

2 a) Chest b) Throat c) Voice box
  d) Windpipe (4)

3 a) Ribs
  b) Diaphragm
  c) Intercostal muscles (3)

4 a) Alveoli (1)
  b) Thin walls, large surface area, moist. (2)
  c) Oxygen is dissolved in the moisture of the lungs. Oxygen is passed from the alveoli across the walls of the blood capillaries into the red cells. Carbon dioxide passes from the plasma out of the blood into the alveoli. (4)

5 Mouth – bronchi – bronchioles – alveoli – blood. (1)

6 a) A inspiration B expiration (1)
  b) More carbon dioxide in A and air is saturated. (2)
  c) Some is taken into the blood (1)
  d) The concentration of carbon dioxide is greater in the alveoli than in the atmosphere (1)
  e) The lungs are moist. (1)
  f) Air must dissolve before it can pass into the blood. (1)
  g) To every cell. Respiration. (1)

❏ 1.6.2 Breathing ❏

1 Breathing (1)

2 As Q2 1.6.1. (4)

3 As Q5 1.6.1 (5)

4 a, d and e are true. (6)

5 a) 21% (1)
  b) 16% (1)
  c) The concentration is higher in the lungs. (2)
  d) The lungs are moist (1)
  e) The air is full of water (2)
  f) Respiration (2)

❏ 1.7.1 Good health ❏

1 Eating the correct amounts of each type of nutrient each day to keep healthy. (1)

2 Sugar, fat and salt. Eating large amounts of these can lead to high blood pressure and being overweight. (4)

3 a) A
  b) E
  c) C
  d) B
  e) D (5)

4 a) The amount of poultry eaten has increased. The consumption of carcass meats has decreased. (2)
  b) People are more health conscious and are eating less fat. (2)
  c) Sausages, pies and foods like burgers. (2)

5 A microbe which causes disease. (2)

6 Any in the text. (3)

7 a) The heart is a muscle and has to be exercised to keep it healthy. Exercise improves circulation and helps to prevent furring of the arteries.
  b) It keeps the muscles and bones healthy, keeps weight down, keeps lungs healthy and reduces stress. (2)

Answers

8 Cancer of the liver and paralysis (2)

9 Own answer (5)

❏ 1.7.2 Good health ❏

1 a) There is less chance of infections of the body and the body does not smell.
b) Being overweight can lead to illnesses like heart disease.
c) Keeps a body healthy, improves circulation, keeps vital organs healthy and reduces stress.
d) There is less risk of lung cancer and other respiratory diseases. (4)

2 i) A ii) B iii) A iv) B v) C
vi) A vii) C viii) C ix) A
x) A (10)

3 a) People are eating less fat. (1)
b) Less (1)
c) Other meats (1)
d) Pies, sausages and burgers (3)

4 Protists, bacteria, viruses and fungi (1)

5 b (1)

6 Any in the text (2)

7 Own answer (6)

❏ 1.8.1 When things go wrong ❏

1 a) Heart disease (1)
b) Eating too much fat (1)
c) Smoking (1)
d) Regular exercise. Eat less fat. No smoking. Less stress. (1)
e) Own graph (5)

2 a) Tar: blocks and irritates the lungs. Causes coughs and infections.
Dust: irritates the lungs, causes coughs.
Carbon monoxide: a poisonous gas, damages cells and narrows arteries.
Nicotine: an addictive drug. It affects the brain and raises blood pressure. (8)
b) Own answer (2)

3 a) Decayed enamel and infected tooth (2)
b) Bacteria (1)
c) Toothache, bad breath or loss of tooth (1)
d) Sweets (1)
e) Brush twice a day. Eat fewer sweet foods. Visit dentist twice a year. (2)
f) Own answer (1)

4 Own answer should comment on how health is affected. (3)

❏ 1.8.2 When things go wrong ❏

1 a) Heart disease (1)
b) Eat too much fat (2)
c) Smoking (2)
d) As 1.8.1 Q1d (2)

2 a) Tar, carbon monoxide, nicotine, dust. (2)
b) i) Carbon monoxide ii) Nicotine
iii) Tar, dust iv) Nicotine (5)
c) Own answer (2)

3 a) Bacteria (2)
b) They turn sugar into acid which dissolves teeth. (2)
c) Toothache (2)
d) Sweets (2)
e) Sugary drinks (2)
f) As 1.8.1 Q3e (2)
g) Own answer (2)

❏ 1.9.1 Micro-organisms ❏

1 Any organism which can be seen under a microscope. (1)

2 Protozoa, virus, fungi and bacteria. (1)

3 a) A is the fungus, B is the bacteria. (2)
b) B will make you ill. (2)
c) Everywhere, in air soil and water. (1)
d) Round, rod shaped and spiral. (1)
e) Mould, yeast or mushrooms. (1)
f) By splitting into two. (2)
g) Cold, dry and high temperatures. (1)

Answers

4 a) Cream (1)
b) Pasteurised (1)
c) Bacteria (1)
d) 'Use by' date in fridge (1)
e) Between 2 to 4 degrees (1)
f) Slows bacterial growth. (1)

5 a) Cook thoroughly, do not store with or prepare with cooked food. Wash hands and utensils after use.
b) Wash feet each day. Do not share towels.
c) Do not mix sewage and drinking water. Sewerage sytems should be built.
d) Having injections. Killing breeding grounds. (4)

6 Own answer (3)

❑ 1.9.2 Micro-organisms ❑

1 As 1.9.1 (2)

2 Athlete's foot and flu (2)

3 a) Fungi (1)
b) Warm and moist (2)
c) Spores (1)
d) Athlete's foot (1)
e) Penicillin (1)

4 a) Fresh cream (1)
b) Pasteurised (1)
c) 2 to 4 degrees (1)
d) Bacteria (1)
e) In the fridge (1)
f) It has a sell by date. (1)

5 a) Micro
b) Multiply
c) Saprophytes
d) Humus
e) Decay. (5)

6 Own answer (4)

❑ 1.10.1 Soil, decay and compost ❑

1 The top layer of the earth's surface. (1)

2 a) The action of rain, heat and snow on rocks. (1)
b) Sandy soil (1)
c) Grains are flat with small spaces in between them. (1)
d) Layers of different rock types in a section of the earth, down to the parent rock (1)
e) Rotted vegetation formed by the decomposition of dead plant material (2)
f) Loam (1)

3 a) Sandy (1)
b) The biggest layer is sand. (2)
c) Gravel. (1) They fall to the bottom first. (1)
d) Dead plants. (1)

4 A: sandy soil. It contains little humus. It drains well so does not get too wet. It contains few nutrients because they get washed through the soil.
B: loam. Because it contains nutrients and retains moisture well. (6)

❑ 1.10.2 Soil, decay and compost ❑

1 a) Weathering (1)
b) Sandy (2)
c) It contains small flat particles with small spaces between them. (1)
d) Soil horizon (1)
e) Rotted vegetation (2)
f) Loam. It contains nurients in the humus and retains water. (1)

2 A: likes sandy soil because it has large particles in it. It does not get too wet and does not contain many nutrients.
B: likes loam because it needs the humus which retains moisture and nutrients. (7)

3 a) Soil profile
b) Leaching
c) Humus
d) Fertile
e) Sandy (5)

Answers

❑ 1.11.1 Reproduction in flowering plants ❑

1 a) Support and transport of nutrient and water.
  b) To make food
  c) To take up nutrients and water.
  d) Houses the reproductive organs. (4)

2 By sexual or asexual reproduction. (1)

3 a) Diagram (7)
  b) i) Stamen (1)
   ii) Carpel (1)
   iii) Petals (1)
   iv) Sepals. (1)

4 a) A: stamen B: carpel (2)
  b) E: stigma F: style G: ovary
   H: ovule (4)
  c) i) Stigma (1)
   ii) Anther (1)
   iii) Ovule (1)
   iv) Style (1)
   v) Ovule (1)
   vi) Ovary (1)

5 The pollen is taken from the stamen of a plant to the carpel of a plant by an animal or by the wind. (2)

6 Pollen is carried by the wind to a carpel. (1)

7 Use the words in the explanation. (4)

❑ 1.11.2 Reproduction in flowering plants ❑

1 a) Leaves
  b) Flower
  c) Roots
  d) Stem (4)

2 a) Own answer (1)
  b) Own answer (7)
  c) i) Stamen (1)
   ii) Carpel (1)
   iii) Petals (1)
   iv) Sepals (1)

3 a) Stamen (2)
  b) Filament (2)
  c) Stigma (2)
  d) Ovary (2)
  e) Ovary (2)

4 a) Pollination (2)
  b) Fertilisation (2)

5 Runner bean Lupin Peas
  Poppy Pansy. (5)

❑ 1.12.1 Leaves and photo-synthesis ❑

1 To make food (1)

2 Thin to allow the diffusion of gasses and broad and flat to absorb sunlight. (2)

3 A: chloroplasts B: vacuole C: nucleus
  D: air spaces (4)

4 a) Cuticle (1)
  b) Epidermis (1)
  c) Palisade layer (1)
  d) Mesophyll layer (1)
  e) Spongy layer (1)

5 a) Carbon dioxide + water ⟶ glucose + water (2)
  b) Chlorophyll (1)
  c) Chloroplasts (1)
  d) Glucose is used for respiration and growth and some is stored. Glucose molecules are small and soluble in water so they have to be converted into starch to be stored. Oxygen is a waste product and is removed from the plant. (4)

6 a) Stoma (1)
  b) It allows the passage of gases and water vapour. (1)
  c) Carbon dioxide (1)
  d) Oxygen (1)
  e) To conserve water (1)

Answers

❑ 1.12.2 Leaves and photosynthesis ❑

1 It makes food. (2)

2 a) To absorb as much sunlight as possible.
b) To allow the passage of gases in and out of the leaf. (2)

3 a) As 1.12.1 (4)
b) Chloroplasts (1)
c) Chlorophyll (1)

4 a) dioxide + water $\xrightarrow{\text{chlorophyll}}$ oxygen (2)
b) Carbon dioxide and water (2)
c) Glucose and oxygen (2)
d) Starch (2)
e) Respiration (2)

5 a) Stomata (2)
b) Carbon dioxide and oxygen (2)
c) To save water. (2)

❑ 2.1.1 Living organisms ❑

1 Putting organisms into groups. (1)

2 a) Plants and animals (1)
b) Plants make their own food. Plants do not move from place to place. (2)

3 Feeding, growing, moving, respiration, growth, sensitivity, excretion. (1)

4 a) Both have flowers/both have leaves/both have seeds/both have roots (1)
b) Different shaped leaves/one has a bulb. (1)

5 a) B (1)
b) A (1)
c) A: Has scales, fins. Breathes through gills. B: Has wings, three parts to its body. (2)

6 With flowers: rose   daffodil   oak tree
horse chestnut tree   daffodil   grass
Without flowers: seaweed   fern
liverwort (2)

7 a) Honeysuckle
b) Oak
c) Wasp
d) Chlamydomonas
e) Snake
f) Pine tree (3)

8 Own answer. To include fur, scales, etc. (4)

❑ 2.1.2 Living organisms ❑

1 Plants and animals. (2)

2 Animals: shark   wasp   frog   jellyfish
The rest are plants. (2)

3 a) Buttercup is a plant, others are animals.
b) Mouse – others are plants.
c) Dandelion – others reproduce by spores.
d) Crab – others are molluscs. (4)

4 a)   i) A   ii) C   iii) B   iv) A, C, D (2)

5 a) Fur on body/feed young on milk/warm blood (2)
b) Soft body/no skeleton (2)
c) B has antlers. (2)
d) D has a shell. (2)

❑ 2.2.1 Grouping animals and plants ❑

1 Amphibians   Birds   Fish
Mammals   Reptiles (5)

2

A	B	F
toad	parrot	halibut
salamander	penguin	salmon
newt	eagle	eel

M	R
doormouse	chameleon
bat	lizard
elephant	python

(5)

3 Birds and mammals. (1)

Answers

4 a) R
   b) A
   c) F
   d) M
   e) B (5)

5 Any two, molluscs/arthropods (2)

6 Flowering and non-flowering (2)

7 Own answer (2)

8 Own answer. (8)

❏ 2.2.2 Grouping animals and plants ❏

1 Any five vertebrates (1)

2 Any five invertebrates (1)

3 As 2.2.1 Q2 (5)

4 a) Reptiles
   b) Amphibians
   c) Fish
   d) Mammals
   e) Birds (5)

5 Flowering: buttercup   oak   bluebell   rose. (2)

6 Own answer (8)

7 Own answer. Use shapes to divide into groups. (8)

❏ 2.3.1 Variation ❏

1 Variation (1)

2 Discontinuous   Continuous (2)

3 Discontinuous (1)

4 a) Discontinuous (1)
   b) Blood group/eye colour (1)
   c) There is no range of statistics. People are either one or the other. (1)

5 Birth weight is affected by smoking or poor diet. Fruit yield is affected by weather, light and food. (2)

6 In nature organisms best suited to living in an environment survive and breed. (2)

7 a) Orang-utan (1)
   b) Gorilla. He is heavier. (2)
   c) Orang-utan. He can get around the tree tops better. (2)

8 a) No animals living any longer. (1)
   b) A large flightless bird with short legs. (2)
   c) It cannot fly. (1)
   d) It could not fly to escape predators. It could not run to escape predators. It got diseases from man and his pets. There was competition for food. Or other suitable suggestions. (3)

❏ 2.3.2 Variation ❏

1 Variation (1)

2 a) Continuous (1)
   b) Blood type/eye colour (1)
   c) Discontinuous (2)

3 Natural selection (2)

4 a) as 2.3.1 Q8 (2)
   b) as 2.3.1 Q8 (2)
   c) as 2.3.1 Q8 (2)
   d) Over 300 years. (2)
   e) Mauritius (2)
   f) As 2.3.1 Q8d (3)

❏ 2.4.1 Why do we look like our parents? ❏

1 Passing on features from parents to offspring. (1)

2 a) Eye colour/nose shape/ear shape/chin shape (2)
   b) Both parents have the genes for blue eyes. They would pass these on to the offspring. (2)

Science Companion © A Porter, M Wood, T Wood and Stanley Thornes (Publishers) Ltd 1995

Answers

3 a) Chromosomes (1)
   b) In the nucleus of a cell (1)
   c) Genes (1)
   d) 46 (1)
   e) 23 (1)

4 Draw the diagram and explain that the egg and the sperm both contain 23 chromosomes each. These add up to 46 when fertilisation takes place so that features from both parents are passed on to the embryo. (5)

2.4.2 Why do we look like our parents?

1 Heredity (2)

2 a) As 2.4.1 Q2 (3)
   b) Blood type eye colour (2)

3 a) Nucleus (2)
   b) 46 (2)
   c) 23 (2)

4 Mother – hair chin
   Father – ears (2)

2.5.1 Extinction – the end of a species

1 a) No food/ eaten by predators/ environment changes/ Other suitable explanation. (2)
   b) 66 to 220 million years ago. (1)
   c) Reptiles (1)
   d) Hard and scaly. (1)
   e) *T. rex*/ Allosaur (1)
   f) Triceratops (1)

2 a) The hardened remains of an animal preserved in rock. (1)
   b) Palaeontologist (1)
   c) How the animals lives. (2)

3 a) The earth was surrounded by a cloud of dust. This prevented the Sun's rays from getting to the Earth. The cold killed the dinosaurs. (2)
   b) Manicouagan crater in Canada. (2)

4 a) 66–220 million years. (1)
   b) There is a thin layer of this metal between the rocks containing fossils and those not containing fossils. It is rare on Earth but found in meteorites. (2)
   c) A layer of carbon also between the rocks with the iridium. (1)
   d) Because they were extinct. (1)

2.5.2 Extinction – the end of a species

1 a) No longer any living on Earth. (1)
   b) 66 to 220 million years. (1)
   c) Reptiles (1)
   d) *T. rex* (1)
   e) The skin is the same. They are reptiles. They lay eggs. (2)

2 a) i) (2)
   b) What the animals looked like. What they ate. (2)
   c) Palaeontologist (2)

3 a) A meteorite. (2)
   b) Earthquakes, landslides, tidal waves and global fires. (2)
   c) Dust (2)
   d) The lack of heat caused by the dust blocking the Sun. (2)

3.1.1 People – a threat to the Earth

1 Air – to breathe / cutting equipment
   Water – washing / drinking
   Food – to give energy / to grow
   Shelter – protection / to keep warm (8)

2 Increases in population / resources not replaced. (2)

3 a) Electricity production – power stations are ugly / smelly / cause pollution.
   b) Mining – uses up resources / minings are ugly / scars the land / waste tips.
   c) Deforestation – kills trees / increase in carbon dioxide in the air / causes soil erosion. (6)

Answers

4 a) In wetland bogs. (1)
  b) Fuel (1)
  c) Gardening (1)
  d) Machines are used which destroy habitat. (1)
  e) It will be used up. It takes 10 000 yrs to form. (1)
  f) Use peat-free alternatives like coconut fibre. (1)
  g) Own answer. (3)

❏ 3.1.2 People – a threat to the earth ❏

1 Earth – food, water, shelter in any order. (4)

2 a) Oxygen (1)
  b) Wheat (1)
  c) Water (1)
  d) Coal / gas (1)
  e) House (1)

3 Mining – coal
  Deforestation – wood
  Quarrying – limestone (3)

4 a) coal (1)
  b) ugly (1)
  c) acid rain / sulphur dioxide gas (1)
  d) electricity (1)
  e) nuclear power / burning gas (2)

5 a) Fuel (1)
  b) Gardening (1)
  c) Cutting peat destroys the habitat of plants and animals. (2)
  d) Coconut fibre / peat-free compost. (2)

❏ 3.2.1 The problem of pollution ❏

1 a) Pollution – damage to the Earth
  b) Pollutant – a substance which causes damage or pollution
  c) Biodegradable – can be broken down by nature
  d) Non-biodegradable – cannot be broken down by nature. (4)

2 The following are biodegradable:
  cardboard   leaves   woollen jumper
  compost   newspaper   wood
  grass cuttings   sewage (4)

3 a) Fossil fuels produce oxides of sulphur and nitrogen. These combine with water in the air to produce weak acids. This is acid rain. (2)
  b) i) Kills trees and plants / prevents the uptake of nutrients.
     ii) Minerals like aluminium leach into lakes and kill fish and plants. (4)

4 a) A layer of gas above the Earth. (1)
  b) It protects the Earth from harmful ultraviolet rays. (2)
  c) It is wearing thin. It has a hole in it. (2)
  d) Skin cancer. (1)

5 a) Raw sewage is dumped in the sea. Causes pollution of beaches.
  b) Factories dump chemicals into rivers. This kills fish and plants.
  c) Chemicals leach into the ground, this runs into streams and rivers. (3)

6 Acid rain   burning coal
  Global warming   carbon dioxide
  Lead poisoning   in the air
  Depletion of the   burning petrol
    ozone layer   use of C.F.C.'s
  Dead lakes   use of phosphates (2)

❏ 3.2.2 The problem of pollution ❏

1 a) Resource
  b) Toxic
  c) Environment
  d) Biodegradable
  e) Pollutant
  f) Non-biodegradable (6)

2 As 3.2.1 (9)

3 a) Fossil fuels (2)
  b) Oxides of sulphur and nitrogen (2)
  c) Sulphuric acid and nitric acid (2)
  d) Kills them (2)

Answers

4 Bleach – W
   Chlorofluorocarbons – A
   Metals – L
   Paper – L
   Petrol – A
   Plastic – L
   Phosphates – W
   Sewage – W
   Sulphur dioxide – A (2)

3.3.1 Are we doing enough?

1 Prosecute polluters / stop emission of harmful gases / stop raw sewage dumping. (1)

2 90% (1)

3 Oil sewage (1)

4 a) A harmful or poisonous substance
   b) Any two (1)

5 Lime (1)

6 Phosphate-free detergents (1)

7 a) A mixture of smoke and fog (1)
   b) The 1956 clean air act (1)

8 Burning alternatives to coal (1)
   Burning coal which contains less sulphur
   Converting power stations (1)

9 a) Flue gas desulphurisation (1)
   b) Sulphur dioxide (1)
   c) It is washed with limestone (1)
   d) Gypsum (1)
   e) Plasterboard (1)

10 Own answers (4)

3.3.2 Are we doing enough?

1 Q1 to 7 as 3.3.1 (same no. of marks)

8 a) Coal (1)
   b) Sulphur dioxide (1)
   c) Acid rain (1)
   d) Limestone (1)
   e) Gypsum (1)
   f) Plasterboard (1)

3.4.1 Living things and their environment

1 To survive (1)

2 Daily changes – light / temperature / moisture
   Seasonal changes – length of day, temperature. (2)

3 a) Active at night (1)
   b) Less likely to be attacked by predators (2)
   c) They feed on nocturnal animals or plants. (2)

4 Own answer (4)

5 a) To spend the winter sleeping (2)
   b) Lack of food (1)
   c) Metabolism slows down / use little energy / stored body fat is used as energy supply (3)
   d) Aestivation (1)

6 a) To escape cold weather / to breed (2)
   b) Birds / eels / whales / salmon (2)
   c) Sun, stars and magnetic fields (2)

3.4.2 Living things and their environment

1 Sleeping / eating / grooming (3)

2 Too cold / cannot get water from the ground (2)

3 a) It loses its leaves in winter. (1)
   b) Lime / oak / sycamore / ash / elm / or any (2)
   c) They have needles instead of leaves, this reduces water loss. Snow does not weigh on their branches. (2)
   d) The stems die down and plant spends winter underground. (1)
   e) Lily / primula / any other (2)
   f) Seeds (2)

4 a) i) bird (2)
      ii) squirrel (2)
      iii) wolf (2)
      iv) ladybird (2)

Science Companion © A Porter, M Wood, T Wood and Stanley Thornes (Publishers) Ltd 1995

Answers

5 These migrate:
arctic tern
penguin
porpoise
turtle
salmon
whale (2)

❏ 3.5.1 Why do animals and plants live in different places? ❏

1 Different animals and plants have different adaptations. (3)

2 Light   moisture   food   vegetation   temperature (2)

3 a) A arctic   B equator   C desert
      D antarctic (4)
   b) Arctic and antarctic cold and freezing. Equator – hot and wet all year, no seasons. Desert – very hot and dry. Cold at night. Little rain. (4)
   c) A – polar bear   lichens
      B – flying frog   orchid   toucan
      C – camel   head-standing beetle
      D – penguin (2)

4 a) Cactus and camel: deserts (2)
   b) No leaves and small surface area – reduces water loss
      Sharp spines – water drips off on to the roots. Give protection.
      Shallow roots – get water from soil
      Camel – large feet stop from sinking into the soil
      Fur insulates
      No water loss from sweating
      Hump has a store of water
      Eyebrows and lids protect from sandstorms (2)

5 a) Equator (2)
   b) Use trees for fuel / clear trees to grow crops. (1)
   c) Greenhouse effect is increased. (2)
   d) Own answer. (1)

❏ 3.5.2 Why do animals and plants live in different places? ❏

1 As 3.5.1 (3)

2 a)   i) B
        ii) C
       iii) A and D (6)
   b) A   polar bear
      B   orang-utan
      C   camel
      D   penguin (4)

3 a) desert (2)
   b) i) large feet (2)
      ii) small surface area and no leaves (2)
     iii) shallow roots (2)
      iv) bushy eyebrows and eyelids (2)

4 Own answer (1)

❏ 3.6.1 Populations within ecosystems ❏

1 i) ecosystem – a place where a lot of organisms live and interact
      ii) ecology – the study of organisms in their environment
     iii) habitat – the place where individual organisms live
      iv) population – a number of a species living in a habitat. (4)

2 Living – food
   Non-living – light (2)

3 i) Hedgerow
      ii) Damp log
     iii) Forest
      iv) Seashore (2)

4 a) They block out the light. (1)
   b) Shrub layer. (1)
   c) They can grow and flower before the canopy leaves open. (1)
   d) Damp leaves to feed on. (1)
   e) Hawthorne / Holly / Elder (1)

Answers

5 a) Fox (1)
b) Kill them (1)
c) Rabbits increase (1)
d) No food / lack of space / diseases spread (1)
e) Rabbits starve and die. No food for foxes (1)

6 Small lichens grow in polluted areas. Leafy lichens grow in unpolluted areas. (2)

❑ 3.6.2 Populations within ecosystems ❑

1 i) Population
ii) Food web
iii) Habitat
iv) Ecosystem
v) Ecology (5)

2 Frog – pond
Blackbird – forest canopy
Mouse – cornfield
Woodlouse – old tree trunk
Blackberry – forest floor. (5)

3 a) Canopy, shrub, herb, ground (1)
b) Elder / Holly (1)
c) There is less light (1)
d) Bluebells (1)
e) Fungi (1)

4 a) Fox (1)
b) Plants (1)
c) Bigger (1)
d) It would be less. They could starve. (1)

❑ 4.1.1 Recycle it ❑

1 a) To make new products from waste materials. (1)
b) Paper tissues / toilet paper / paper bags (1)

2 a) A resource which gets used up or cannot be used again. (1)
b) Coal / oil / gas (1)

3 a) A hole in the ground where rubbish is tipped. (1)
b) Toxic gases can be given off / animals spread diseases / chemicals leach into streams. (3)

4 Less resources are used and money and energy are saved when they are made. (2)

5 Textiles – taken to charity shops
Glass – recycle bins
Plastic – some can now be recycled
Paper – paper bin or merchant (2)

6 a) Pollutes water / kills fish and plants / destroys birds' feathers.
b) Causes poisonous fumes.
c) Chemicals leach into streams and rivers. (3)

7 a Smaller packs cost less to produce so saving is passed on to the customer. (1)
b) Smaller packs produce less waste packaging and more can be carried in one trip saving petrol and reducing pollution. (1)

8 a) A – drinks B – vegetables C – jam
D – tissue box (2)
b) It tells them where to take packs for recycling and what type of resources are used to produce them. (2)
c) Own answer (2)

9 Own answer (2)

❑ 4.1.2 Recycle it ❑

1 Paper / glass / aluminium (2)

2 a) Ugly to look at / smelly / spread diseases (2)
b) Methane (2)
c) Rats / birds (2)

3 a) Uses valuable resources / causes a lot of litter (2)
b) It is smaller so uses fewer resources. It produces less waste paper. (2)

Science Companion © A Porter, M Wood, T Wood and Stanley Thornes (Publishers) Ltd 1995

Answers

4 a) A – aluminium   B – recycled paper   (2)
   b) A – sell to recycle plant
      B – recycle bin   (2)
   c) Drinks / coffee   (2)
   d) Tissue box / stationery   (2)
   e) B   (1)

5 Own answer   (3)

❑ 4.2.1 Passing on energy ❑

1 Producers are plants which make the food that animals eat.
Consumers eat the food which plants make.
Decomposers break down dead organic material and returns the nutrients back into the soil.   (3)

2 Plants are producers. They convert energy from the Sun into chemical energy which is passed on to animals. Without plants there would be no food for animals.   (1)

3 Any suitable chains. They must be done correctly, i.e.
lettuce   snail   robin   (3)

4 It decreases.   (1)

5 a) It shows how many organisms there are at each trophic level.   (2)
   b) Less energy is passed along.   (1)
   c) e.g. respiration.   (3)

6 A 'pyramid of biomass' shows the mass of organisms and represents the amount of energy available rather than the number of organisms.   (2)

7
PREDATOR	PREY
mouse	owl
hedgehog	ground beetle
lioness	wildebeest
fox	chicken
dingo	lizard
mongoose	snake
aardvark	ant

(7)

8 a) Forests   (1)
   b) Insects and caterpillars   (1)
   c) It eats another animal.   (1)
   d) Protected by laws   (1)
   e) Leaves → insect → wood ant   (1)
   f) Ants would be killed. Trees would be damaged by the large numbers of insects and caterpillars.   (2)

❑ 4.2.2 Passing on energy ❑

1 a) Consumer   (2)
   b) Producer   (2)
   c) Decomposer   (2)

2 Any chain   (6)

3 a), c) and e) are true.   (3)

4 a) In forests, in mounds of earth.   (2)
   b) They eat the insects and caterpillars that eat the trees.   (2)
   c) Because they prevent damage to the trees they must be kept from dying out.   (2)
   d) An animal which feeds on another animal. Usually one level below it in the food chain.   (2)
   e) It means to support and help.   (2)

5 Own answer   (5)

❑ 4.3.1 The carbon and nitrogen cycles ❑

1 a) Non-metallic   (1)
   b) Diamond   (1)

2 Respiration / Combustion / Decomposition   (3)

3 a) A – combustion
      B – photosynthesis
      C – animal respiration   (3)
   b) Coal / oil   (1)
   c) Plants   (1)
   d) They use carbon dioxide from the air. They are the only organisms which can do this.   (1)
   e) Decomposers   (1)

Answers

   f) They break down dead plant and animal material, releasing carbon for recycling. (1)
   g) They eat plants. (1)
   h) Carbon dioxide builds up in the atmosphere, preventing the Sun's radiation from escaping. This warms up the atmosphere. (1)

4 a) Marble chips (calcium carbonate) and hydrochloric acid. (1)
   b) Turn limewater milky. (1)
   c) Acid (1)
   d) Drinks / fire extinguishers. (2)

4.3.2 The carbon and nitrogen cycles

1 a) non-metallic (2)
   b) CO_2 (2)
   c) Diamond (2)
   d) Oxygen (2)

2 a) Combustion (2)
   b) Respiration (2)
   c) Photosynthesis (2)
   d) Decomposers (2)
   e) Plants (2)

3 To turn limewater milky (2)

5.1.1 Types of materials

1 a) Found in the world and unchanged by people. (2)
   b) Made in the factories by people from raw materials. (2)
   c) The starting materials for making items in factories. (2)
   d) Made from clay. The clay is heated in an oven. Also known as pot. (2)
   e) A solid, often transparent and brittle, made by melting sand, limestone and sodium carbonate. (2)

2 Many examples, for instance:

Item	Old material
car bumper	chrome-plated steel
window frame	wood
carrier bag	paper (brown)
suitcase	leather or cardboard

(5)

3 a) nylon, acrylic (2)
   b) 60% in A, 35% in B (2)
   c) To try to give the final garment some of the best characteristics of each fibre. (1)
   d) So you know how to wash it, and decide whether the final product is suitable for winter or summer wear. (2)
   e) Wool traps the warm air around the body better than acrylic. (1)
   f) A wash by hand in hand-warm water (1)
   B wash in water at 40°C (1)

5.1.2 Types of materials

1 natural – E; manufactured – D; raw – A; ceramic – B; glass – C. (5)

2 a) It is lighter. (1)
   b) Cheaper, can be thrown away without washing. (2)
   c) Cheaper, can be moulded rather than fitted together. (2)
   d) Don't break when you drop them. Don't chip. Cooler to the touch. (2)
   e) Plastic doesn't corrode and is slightly flexible. (1)

3 a) Sirdar PLC, Wakefield (2)
   b) Patons and Baldwins Ltd., Darlington (2)
   c) 60% (1)
   d) 32% (1)
   e) B (1)
   f) nylon (1)
   g) cotton (1)
   h) 50 g (1)
   i) hand wash (2)

5.2.1 Properties, uses and development of materials

1 The surface of the tin is flat because it is a liquid. Gravity forces the glass down on to it making it flat too. (2)

Answers

2 Tensile strength is the resistance of a material to a pulling force. Compressive strength is the material's resistance to a squeezing force. (2)

3 It needs both good tensile and good compressive strengths. (2)

4 Beam A is strong because it is deep so will not bend easily under a load. However it is not wide enough to support much. (1)
Beam B is wide and will support a good area and support joints. However it is shallow and lacks strength. (1)
Beam C combines the strengths of both beams and so is widely used being strong but as light as possible. (1)

5 Metals are strong, easily pressed into shapes and can give a smooth finish. (3)

6 Research. (8)

❏ 5.2.2 Properties, uses and development of materials ❏

1 Diamond, corundum, apatite, calcite, talc. (5)

2 Talc is the softest thing known, calcite is harder, so we use the softest thing to be kind. (1)

3 a) yes (1)
  b) no (1)
  c) no (1)
  d) yes (1)

4 a) Z (2)
  b) Y (2)
  c) Z (2)
  d) X (2)
  e) Y (2)

❏ 5.3.1 Separating and purifying ❏

1 The particles are close together (1) but are free to move about. (1) (2)

2 In hot water the particles are moving faster than in cold water. (1)

3 Particles in steam are further apart than particles in water. (1)

4 a) point 3 (1)
  b) This is where the steam cools down below 100°C and condenses into liquid water. (1)

5 When the water evaporates (1), the solid substances remain behind on the element (1). (Pupils need to show they understand that the solid and the liquid have been separated when the water boils.) (2)

6 Distilled water is water which has been boiled into a gas (1) and then condensed into a liquid. (1) (2)

7 Solid substances will dissolve (1), in this case in the water. Insoluble means will not dissolve. (1) (2)

8 The soluble part of the coffee dissolves. (1)

9 In solution, the particles of the coffee are small enough to pass through the holes in the filter paper. (1) The insoluble particles are too big to go through the holes. (1) (2)

❏ 5.3.2 Separating and purifying ❏

1 The water boils/evaporates. Allow 'heats up'. (1)

2 100°C (1)

3 It has boiled (turned into a gas) (1) and then condensed (turned into a liquid). (1) (2)

Answers

4 a) Something impure is a mixture containing unwanted material. (1)
   b) The impurities are left behind when the water boils, they are left on the element. (1)

5 Condensed means has turned from a gas (1) to a liquid. (1)

6 a) colourless (1)
   b) brown (1)
   c) The water dissolved some of the colour in the coffee. (1)

7 The holes in the filter paper are big enough to let the water through (1) but small enough to trap the solids. (1) (2)

8 The high quality paper has smaller holes or is waterproof or won't let the water through it. (1)

9 Solid coffee particles dissolved in the water would be left behind. (1)

❑ 5.4.1 Acids, alkalis and indicators ❑

1 a) Caustic soda (1)
   b) Sodium hydroxide (1)
   c) Alkaline (1)
   d) They said caustic soda was an acid-like substance when in fact it is alkaline. (1)
   e) Dissolve some "Wonderwhite" in water. Add Universal Indicator Solution. The alkaline caustic soda should turn the indicator blue/purple. If it had been an acid it would have gone red. (3)

2 Mix equal amounts so that they balance each other. Have an indicator in the first one, add the other, with stirring, until the indicator just goes to its neutral colour. (3)

   NOTE: NOT equal volumes unless the acid and alkali are of equal CONCENTRATION.

3 a) acids (1)
   b) Toothpastes are often slightly alkaline to neutralise any acids in the mouth. (1)

4 One at a time, place a little of the contents of a bottle in a test tube. Add a few drops of UIS. (1)
   Results: sulphuric acid – red (1)
   pure water – green (1)
   ethanoic acid – orange/yellow (1)
   calcium hydroxide – blue/purple (1)

5 Varied, but possible answers are:
   Coca-cola – phosphoric acid
   orange-jelly marmalade – citric acid
   mayonnaise – vinegar (ethanoic acid) (3)

❑ 5.4.2 Acids, alkalis and indicators ❑

1 a) Wonderwhite (1)
   b) Washing clothes (1)
   c) 8 (1)
   d) West Coventry Hospital (1)
   e) Sodium hydroxide (1)
   f) An alkali (1)
   g) Sodium hydroxide is an alkali not an acid. (1)
   h) Add Universal Indicator Solution, if it goes red it is an acid, if it goes blue/purple it is an alkali. (3)

2 a) Sodium hydroxide (1)
   b) Burn it (2)
   c) Long rubber gloves (2)

3 Take a test tube. (1) Carefully pour a little liquid into it from 1 bottle. (1) Add a few drops of Universal Indicator Solution. (1) If the liquid is an acid the UIS goes red, (1) if an alkali (calcium hydroxide) it goes blue/purple. (1) (5)

❑ 5.5.1 The Periodic Table ❑

1 a) carbon (1)
   b) silicon (1)
   c) Group IV (1)
   d) Germanium, Group IV next element, and group IV is providing the peaks. (2)
   e) Group 0, Noble Gases. (2)
   f) Higher. So far Group II is always higher than Group I. (2)

Answers

2
a) sodium, Na (2)
b) sodium, Na OR aluminium, Al (2)
c) hydrogen, H (2)
d) fluorine, F (2)
e) mercury, Hg (2)

3
a) metal (2)
b) softer (1)
c) more (1)
d) one (2)

❑ 5.5.2 The Periodic Table ❑

1
a) E (1)
b) A (1)
c) F (1)
d) B (1)

2
a) carbon (1)
b) C (1)
c) 6 (1)
d) 4 (1)
e) 4 (1)

3
a) II (1)
b) Portium (1)
c) Portium (1)
d) 2.2 (1)

4
a) mercury, bromine (2)
b) SIX of: hydrogren, helium, neon, argon, krypton, xenon, radon nitrogen, oxygen, fluorine, chlorine (6)
c) 96 (1)
d) America, California, Einstein (3)

❑ 5.6.1 Reactivity of metals ❑

1
a) Gold does not react with air or water (1)
b) Iron reacts with air and water to form compounds in rocks (1)

2
a) The metal reacts with the air (1)
b) Sodium oxide (1)
c) Sodium is stored in oil (1)

3
a) A (1)
b) Hydrogen (1)
c) B (1)
d) Copper (silver, gold, platinum) (1)
e) Bubbles of gas from metal D (1)
Less than eight bubbles (1)

4
a) Lead is waterproof and malleable (2)
b) Lead is poisonous (1)
c) The less reactive the metal is, the less it will dissolve in the water. (1)

❑ 5.6.2 Reactivity of metals ❑

1
a) Gold does not react with air or water (1)
b) The shine of gold lasts a long time because it does not easily react. (1)

2
a) air and water (2)
b) Iron can be painted, greased, kept dry, etc. (1)

3
a) Magnesium (1)
b) Hydrogen (1)
c) Copper (1)
d) Potassium, sodium, lithium, calcium (1)
e) Bubbles of gas from metal D (1)
Less than eight bubbles (1)

4
a) Lead is waterproof and malleable (2)
b) Lead is poisonous (1)
c) The less reactive the metal is, the less it will dissolve in the water. (1)

❑ 6.1.1 Solids, liquids and gases ❑

1
a) Porterite – solid (1)
b) Batlium – gas (1)
c) Woodite – liquid (1)
d) Minerite – gas (1)
e) Glenium – liquid (1)

2 Vapour particles are much smaller than liquid particles. (1)
The holes are small enough to stop liquid particles getting through but large enough to let vapour particles through. (1)

3 Solid (ice) – least energy
Gas (steam) – most energy
Water (liquid) – intermediate energy (2)

4 In a suspension tiny particles of solid remain spread out throughout the liquid. In a solution the solid dissolves into the solvent and no solid remains, it is all in the liquid state. (3)

Answers

5 a) Under pressure (1)
b) The gas which causes the fizz is less soluble in hot water than cold and so escapes. (2)

6 a) A pair of liquids which do not mix. (1)
b) A liquid which can dissolve solids or gases. (1)
c) A solute is a solid which dissolves in a liquid to form a solution. (1)
d) Alcohol and water will mix completely in any proportion. (2)

❑ 6.1.2 Solids, liquids and gases ❑

1

SOLIDS	LIQUIDS	GASES
apple	petrol	air
paper	vinegar	steam
ice	milk	

(8)

2 a) immiscible (1)
b) bottom (1)
c) tap funnel OR separating funnel (1)

3 Energy OR heat. (1)

4 A – gas (1)
B – liquid (1)
C – solid (1)

5 A – liquid (1)
B – solid (1)
C – gas (1)
D – gas (1)
E – liquid (1)

❑ 6.2.1 Atoms and temperature scales ❑

1 −273°C is the lowest possible temperature so far as we know. With zero kelvin at this point there are no negative kelvin temperatures. (2)

2 a) These atoms cannot be split, made or destroyed. Scientists can split atoms in atomic reactors, new atoms can be made from old ones and atoms can be destroyed. (2)
b) Electron microscope (1)

3 a) There is a time lag because the water has to heat the glass of the thermometer bulb and heat then has to be transferred to the liquid in the bulb which then starts to expand. (2)
b) Capillary tube. (1)
c) The bottom of the beaker will be slightly hotter than the body of water because it is in contact with the bunsen flame and hot gauze. (2)
d) It is only liquid over this range. Below −39°C it is a solid. Above 360°C it is a gas. (2)
e) Alcohol filled, because alcohol remains a liquid at the low temperatures found there. (2)

4 Research (6)

❑ 6.2.2 Atoms and temperature scales ❑

1 a) 100°C
b) 0°C
c) −13°C
d) 7.5°C
e) 47.5°C (5)

2 a) iron 1540°C (2)
b) gold 3080°C (2)
c) aluminium 1810°C
copper 1486°C
gold 2016°C
iron 1210°C
lead 1413°C
silver 1251°C
tin 2038°C (7)
d) iron (1)

3 a) 2 (1)
b) We can split atoms in atomic reactors, new atoms can be made from old ones and atoms can be destroyed. (1)
c) Electron microsoope (1)

Answers

❑ 6.3.1 Particles and changes of state ❑

1 As solids we would be holding hands and swaying a little as though we were vibrating. (1)
As we got warmer we would sway more, and some would stop holding hands as bonds broke. (1)
Gradually all hands would stop holding as all bonds broke (1) and we would start moving about as the solid was now a liquid and particles were free to move. (1)

2 Drying is the evaporation of the water from the clothes. (1)
The Sun supplies heat energy and the wind blows away the particles with most energy. (1)
Without sun and wind evaporation is much slower as energy is not supplied and the particles then move off on their own rather than with help from the Sun and wind. (1)

3 The energy is being used to change particles from the solid state into the liquid state. (2)

4 They vibrate around a fixed point. (1)

5 Liquids are closer together and particles have less energy so move more slowly and have to 'fight' their way through more particles. (2)

6 Meet at C because the lighter ones (B) will move twice as far in the same time. (2)

7 a) solids (1)
b) solids and liquids (2)
c) gases (1)
d) solids and liquids. (2)

❑ 6.3.2 Particles and changes of state ❑

1 B is time zero because all the ink is together. (2)
C is after 5 minutes because the ink is starting to spread out. (2)
A is after 10 minutes because the ink is spread out the most. (2)

2 a) sublimation (1)
b) condensation (1)
c) melting (1)
d) boiling (1)
e) freezing (2)

3 3 marks for the general shape of the curve. 1 for boiling point. 1 for melting point.

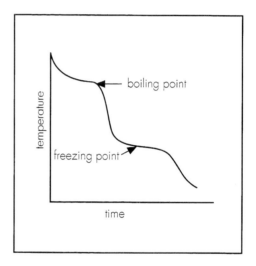

4 The heat energy is needed to change the state of the substance from a solid into a liquid instead of raising the temperature. (3)

❑ 6.4.1 Atoms, ions and molecules ❑

1 a) C (1)
b) A (1)
c) B (1)
d) D (1)
e) A (1)

Answers

2 a) a single pure substance that cannot be split into anything simpler. (1)
b) the smallest part of an element that can exist and still behave like the element. (1)
c) a substance which is made when two or more elements team up in fixed amounts. (1)
d) does not contain fixed amounts of each substance and the individual substances keep their own properties. (1)
e) the smallest part of a compound which can behave as the compound. (1)
f) an atom which has gained or lost electrons. (1)

3 All atoms contain equal numbers of protons and electrons. Protons and electrons have equal and opposite charges and so cancel each other out. (2)

4 Sodium is positive, so the negative oxygen on the water is attracted to it. (1)
Chloride is negative, so the positive hydrogen on the water is attracted to it. (1)

5 Compounds have new properties different from their components. (1)
Mixtures retain the individual properties of their components. (1)
Hydrogen is a gas which is explosive. (1)
Chlorine is a yellow/green poisonous gas. (1)
Hydrochloric acid is a colourless liquid which is acidic, it has new properties, not those of its components. (1)

❏ 6.4.2 Atoms, ions and molecules ❏

1 a) element (1)
b) atom (1)
c) compound (1)
d) mixture (1)
e) molecule (1)
f) ion (1)

2 A +1
B 0
C −1
D 1
E 1 (5)

3 Six protons of charge +1 each. Six electrons of charge −1 each. The charges are equal and opposite and so cancel. (3)

4 The attraction of positive ions for negative ions is very large. A very large force is needed to break this attraction so a lot of heat is needed, therefore a high melting point. (3)

5 Water has new properties and not those of hydrogen and oxygen. So it must be a compound because in a mixture the components keep their own properties. (3)

❏ 6.5.1 Radioactivity ❏

1 a) Alpha radiation, because it was stopped by the paper. (2)
b) 3 hours. (1)
c) The half-life is the time it takes for the radioactivity of a radioisotope to decay to half its original level. (2)
d) 211 counts per minute. (1)

2 a) Isotopes are atoms of the same element that differ only in the number of neutrons in the nucleus. (2)
b) 14 (1)
c) (2)

(2)

d) 1120 years, i.e. 100 → 50 → 25 is 2 half-lives equals 2 × 5600. (2)
e) The radioactive decay is a random process which will bring in greater errors if the time scale is small. (2)

3 Due to their ability to knock electrons out of atoms. (2)

Science Companion © A Porter, M Wood, T Wood and Stanley Thornes (Publishers) Ltd 1995

Answers

4 In 1896 Henri Becquerel discovered radiation. He placed a uranium compound on a wrapped photographic plate and the plate darkened owing to the radiation. (3)

5 a) A large atomic nucleus breaks down into two smaller nuclei, releasing neutrons and large amounts of energy. (2)
b) It is carried away by a cooling system. (1)
c) Use rods of cadmium or steel with a high boron content. (1)
d) They are slowed down by using graphite rods. (1)

❑ 6.5.2 Radioactivity ❑

1 a) Atoms of the same element with different masses. They have different numbers of neutrons but the same numbers of protons and electrons. (3)
b) Deuterium and tritium. (2)
c) $^{3}_{1}H$ (1)
(1)
(1)

2

Isotope	Number of protons	Number of neutrons	Number of electrons	
$^{4}_{2}He$	2	2	2	(2)
$^{12}_{6}C$	6	6	6	(2)
$^{41}_{19}K$	19	22	19	(2)
$^{238}_{92}U$	92	146	92	(2)

3 It is naturally present radiation all over the Earth. It can come from space, food, air or rocks by natural breakdown of the nuclei of unstable atoms. (3)

4 a) Nuclear fission makes heat. (2)
b) The heat is carried away by the cooling system. (1)
c) It is used to boil water, and this steam drives turbines to make electricity. (3)

❑ 6.6.1 Solvents, glues and hazards ❑

1 a) This substance can cause serious skin burns and also damage furniture, so wear protective clothing and handle with care. (1)
b) There is a danger of this substance exploding. (1)
c) General warning telling users to be careful. (1)
d) Will catch fire if near a flame. (1)
e) This substance causes the skin or eyes to become sore and itchy. (1)
f) This type of substance provides oxygen and helps things to burn. (1)
g) Toxic means poisonous. This substance may kill us if we swallow, or even touch it – some substances can kill by soaking into the skin. (1)

2 a) Not in control of the car and cannot react quickly to problems. (2)
b) Wet so cannot stop quickly. So will crash into the other car if it has to stop quickly. (2)
c) If you fall off or are in a crash there is likely to be damage to your head and it could kill. (2)
d) If cars are near they might not be able to stop and you could be killed. (2)
e) Someone could trip over the cable and pull the kettle on them and get badly burned. (2)

3 common salt + water → salt water (1)
copper sulphate + water → copper sulphate solution (1)
potassium nitrate + water → potassium nitrate solution (1)

Answers

4 a) Substances often jump out when heated and they are hot so could blind us if they get in our eyes. (1)
b) There are many poisonous things used in labs, and so we must never eat in case we get one of the poisonous things in our mouths. (1)
c) This is to stop someone tripping over them when moving round in a practical lesson and perhaps causing an accident by spilling a hot substance. (1)
d) If something does jump out it will land on the bench and not hit someone. (1)
e) Rapid movement can cause things to be knocked, and spills can happen to cause burns and other injuries. (1)

❏ 6.6.2 Solvents, glues and hazards ❏

1 a) concentrated sulphuric acid – A
b) gunpowder – B
c) rat poison – D
d) pure alcohol – C
e) concentrated nitric acid – A
f) petrol – C
g) arsenic – D (7)

2 a) Traffic has no time to stop, you are being a nuisance. (2)
b) Hot substance can jump and go in the eye causing blindness. (2)
c) It could go off at any time and cause death or serious injury. (2)
d) The tide could come in and carry you out to sea. (2)
e) On stopping suddenly you could be thrown through the glass, on to the road and badly hurt or killed. (2)

3 Glues contain harmful solvents. They can kill or make you very ill. Young children cannot read and understand hazard signs. They like to explore, open things, smell and taste things. They could die. (2)

4 Own idea. (3)

5 The bleepers help blind people who hear a sound saying in effect that it is safe to cross. The lumps on paving stones also help blind people feel through their shoes that they are at a crossing point. (3)

❏ 7.1.1 Physical and chemical changes ❏

1 The particles vibrate faster as the temperature rises. (1)

2 The melting point of the fat is lower than the melting point of the aluminium, which must also be higher than the temperature of the cooker. (2)

3 It becomes a liquid which takes the shape of the container. (1)

4 The particles will slow down and as they join together will stop moving around. (2)

5 Physical – no new material is made and the change is temporary. (2)

6 The egg changes from liquid to solid (solidifying). (1)

7 Chemically – the change is permanent, the egg does not return to liquid when it cools down. (2)

8 a) A
b) B
c) B
d) A (4)

❏ 7.1.2 Physical and chemical changes ❏

1 The melting point of aluminium is higher than the temperature of the heating ring. (1)

2 It has a low melting point. (1)

3 Physical (1)

4 It will turn back to a solid. (1)

Answers

5 65°C (1)

6 Raw egg is liquid because it is runny and will take up the shape of its container. (2)

7 Solid (1)

8 The egg is chemically changed because it does not return to its original state when it cools down. (2)

9 a) B
b) A
c) C
d) A
e) D (5)

❑ 7.2.1 Changes for the better from land and air ❑

1 sulphur + oxygen → sulphur dioxide (1)

2 oxygen (1)

3 It becomes water. (1)

4 hydrogen sulphide + oxygen → water + sulphur (2)

5 It would turn to sulphur dioxide. (1)

6 When the gas burns the hydrogen sulphide becomes sulphur dioxide which can cause acid rain. (2)

7 a) Ports are used to import the sulphur from abroad. (2)
b) Rail and motorways are needed to distribute the sulphuric acid. (2)
c) The plant needs workers to operate the machinery. (2)

8 Car battery acid. (1)

❑ 7.2.2 Changes for the better from land and air ❑

1 Use an indicator and watch for a change in colour (or a named indicator and its correct colour change). (2)

2 CORROSIVE (1)

3 There is no volcanic activity in the UK. (1)

4 Oxygen (1)

5 North Sea (1)

6 The sulphur comes from abroad and if a plant is near to a port it cuts the cost of transporting it. (2)

7 a) to make detergent
b) to make paint
c) to make dyes (3)

8 superphosphate, paints and pigments, fibres, soaps and detergents, plastics, other fertilisers, dyes, pickling metals (1)

9 Acid rain – formed when emissions of sulphur dioxide dissolve in water in the atmosphere. (2)

10 Car battery acid. (1)

❑ 7.3.1 Changes for the better from water and living things ❑

1 a) Steel (1)
b) Extracted means removed from. (1)
c) 1 000 000 (1)
d) 1 000 kg (1)
e) 6.8 g per cm^3 (1)
f) salt, sodium chloride (1)
g) Evaporation of the water leaves mostly salt. (1)

2 Chopping up – the leaves are made small to fit into a beaker. (2)
Boiling – the leaves and water are heated to help remove the dye from the leaves quickly. (2)
Filtering – this separates the leaves from the dye solution. (2)
Evaporating – this removes the water from the dye solution leaving only the dye. (2)

Answers

❑ 7.3.2 Changes for the better from water and living things ❑

1 a) Steel (1)
b) Magnesium (1)
c) Solid metal would be too heavy, tubing reduces the weight of the bicycle. (1)
d) Extracted (1)
e) Alloys (1)
f) Aluminium (1)
g) 6.8 g per cm^3 (1)

2 i) Bunsen burner, tripod, gauze, heat mat, beaker (5)
ii) Filter funnel, filter paper (2)
iii) Evaporating dish (1)

❑ 7.4.1 Chemicals from oil ❑

1 Hydrogen and carbon (2)

2 Crude oil is heated (1) and the fractions condense (1) in different compartments of a tower depending on their boiling points. (1) (3)

3 Under pressure (1)

4 The fractions are not pure liquids but mixtures (1)

5 As the number of carbon atoms increases so the boiling point increases. (2)

6 a) The molecule is broken up into smaller ones. (1)
b) Molecules are cracked using heat and/or a catalyst. (1)
c) The small molecules are more useful – they are made into polymers. (1)

7 a) The double bond opens up allowing monomers to link together and still keep intact. (1)
b) Polymers (or plastics) (1)

8 a) bags, bottles, wrapping, packaging etc. (2)
b) records, drainpipes, synthetic fabrics etc. (2)
c) ceiling tiles, packaging, egg boxes etc. (2)

❑ 7.4.2 Chemicals from oil ❑

1 Crude oil is heated (1) and the fractions condense (1) in different compartments of a towers depending on their boiling points. (1) (3)

2 Their boiling points are below room temperature. (1)

3 Bitumen and wax/grease (2)

4 It increases (1)

5 The boiling points are over a range of temperatures (1) whereas pure liquids have specific boiling points. (1) (2)

6 Hydrogen and carbon (2)

7 a) monomers (1)
b) cracking (1)
c) heat and/or catalyst (1)

8 a) bags, bottles, wrapping, packaging, etc. (2)
b) records, drainpipes, synthetic fabrics, etc (2)
c) ceiling tiles, packaging, egg boxes, etc. (2)

❑ 7.5.1 Changing our food ❑

1 a) Curds and whey (2)
b) Curds (1)
c) Enzymes are natural (biological) catalysts which increase the rate of reactions. (2)
d) Rennet (1)

2 a) They are both fungi. (1)
b) Water, warmth, food, oxygen (2)
c) Alcohol and carbon dioxide (2)
d) Fermentation is the process of changing sugar into alcohol by the use of yeast. (2)

Answers

3 a) Bacteria reproduce more slowly in the cold. (1)
b) Too low and the food will freeze (1), too high and bacteria will grow rapidly. (1) (2)
c) The chicken should not be stored above the cake. (1)
Uncooked juices containing bacteria could drip down on to the cake. (1)
d) *Salmonella* (1)
e) Harmful bacteria (*Salmonella*) can live in the under-cooked chicken. (1)
f) These bacteria are killed off at high temperatures. (1)

4 a) Bacteria need water to grow well. (1)
b) Food is sterilised by heat and then sealed so no bacteria can get in. (1)
c) The acid in vinegar stops the growth of bacteria. (1)

❏ 7.5.2 Changing our food ❏

1 a) Milk (1)
b) Curds are the solids and whey is the liquid which separate when milk curdles (2)
c) Rennet is an enzyme which increases the speed at which milk separates (2)
d) Mould (1)

2 a) Alcohol and carbon dioxide (2)
b) Yeast turns sugars into alcohol for drink like beer and wine (2)
c) Yeast turns sugars into carbon dioxide to make the bread rise (2)
d) The fermentation takes too long (1)
e) The yeast could die at high temperatures (1)

3 a) The temperature is lower (1)
b) It could be frozen solid (1)
c) Bacteria could grow rapidly (1)
d) The chicken should not be stored above the cake. (1)
Uncooked juices containing bacteria could drip down on to the cake (1)
e) Chicken, eggs, etc. (1)
f) *Salmonella* makes toxins which poison our bodies (1)

g) *Salmonella* is killed at high temperatures (1)

4 a) canning
b) using vinegar
c) drying (3)

❏ 7.6.1 Oxygen – the gas of life ❏

1 a) During the day the carbon dioxide enters the leaf and oxygen is put into the air. (1)
During the night the carbon dioxide is put into the air as the oxygen is used up. (1)
b) Through small holes (stomata) (1)
c) photosynthesis (1)
d) carbon dioxide, water and energy (3)

2 a) gills (1)
b) lungs (1)
c) skin (1)

3 a) oxygen (1)
b) 21% (1)
c) 16% (1)
d) carbon dioxide (1)
e) moisture in the lungs (product of respiration) (1)

❏ 7.6.2 Oxygen – the gas of life ❏

1 a) Small holes in the surface of the leaf (1)
b) Carbon dioxide enters the leaf. (1)
c) Oxygen (1)
d) Photosynthesis (1)
e) Oxygen enters the leaf and carbon dioxide is emitted (2)

2 any three suitable examples (3)

3 a) nitrogen (1)
b) oxygen (1)
c) nitrogen (1)
d) oxygen (1)
e) carbon dioxide (1)
f) water vapour (1)

Answers

❑ 7.7.1 Oxidation ❑

1 a) oxygen (1)
b) carbon dioxide (1)
c) water (steam) (1)
d) carbon monoxide (1)
e) carbon monoxide is a poisonous gas (1)

2 a) Tube 1 – air and water
Tube 2 – air
Tube 3 – water
Tube 3 – air, water and salt (4)
b) Tube 1 – rusty
Tube 2 – no rust
Tube 3 – no rust
Tube 3 – very rusty (4)
c) water and air (2)
d) In winter there is more rain and salt is often put on the roads (2)
e) paint, grease, electroplating, keep dry (3)

❑ 7.7.2 Oxidation ❑

1 a) oxygen (1)
b) No oxygen on the Moon (1)
c) Oxygen is supplied from one tank to the fuel from the other tank. It carries its own supply of oxygen. (2)
d) i) carbon dioxide (monoxide) (1)
ii) magnesium oxide (1)
iii) water (steam) (1)

2 a) orange-brown colour (1)
b) 1 and 2 (2)
c) 1 and 3 (2)
d) Air and water were present. (2)
e) In tube 2 there was no water. (1)
In tube 3 there was no air. (1)
f) Salt (sodium chloride) (1)
g) paint, grease, electroplating, keep dry (3)

❑ 7.8.1 Chemical energy ❑

1 a) Carbon dioxide and water (steam) (2)
b) Temperature increases (1)
c) Reactants (1)
d) Enthalpy change (1)
e) ΔH (1)

2 The energy is taken in from the surroundings (1) and so the temperature falls. (1) (2)

3 a) Energy is supplied (1)
b) electricity (1)
c) products (1)
d) copper and chlorine (2)
e) copper (1)
f) exothermic (1)

❑ 7.8.2 Chemical energy ❑

1 a) exothermic (1)
b) carbon dioxide and water (steam) (2)
c) Temperature rises (1)
d) enthalpy (1) change (1) (2)
e) 2252 kJmol^{-1} (1)

2 a) Energy is supplied. (1)
b) battery or power pack (1)
c) electricity (1)
d) products (1)
e) copper and chlorine (2)
f) copper (1)
g) exothermic (1)

❑ 7.9.1 Rates of reaction ❑

1 a) Increasing the temperature (1)
increases the rate of reaction (1)
b) Particles move faster at higher temperatures and therefore there will be more collisions. (2)

2 a) carbon dioxide (1)
b) Industry produces sulphur dioxide (1)
which adds to the acidity of rain. (1)
c) Increasing concentration will increase the number of particles (1) which will increase the number of collisions. (1) (2)
d) Use a pH meter or universal indicator. (1)

3 Smaller particles means a larger surface area (1) which increases the rate of reaction/cooling (1) (2)

Answers

4 Catalysts speed up the rate of reaction (1) and enable reactions to be carried out at a lower temperature (1) which saves money (1)   (3)

❏ 7.9.2 Rates of reaction ❏

1. a) Increasing the temperature (1) increases the rate of reaction. (1)
   b) A fridge will slow down the rate of decay (1) and a freezer will slow it down even further (1)   (2)

2. a) carbon dioxide (1)
   b) More sulphur dioxide has been made by more power stations. (2)
   c) It increases the rate of reaction. (1)
   d) pH meter or universal indicator (1)

3. Smaller particles means a larger surface area (1) which increases the rate of reaction/cooking. (1)   (2)

4. Raise the temperature, increase the concentration of acid, use magnesium with a larger surface area.   (3)

❏ 8.1.1 The weather and weather forecasting ❏

1. a) March (1)
   b) January (1)
   c) January (1)
   d) 247.5 mm (1)
   e) 498 hrs (1)
   f) 8 hrs (1)
   g) 1 mm (1)

2. a) warm front (1)
   b) cold front (1)
   c) occluded front (1)
   d) depression (1)
   e) anticyclone (1)
   f) isobar (1)

3. a) They blow from high pressure to low pressure and pressure is greater at the centre. (1)
   b) rain (1)
   c) south west (1)
   d) anti-clockwise (1)

e) Two of: satellite, radar stations, weather ships, aircraft, instruments on rockets and balloons.   (2)

❏ 8.1.2 The weather and weather forecasting ❏

1. a) E (1)
   b) B (1)
   c) C (1)
   d) A (1)
   e) D (1)

2. N   North (1)
   NE North-east (1)
   E   East (1)
   SE South-east (1)
   S   South (1)
   SW South-west (1)
   W   West (1)
   NW North-west (1)

3. a) South-west (1)
   b) Atlantic Ocean (1)
   c) North-east (1)

4. The spinning of the Earth (1)

5. a) (1)
   b) (1)
   c) (1)

❏ 8.2.1 Weather science ❏

1. a) i) 45% (1)
      ii) 40% (1)
      iii) 15% (1)
   b) Heat radiation (2)
   c) They show the accurate air temperature because they are not being heated by radiation from the Sun. (2)
   d) Heat is conducted through a material by the vibration of its particles. Space is a vacuum and there are no particles to pass on the energy. (2)

Answers

2 a) Particles are closer together in a solid than in a liquid and conduction is heat transfer by vibrating particles giving energy from those particles with a lot of energy to ones with less. (2)
b) Particles need to move to transfer heat by convection. Particles are not free to move in a solid but they are in gases and liquids. (2)

3 a) Fog is tiny droplets of water in the air close to the ground. (1)
b) Mist – thin fog with visibility greater than 1 km.
Fog – visibility 0–1 km.
Smog is fog containing smoke and general industrial pollution. (3)
c) Frontal fog forms when two masses of air meet which are at different temperatures and hold different amounts of water vapour.
Cold air cannot hold as much water vapour as hot air. (2)

4 a) Cumulonimbus (1)
b) Rising warm air currents. (1)
c) Water droplets and ice crystals bumping together in the high wind inside the cloud. (2)
d) Lightning is very hot and so it expands the air through which it travels, causing a pressure wave which we hear as thunder. (2)

❏ 8.2.2 Weather science ❏

1 a) i) 45% (1)
ii) 40% (1)
iii) 15% (1)
b) They show the accurate air temperature because they are not being heated by radiation from the Sun. (2)

2 a) Radiation (1)
b) Convection (1)
c) Conduction (1)

3 a) Cold (1)
b) It causes the water to come out of the air and make fog. (1)
c) Tiny droplets of water in the air close to the ground (2)
d) Fog containing smoke and general industrial pollution (2)
e) Mist (2)

4 a) 30 000°C (1)
b) Rising warm air currents (2)
c) Lightning is very hot and so it expands the air through which it travels, causing a pressure wave which we hear as thunder. (2)
d) Positive (2)
e) The top of the cloud, and the ground are positive, the bottom of the cloud negative, and lightning flashes between points of opposite charge. (2)

❏ 8.3.1 Climate, farming and catastrophies ❏

1 Weather is the changes which happen in the air around you every day. Climate describes the weather pattern at a place over a period of 30 years. (2)

2 Four of:
energy from the Sun, the Earth's shape, position of the Earth in space, the spin of the Earth, the atmosphere, the oceans. (4)

3 (4)

4 a) The air cools as it rises. Cool air cannot hold as much moisture as hot air so it lets it go as rain. (2)
b) The air is dry, so no rain, because it lost water rising at the equator. (2)
c) Nomadic means wander from place to place. Oases are areas where water is found. (2)
d) Nutrients in the soil are leached (washed away). (1)

Answers

5 a) Caribbean (1)
b) warm (1)
c) wet (1)

6 In our summer, areas north of the Equator are closer to the Sun so are warmer. (2)

7 a) The gravitational pull of the Moon, and to a lesser extent, the Sun. (2)
b) When both Moon and Sun pull in the same direction and cause a very high tide. (1)

❏ 8.3.2 Climate, farming and catastrophies ❏

1 At the equator the heat is concentrated over a smaller area than the same amount of heat at the North Pole owing to the curve of the Earth. (4)

2 a) The change which happens in the air around you every day. (2)
b) 30 years (1)

3 a)

```
      N
      ↑
W ————+———— E
      ↓
      S
```
(3)
b) over 750 mm (1)
c) Caribbean (1)
d) warm (1)
e) Warm moist air blows in from the south-west. It rises as it reaches the mountains. It cools and loses its moisture as rain. Once over the hills the air is drier so the land below gets less rain. (5)

4 a) The Sun heats the air, makes it less dense, so it rises. (3)
b) The air is dry so there is no rain. (2)
c) The rising air pulls air away from the surface lowering the air pressure. Falling air 2000 miles away raises the pressure. (2)

❏ 8.4.1 Air masses and Charles' law ❏

1 1 – arctic; 2 – polar continental;
3 – tropical continental;
4 – tropical maritime;
5 – polar maritime (5)

2 a) It has come from hot tropical regions. (1)
b) It has travelled over many thousands of miles of water. (1)

3 The air rises as it meets the land, cools slightly, and deposits what little moisture it contains. (2)

4 a) Concentrated sulphuric acid. (1)
b) It expands, and pushes up the acid bubble. (1)
c) Gas has zero volume at zero kelvin and so this means the temperature units increase in line with the mathematical formula. Celsius would mean negative values for many results and this would not fit the maths formula. (3)
d) Add 273 on to the Celsius figure. (1)

5 a) neatness (3); line A points (5); line B points (5)
b) line B (1), because it is a straight line which cuts the x-axis at –273°C. (1)

❏ 8.4.2 Air masses and Charles' law ❏

1 1 – arctic; 2 – polar continental;
3 – tropical continental;
4 – tropical maritime;
5 – polar maritime (5)

2 a) It has come from the equator where it is hot. (1)
b) It comes over land for most of the way, particularly the Sahara Desert. (1)

3 a) It comes from cold polar regions. (1)
b) Air mass 5 travels over a lot of water so is moist, but air mass 2 comes over a lot of land so is much drier. (2)

Answers

4 a) V_1 = initial volume (1)
V_2 = final volume (1)
T_1 = initial temperature (1)
T_2 = final temperature (1)
b) Add 273 on to the Celsius figure (1)

5 a) neatness (3); line A points (5);
line B points (5)
b) line B (1), because it is a straight line which cuts the x-axis at –273°C. (2)

❏ 8.5.1 Weathering and landscaping ❏

1 a) USA (1)
b) In the south-west (1)
c) Utah and Arizona (2)
d) Navajo (1)
e) Rocky, table-like structures formed by wind and rain over millions of years, which have eroded the soft rock. (3)
f) A narrow column of rock. (3)
g) Westerns (1)

2 Water runs into narrow cracks in the rocks. In low temperatures it freezes and expands forcing the rock apart. Eventually this breaks up the rock into smaller pieces. (2)

3 The roots of trees, plants and grass will bind the soil together to stop erosion. (2)

4 The rock under the fall is soft. The force of the water pouring over has eaten away the soft rock to this depth. (2)

5 a) The burning of fossil fuels. (1)
b) sulphur dioxide (1)
c) It is a product of the Industrial Revolution. (1)
d) One of:
Sandstone, limestone (1)

❏ 8.5.2 Weathering and landscaping ❏

1 Water expands when it freezes. When water gets into cracks in rocks and freezes, it opens the cracks. Over several winters this causes pieces of rock to fall away. These form a pile of scree at the bottom of a rock face. (4)

2 By binding the soil. (1)

3 a) Sandstone and limestone. (2)
b) It is found in the ground. (1)
c) It is soft and easy to carve. (2)
d) The Industrial Revolution has meant the burning of more fossil fuel so more acid rain. (2)

4 a) It flows too gently to dig too deeply into the land. (2)
b) Meandering (1)
c) It wears away the outside of bends, depositing the material further downstream on the inside of bends. (2)

5 The rock under the fall is soft. The force of the water pouring over has eaten away the soft rock to this depth. (3)

❏ 8.6.1 The water cycle ❏

1 A – energy from the Sun;
B – evaporation;
C – respiration;
D – transpiration;
E – rain or snow;
F – stream/river returns water to the sea. (6)

2 Percolation (1)

3 3% (1)

4 Polar ice caps (1)

5 Water table (1)

6 Removes large objects like fish, trees, twigs, etc. (1)

Answers

7 They make tiny solid particles of dirt etc. stick together and settle out. (1)

8 To kill any bacteria. (1)

9 In an underground service reservoir. (1)

10 The screw just fits into a cylinder. As it turns, trapped water is forced to move upwards. (2)

11 ONE of: fertilizer on fields or making methane gas. (1)

12 Microbes (1)

13 A network of underground pipes needed to transport domestic and industrial sewage to the treatment works. (1)

14 ONE of: typhoid, cholera and types of dysentery. (1)

❑ 8.6.2 The water cycle ❑

1 a) A – energy from the Sun;
B – evaporation;
C – respiration;
D – transpiration;
E – rain or snow;
F – stream/river returns water to the sea. (6)
b) It gets colder higher up so it may rain low down then snow high up if it is cold enough. (1)
c) The Sun. (1)

2 a) It removes large objects like fish, trees, twigs, etc. (1)
b) They make tiny solid particles of dirt etc. stick together and settle out. (1)
c) To kill any bacteria. (1)

3 a) Insoluble (1)
b) More dense than the water. (1)
c) Spread on fields as fertilizer and used to make methane. (2)

4 a) 3% (1)
b) Polar ice caps. (1)

5 a) A network of underground pipes needed to transport domestic and industrial sewage to the treatment works. (1)
b) TWO of: typhoid, cholera and types of dysentery. (2)

❑ 8.7.1 The rock cycle ❑

1 a) Weight of new layers above. (1)
b) sedimentary – random;
metamorphic – layers (2)
c) The layers mean it can be split into thin sheets and make the roof light. (2)

2 a) They are formed in layers from sediments carried to one place and allowed to settle. The weight of new layers compresses old ones and rocks are formed. (2)
b) Old ones are lower into the ground. (1)

3 a) A complete change of appearance and character. (2)
b) Pressure and heat. (2)

4 a) Dead creatures fall to the bottom of seas and lakes and become part of the sediment to be compressed into rock. (2)
b) The forces needed to create them destroy the fossils. (2)

5 a) Hot molten rock deep inside the Earth. (2)
b) Large crystals – slow cooling deep inside the Earth. Small crystals – rapid cooling at the Earth's surface. (4)
c) lava (1)
d) It is less dense than water being full of holes caused by expanding gases which were given off as it was formed. (2)

6 a) limestone (1)
b) smaller and more uniform (2)
c) smoother and harder (2)

Answers

❑ 8.7.2 The rock cycle ❑

1 a) Weight of new layers above. (1)
  b) sedimentary – random;
    metamorphic – layers (2)
  c) The layers mean it can be split into thin sheets and make the roof light. (2)

2 a) lakes and seas (2)
  b) old (1)
  c) mudstone (1)
  d) sandstone (1)
  e) Creatures die and are buried with sediments over millions of years and are compressed into fossils. (2)

3 a) Hot molten rock deep inside the Earth. (2)
  b) large (1)
  c) lava (2)
  d) at the surface (2)
  e) smaller (1)
  f) Expanding gases given off as the magma (lava) cooled at the surface. (2)

4 a) heat and pressure (2)
  b) calcium carbonate (2)
  c) The crystals become smaller and more uniform. (2)
  d) Marble is smoother and harder than limestone. (2)

❑ 8.8.1 Earthquakes and volcanoes ❑

1 a) To the left. (1)
  b) The temperature varies, there are hotter and cooler areas. (1)
  c) They are moving together. (1)
  d) They are moving apart. (1)
  e) Point C (moving apart). (1)

2 a) A line where two of the Earth's plates meet. (1)
  b) Gas mains break and the gas catches fire. (1)
  c) Wind fanned the flames to spread them. In 1989 the lack of wind prevented the fires spreading. (1)

3 a) It melts owing to friction. (2)
  b) Slow. It is thick so cannot move quickly. (2)
  c) Batholith (1)
  d) Slow moving magma forms steep sided cones because it solidifies quickly as it leaves the crater. (2)

❑ 8.8.2 Earthquakes and volcanoes ❑

1 a) To the left. (1)
  b) They are moving together. (1)
  c) They are moving apart. (1)
  d) There are areas of different temperatures causing the currents to flow. (1)

2 a) Gas mains were broken and the gas caught fire. (1)
  b) Water mains are also burst. (1)
  c) They've been built on a suspension system rather like a car, and are designed to rock in an earthquake. (2)
  d) The strong wind helped to spread the fires in 1906. Lack of wind in 1989 helped to stop the fires spreading. (2)

3 a) It melts. (1)
  b) Slowly. It is thick so it cannot move quickly. (2)
  c) It solidifies quickly. (1)
  d) lava (1)

❑ 9.1.1 How things work ❑

1 a) A system of cables which distributes electricity through the country. (1)
  b) Electricity is flowing through the wire. (1)
  c) To prevent electricity passing through the workmen. (1)
  d) Electricity from the cables would pass through their bodies and kill them. (2)
  e) To avoid conducting electricity – plastic is an insulator. (1)
  f) The cables are insulated from the pylons using separators made from ceramics. (2)
  g) Birds do not make contact with the ground so electricity will not flow through them. (1)

Answers

2 a) The cell will quickly use up its
   energy because there is nothing
   to stop it flowing from one end to
   the other. (2)
   b) A gap in the circuit would mean that
   electricity would not flow because
   the air is a poor conductor of
   electricity. (2)
   c) Cells facing each other will try to
   push the electricity in opposing
   directions and so no flow will be
   possible. (2)

❑ 9.1.2 How things work ❑

1 a) A system of cables which distributes
   electricity through the country. (1)
   b) metal (1)
   c) It is a conductor of electricity. (1)
   d) The electricity will flow through their
   suits because of the metals woven
   into the fabric. (2)
   e) plastic (1)
   f) It is an insulator of electricity. (1)
   g) ceramic (1)
   h) It is an insulator of electricity. (1)

2 a) There is a direct connection between
   the two ends of the cell. (1)
   b) The cell is quickly drained of energy. (1)
   c) There is a gap (switch) in the second
   circuit (1) and the cells are in opposing
   directions in the third circuit. (1) (2)
   d) Close the switch in the second circuit (1)
   and reverse one of the cells in the
   third circuit. (1) (2)

❑ 9.2.1 Go with the flow ❑

1 a) It is the flow of electrical charges. (1)
   b) A circuit where bulbs are connected
   to the power supply in turn. (1)
   c) A circuit where each bulb is
   connected to the power supply
   directly. (1)
   d) It is a device for measuring electric
   current. (1)
   e) Any device which will slow down
   the flow of electricity. (1)

2 a) cells
   resistor
   voltmeter (V)
   ammeter (A) (4)
   b) As the voltage goes down so does the
   current (1) but – when the voltage
   halves, the current halves (1) but –
   the voltage is proportional to the
   current (1) (3)

3 a) 3 ohms (1)
   b) 1 ampere (1)
   c) 3 volts (1)

❑ 9.2.2 Go with the flow ❑

1 a) ammeter (1)
   b) series (1)
   c) current (1)
   d) parallel (1)
   e) resistor (1)

2 a) cells
   resistor
   voltmeter (V)
   ammeter (A) (4)
   b) 0.3 ampere (1)
   c) 5 volts (1)
   d) 10 ohms (1)

3 a) It is where a wire connects across
   a power supply directly. (1)
   b) A short circuit drains the energy
   in a cell very quickly. (1)
   c) No. (1)

❑ 9.3.1 Electricity in your home ❑

1 a) 531 units (1)
   b) 7.67p (1)
   c) 4073p = £40.73 (1)

Science Companion © A Porter, M Wood, T Wood and Stanley Thornes (Publishers) Ltd 1995

Answers

2 a) 0.5 units (1)
b) 0.5 units (1)
c) 0.25 units (1)
d) 0.462 units (1)
e) 4p, 4p, 2p, 45p (4)

3 earth wire – returns electricity safely
to earth (1)
if the appliance should become live (1)
flex grip – stops the bare wires from
coming loose inside the plug (1)
fuse – melts if the current gets too high (1)

❏ 9.3.2 Electricity in your home ❏

1 a) 500 units (1)
b) 8p (1)
c) 4000p = £40.00 (1)

2 a) electric fire, iron, microwave oven,
hair dryer, light bulb, television (6)
b) 2 hours (1)
c) 4 units (1)

3 a) yellow and green (1)
b) The flex grip stops the bare wires
from coming loose inside the plug. (1)
c) 3 A for the lamp and 13 A for an iron (2)

❏ 9.4.1 The life of Michael Faraday ❏

1 Some of the energy from the cyclist
turning the pedals goes into turning
a magnet within a coil of wires which
produces an electric current in the wires. (3)

2 An ammeter measures the electric
current. (1)

3 e.g. hair dryer, hover lawn mower,
fan, electric heater, washing machine
(any three) (3)

4 a) 1791–1867 (2)
b) Newington, Surrey (1)
c) blacksmith (1)
d) Humphrey Davy (1)
e) Faraday sent Davy a hand-bound
book of his lectures. (1)
f) 1826 (1)
g) benzene and stainless steel (any one) (1)

❏ 9.4.2 The life of Michael Faraday ❏

1 a) from kinetic energy of the bicycle (1)
b) a magnet (1)
c) the lights on the bicycle (1)
d) An ammeter measures the electric
current (1)

2 a) The motor rotates wet clothes to
drive the water out of them. (1)
b) The motor turns the blades of the
mower which cut the grass. (1)
c) The motor spins the blades of the
fan making the air move. (1)

3 a) 1791–1867 (2)
b) Newington, Surrey (1)
c) blacksmith (1)
d) Humphrey Davy (1)
e) Faraday sent Davy a hand-bound
book of his lectures. (1)
f) 1826 (1)
g) benzene and stainless steel (any one) (1)

❏ 9.5.1 Electromagnetism ❏

1 a) electrons (1)
b) alternating current and direct current (2)
c) In alternating current the direction
of the current changes. (1)
In direct current the direction of the
current is only one way. (1)

2 a) The compasses are aligned to the
Earth's magnetic field. (1)
b) north (1)
c) There is a gap in the circuit so no
electricity flows. (1)
d) the cell (1)
e) The needles of the compasses are
deflected and point in a circle round
the wire. (2)
f) The electricity in the wire produces
a magnetic field. (1)

3 a) iron (1)
b) Two magnets repel each other when
the two north poles or the two south
poles are facing each other. (2)
c) Two magnets attract each other
when the facing poles are opposite. (1)

Science Companion © A Porter, M Wood, T Wood and Stanley Thornes (Publishers) Ltd 1995

Answers

4 a) The electromagnet in a loudspeaker can transfer the electric current in a wire into movement of the paper cone inside the loudspeaker which vibrates to make a sound. (2)
b) In a doorbell the electromagnet forces a small hammer to hit the chimes of the bell when the push switch is pressed. (2)

❑ 9.5.2 Electromagnetism ❑

1 a) direct current (1)
b) The current flows one way in the circuit. (1)
c) alternating current (1)
d) The current changes direction. (1)

2 a) A magnetised needle is balanced on a pin so that it is free to turn. (2)
b) north (1)
c) the Earth's magnetic field (1)
d) The needle of the magnet would be deflected (1) anticlockwise (1) (2)

3 a) iron (1)
b) i) attraction
ii) repulsion
iii) attraction
iv) repulsion (4)

4 a) electrons (1)
b) the cell (1)
c) a magnetic field in the wire (1)
d) The current changes direction and so does the magnetic field – too quickly for the compasses to change. (2)

❑ 9.6.1 Numbers and words ❑

1 a) They could have kept count of animals they owned or the size of fields etc. (1)
b) The knots could easily be added to or taken away to deceive people. It would be easy to forget what the knots meant or to get them confused with others. (2)

2 a) Charles Babbage (1)
b) The computer worked mechanically using cogs and levers. (1)

3 a) on magnetic tape (1)
b) You have a personal identification number (PIN) which only you know. (2)

4 a) bar code (1)
b) lasers (1)
c) Supermarkets don't have to label each item on the shelves with the price. They can keep an instant check on their stock. (2)
d) Shoppers get an itemised bill. (1)

5 Hieroglyphics are the written language of the ancient Egyptians. (2)

❑ 9.6.2 Numbers and words ❑

1 a) 10 (1)
b) It is how many fingers we have on our hands (1)

2 An abacus is a counting device which uses beads strung on to wires. (2)

3 a) magnetic tape (1)
b) You have a personal identification number (PIN) which only you know. (2)

4 a) bar code (1)
b) lasers (1)
c) Supermarkets don't have to label each item on the shelves with the price. They can keep an instant check on their stock. (2)
d) Shoppers get an itemised bill. (1)

5 a) 26 (1)
b) The Chinese have an alphabet based on characters which represent words. (1)
c) Ancient Egyptians (1)

Answers

❑ 9.7.1 Microelectronics ❑

1 a) i) 5
  ii) 9
  iii) 12
  iv) 15
  v) 21 (5)
 b) i) 110
  ii) 1011
  iii) 10010
  iv) 10100
  v) 11111 (5)

2 a) a miniaturised circuit in which all the components are etched on to silicon (1)
 b) read only memory – permanent memory (1)
 c) random access memory – temporary memory (1)

3 a) with a word processor (1)
 b) with a scanner (1)
 c) with a sampler (1)

4 a) through the keypad (1)
 b) silicon (1)
 c) display screen (1)
 d) display screen (1)

❑ 9.7.2 Microelectronics ❑

1 a) integrated circuit (1)
 b) read only memory (1)
 c) random access memory (1)
 d) RAM (1)

2 a) allows words to be stored in a computer and then alter them (1)
 b) allows pictures and photographs to be stored on computer (1)
 c) allows sounds to be stored in a computer (1)

3 a) work out arithmetical sums (1)
 b) through the keypad (1)
 c) silicon (1)
 d) display screen (1)
 e) display screen (1)

4 a) binary (1)
 b) 1 and 0 (2)
 c) 10 (1)

❑ 9.8.1 Logic gates ❑

1 a) changes in the environment (1)
 b) i) A push switch comes on when something presses down on it. (1)
  ii) A temperature sensor comes on when the temperature reaches a certain level. (1)
  iii) A light sensor comes on when light shines on it. (1)

2 a) NOT – 1 input, OR – 2 inputs, AND – 2 inputs (3)
 b) NOT – input off (1)
  OR – either input on or both inputs on (3)
  AND – both inputs on (1)

3 a) NOT (1)
 b) OR (1)
 c) AND (1)

❑ 9.8.2 Logic gates ❑

1 a) i) pressure on the switch
  ii) a rise in temperature
  iii) light shining on the sensor (3)
 b) i) light sensor
  ii) temperature sensor
  iii) push switch (3)

2 a) one (1)
 b) off (1)
 c) two (1)
 d) both inputs off (1)
 e) two (1)
 f) both inputs on (1)

3 a) NOT (1)
 b) OR (1)
 c) AND (1)

Answers

❑ 10.1.1 Logic gates ❑

1 a) Non-renewable fuels will run out before they can be made again. (1)
b) Fossil fuels are formed from the decayed remains of plants and animals. (1)
c) Dead plants or animals (1) decayed without oxygen (1) and were compressed and heated under rock layers for millions of years. (1)
d) oil, gas, coal, peat (1)

2 a) Petroleum is a mixture of oils and gases. (1)
b) Fractional distillation (1)
c) The oils have different boiling points. (1)

3 a) Generating electricity (1)
b) Coal was formed from fern-like plants. (1)
c) smog (1)
d) It made smokeless fuels compulsory in towns and cities. (1)
e) Coke is a smokeless fuel. (1)

4 a) water (1)
b) no pollution (1)
c) hydrogen is potentially explosive. (1)

5 a) uranium (plutonium) (1)
b) Atoms break up (fission). (1)
c) The heat boils water into steam which drives the turbines. (1)

❑ 10.1.2 Fuels ❑

1 a) millions (1)
b) They are used up before they can be made again. (1)
c) oil, coal, gas (3)
d) coal (1)
e) The smoke caused pollution. (1)
f) coke (1)

2 a) North Sea (1)
b) pipeline (1)
c) Impermeable rock above them (2)
d) Dead sea plants and animals (2)
e) Fractional distillation (1)

3 a) oxygen (1)
b) water (1)
c) no pollution (1)
d) The hydrogen exploded. (1)
e) Helium is not explosive. (1)

❑ 10.2.1 Making use of energy ❑

1 Sun's energy stored in plants (1)
Plants converted to coal (1)
Stored energy released as coal burns (1)
Turned into electricity in a power station (1)

2 a) lead (1)
b) sulphuric (1)
c) chemical (1)
d) turning over the engine (1)
e) kinetic energy of the engine (1)

3 a) electrical and chemical (2)
b) body heat, hot water, other appliances, solar, cooking and lighting (1)
c) More people mean less bought in energy. (1)
d) They would trap more solar energy. (1)
e) Lights get hot as the filament inside them glows. (1)

❑ 10.2.2 Making use of energy ❑

1 a) photosynthesis (1)
b) Plants decay over millions of years. (1)
c) chemical (1)
d) Coal is burned to make steam which drives turbines. (1)
e) heat (1)

2 a) In the elastic band (1)
b) The elastic band is twisted. (1)
c) rotating the propeller (1)

3 a) bought in energy (1)
b) body heat, hot water, other appliances, solar, cooking and lighting (1)
c) Body heat is greater than solar heat; (1) it is twice as much. (1)
d) TV, video, iron, hair dryers etc. (2)
e) more in summer and less in winter (1)

Answers

❏ 10.3.1 On the move ❏

1 a) by stretching the band (1)
   b) in the chemicals inside the cell (1)
   c) in the chemicals of the gas (1)

2 a) The teeth of the first cog push on to the teeth of the cog A. (2)
   b) They turn clockwise and anti-clockwise. (1)
   c) B turns faster. (1)
      C turns more slowly. (1)

3 a) A (1)
   b) A (1)
   c) B (1)

4 a) stored energy (1)
   b) movement energy (1)
   c) increases (1)
   d) as the child goes down the slide (1)

❏ 10.3.2 On the move ❏

1 a) gas fire (1)
   b) cell (1)
   c) elastic band (1)

2 a) clockwise (1)
   b) anti-clockwise (1)
   c) 60 times a minute (1)
   d) It would turn more slowly. (1)

3 a) Energy comes from muscles of the legs which get their energy from the food the person eats. (2)
   b) A chain connects the two (1)
      and pulls on cogs in each gear wheel. (1)
   c) A (1)

4 a) potential (1)
   b) kinetic (1)
   c) at the top of the hill (1)

❏ 10.4.1 Energy conservation ❏

1 a) They have to be imported. (1)
   b) diesel (1)
   c) crude oil (petroleum) (1)
   d) This reduces the cost of transporting the fuel. (1)
   e) hydroelectric power (1)
   f) fast flowing water (1)
      is used to drive electric turbines (1)

2 a) Wind can be used to turn propellers (1)
      which in turn drive electric turbines (1)
   b) hot rocks (1)
   c) in solar cells (1)

3 a) double glazing (1)
   b) cavity wall insulation (foam) (1)
   c) loft insulation (fibre) (1)
   d) carpets/underlay (1)

❏ 10.4.2 Energy conservation ❏

1 a) Isle of Man (1)
   b) diesel (1)
   c) brought into the island (1)
   d) transportation costs (1)
   e) hydroelectric (1)
   f) flowing water (1)
   g) wave/tidal energy (1)

2 wind = wind turbines
   lakes = hydroelectric
   hot rocks = geothermal
   Sun = solar (4)

3 cavity insulation = walls
   double glazing = windows
   carpet underlay = the floor
   loft insulation = the roof (4)

❏ 10.5.1 Human life and activity – the energy we need ❏

1 Energy is used to keep war, carry out chemical reactions, work muscles and pump blood. (1)

Answers

2 a) Protein, fat and carbohydrate (3)
b) Protein – fish / meat
Carbohydrate – bread / potatoes
Fat – butter / oil (3)
c) Protein (1)

3 a) kJ and kcal (1)
b) 142 (1)
c) 15.7 g (1)
d) It helps people to know what the foods are made of and how much fat they contain. (1)

4 a) i) chocolate (1)
ii) chocolate (1)
iii) crisps (1)
iv) cola (1)
b) Because they contain a lot of fat and sugar which can be unhealthy in large amounts (1)
c) Apple / fruit / yoghurt (1)
d) Any five – bread, rice, pasta, potatoes (1)

5 a) Own graph (4)
b) Different people do different activities. The more active need more energy. Growing children need more energy than adults. (1)
c) The body is still working so needs energy. (1)
d) Own answer (5)

❑ 10.5.2 Human life and activity – the energy we need ❑

1 As 10.5.1 (2)

2 P – oil / cream
C – flour / potatoes
F – cheese / meat (6)

3 a) proteins (2)
b) carbohydrates (2)
c) fats (2)

4 a) biscuits (1)
b) 470 kcal/1984 kJ (1)
c) carbohydrate (1)
d) Kilocalories / kilojoules (2)
e) Own answer (5)

5 a) chocolate (1)
b) Cola (1)
c) Chocolate (1)
d) They contain a lot of fat and sugar so are unhealthy. (2)
e) Apple / fruit / yoghurt (1)

❑ 11.1.1 What is a force? ❑

1 a) 2 N
b) 12 N
c) 0.1 N
d) 0.06 N
e) 0.001 N (5)

2 a) gravity (1)
b) weight (1)
c) weight (1)

3 a) gravity (1)
b) the ground (1)
c) balanced (1)
d) The bag sinks because the downward force of gravity is greater than the upward force of the road. (2)

4 a) The forces are balanced. (1)
b) Forwards because there is a decrease in the backward force. (2)
c) The increased force of "the seat" pushing you forward. (2)

5 It is due to the recoil of the shell. Firing the shell produces a reaction in the tank. (2)

❑ 11.1.2 What is a force? ❑

1 a) 2 N
b) 7 N
c) 5 N
d) 8 N
e) 10 N
f) 400 g
g) 800 g
h) 50 g
i) 10 g
j) 110 g (10)

Answers

2 a) gravity (1)
  b) the ground (1)
  c) balanced (1)
  d) The bag sinks because the downward force of gravity is greater than the upward force of the road. (3)

3 a) Forwards because there is a decrease in the backward force. (2)
  b) The increased force of "the seat" pushing you forward. (2)

❑ 11.2.1 Going down and slowing down ❑

1 a) Gravity squeezes the cartilage discs as you stand and you get shorter. (2)
  b) Long spaceflights allow the discs to expand owing to lack of gravity. (2)

2 a) The amount of matter in an object. (1)
  b) kilograms (1)
  c) weight (1)
  d) newtons (1)
  e) In space the astronauts still have mass but there is no force of gravity pulling on them so they have no weight. (2)
  f) Gravity is the pull force of a large body. The bigger the mass of the pulling body the bigger the gravity. The Earth is bigger than the Moon. (3)

3 The air particles are forced to move around the parachute thus slowing its fall. (2)

4 a) Cone and bullet because they are pointed and will allow the liquid particles to slide past them easily. (3)
  b) Irregular because the sides were not smooth and the particles would move in and out slowly as they went round the object. (2)

❑ 11.2.2 Going down and slowing down ❑

1 a) You need to see through it. You can see through clear glass and plastic but you cannot see through pot. (1)
  b) This makes them all the same mass, so all have the same weight. (2)
  c) It is thicker so will slow down the movement of the objects, making timing easier. (2)
  d) Gentle release gives them all a slow start, dropping would give some shapes a faster start than others. (2)
  e) Cone and bullet because they are pointed and will allow the liquid particles to slide past them easily. (3)
  f) Irregular because the sides were not smooth and the particles would move in and out slowly as they went round the object. (2)

2 a) The amount of matter in an object. (1)
  b) kilograms (1)
  c) weight (1)
  d) newtons (1)
  e) In space the astronauts still have mass but there is no force of gravity pulling on them so they have no weight. (2)

3 The metal skin gets hot as air passes quickly over it, so hot, it would melt aluminium. (2)

❑ 11.3.1 Starting and stopping ❑

1 a) 9.2 m (1)
  b) 15.2 m (1)
  c) 6 m (2)
  d) 24.4 m (2)
  e) 30.4 m (2)
  f) 30 mph   27.4 m
     50 mph   76.2 m
     70 mph   149.4 m (3)
  g) 30 mph   36.5 m
     50 mph   76.2 m
     70 mph   170.7 m (3)

Science Companion © A Porter, M Wood, T Wood and Stanley Thornes (Publishers) Ltd 1995

Answers

2 Friction between the pads and the rim of the wheel creates heat as breaking occurs. (2)

3 a) synovial fluid (1)
b) cartilage (1)
c) The cartilage in the joints becomes worn and rough after years of use. (2)

❑ 11.3.2 Starting and stopping ❑

1 a) (i) 40 km/h, (ii) 80 km/h, (iii) 113 km/h. (3)
b) 9.2 m and 13.7 m (2)
c) 27.4 m (2)
d) 36.6 m (2)
e) 15.2 m and 38.1 m (2)

2 Oil (2)

3 a) synovial fluid (2)
b) cartilage (2)

4 a) non-stick (2)
b) PTFE (1)

❑ 11.4.1 Sinking and floating ❑

1 a) The left-hand one containing air. (1)
b) When full of air they give the submarine a density less than that of water and so it floats. When full of water its density becomes greater than that of water and so the submarine sinks. (4)
c) 1 cm^3 of pure water has a mass of 1 g. (1)

2 a) displaced water (1)
b) It equals the object's own weight. (1)

3 a) To give upthrust. (1)
b) Much cheaper/more available. (1)
c) flammable gas (1)
d) Unreactive, so puts out rather than causes fires. (1)
e) Upthrust not so good/more expensive. (1)
f) Ballonets (1)
g) Push air out, the mass of the airship falls so the airship goes up. Suck air in, mass increases so airship goes down. (4)
h) Saves weight. (2)

❑ 11.4.2 Sinking and floating ❑

1 a) Left-hand one. (1)
b) less (1)
c) air (1)
d) water (1)
e) Regulate the mixture of air and water in its ballast tanks. (2)

2 a) Upthrust (1)
b) Flammable (1)
c) This lowers the total mass of the airship so the helium can lift it. (2)
d) Sucking in air increases the mass of the airship so the helium cannot lift it, so it falls. (2)
e) Fabric is less dense so it does not make as big a contribution to the weight. (2)

3 aluminium 2.7 g/cm^3; calcium 1.5 g/cm^3; gold 19.3 g/cm^3; lead 11.3 g/cm^3; silver 10.5 g/cm^3; tin 7.3 g/cm^3 (6)

❑ 11.5.1 Some forces in action ❑

1 a) It will seem very heavy as its weight will act against the direction you want it to move. (1)
b) Its weight will help it to move forward and make it easier to push. (2)
c) The pushing force gets bigger as more people join but the mass of the car stays the same. (1)
d) the gaining of speed (1)
e) brakes (1)

2 1.54 mm (1)

Answers

3 a) As the balloon deflates, the air rushes down the tube and the wood floats on the cushion of air. (2)
b) A hole in the tube from the balloon would allow a jet of air out in one direction and the hovercraft would move in the opposite direction. (2)

4 a) The left-hand one with the flat top. (1)
There is less force upwards to support the middle than in the curved one. (1)
b) Light books could be pushed aside more easily and give less support. (1)
c) Great strength but it cannot be bent without destroying its corrugated nature. (1)

❑ 11.5.2 Some forces in action ❑

1 a) iii) 2 friends (1)
b) More people to push the same mass which makes it less effort from each person (2)
c) Applies the brakes. (1)
d) The heavier a car, the more force is needed to make it move, this means a bigger engine if it is to move as fast as a smaller, lighter car. (2)
e) Fit a much bigger engine. (1)

2 a) Blow it up. (1)
b) As the balloon deflates, air rushes down the tube and the wood floats on a cushion of air. (1)
c) Air will come out in a little jet and help the hovercraft to move in one direction. (2)

3 a) The flat one on the left. (1)
b) To make it a fair test. (1)
c) To give good support. Light books might move as the forces increased. (1)
d) Great strength but it cannot be bent without destroying its corrugated nature. (1)

❑ 11.6.1 Turning forces ❑

1 Turning effect is called the moment. This is the size of the force times the perpendicular distance from the hinge. Since the turning effect is the same, if the distance falls, the force must increase; if the distance increases the force increases. (4)

2 a) 40 N m (1)
b) Anti-clockwise (1)
F is to the right, the force is moving down from '9' to '6' on a clockface so it is anti-clockwise. (1)
c) 8 cm (1)
d) 16 cm (1)
e) $W = 4$ cm; $X = 9$ N; $Y = 3$ cm; $Z = 3$ N. (4)

3 $S = 10$ cm; $T = 2$ N (2)

❑ 11.6.2 Turning forces ❑

1 a) A (1)
b) $10 \text{ N} \times 0.5 \text{ m} = 5 \text{ N m}$ (2)
c) Twice the distance, half the force so 5 N (2)

2 a) $200 \text{ N} \times 4 \text{ m} = 800 \text{ N m}$ (2)
b) $400 \text{ N} \times 2 \text{ m} = 800 \text{ N m}$ (2)
c) yes (1)
d) The moments are equal. (2)
e) B (1)
f) The clockwise moment becomes larger than the anti-clockwise one, so balance is lost and the right side falls. (2)

❑ 11.7.1 Work and power ❑

1 work done = force × distance moved in the direction of the force (1)

2 Work done requires force to move a given distance and if everything is still, there is no movement. (1)

3 a) kinetic energy (1)
b) gravitational potential energy (1)

Answers

4 a) 100 N × 1.5 m = 150 N m (1)
b) The weight and that of the shopping act vertically downwards. The force must act in the opposite direction, so use the vertical height of 1.5 m in the calculation. (2)

5 power = work done/time taken (1)

6 More power doing the job quickly (1)
Same work done but divide by less time gives a bigger power figure. (2)

7 a) 80 J of energy every second. (1)
b) Higher initial cost but last 8 times longer and use only 20% of the electricity so over the years they should be in use they will more than pay for themselves in the cost of electricity and ordinary bulbs. (3)

❑ 11.7.2 Work and power ❑

1 a) B (1)
b) A and C (2)
c) A and C (2)
d) B (1)

2 a) 20 N × 2 m = 40 J (1)
b) 20 N × 15 m = 30 J (1)
c) 1 N × 15 m = 15 J (1)

3 a) 10/2.5 = 4 N (1)
b) 450/45 = 10 N (1)
c) 90/100 = 0.9 N (1)

4 a) 100/2 = 50 watts (1)
b) 100/4 = 25 watts (1)
c) slowly (1)

❑ 11.8.1 Under pressure ❑

1 a) 2 N (1)
b) 0.4 N per cm^2 (2)
c) 0.02 N per cm^2 (2)
d) The weight is spread over a larger area. (1)

2 The pressure on the snow is reduced (1) by the larger area of the snow shoes (1). The pressure with shoes is ten times greater than the pressure with snow shoes. (1) (3)

3 a) One mark for each point plotted. One mark for each scale on each axis. One mark for a straight line joining the points. (7)
b) The pressure stayed the same. (1)
c) The gas particles moved faster. (1)
d) The pressure would increase. (1)
e) The volume becomes zero. (1)

❑ 11.8.2 Under pressure ❑

1 a) To increase the area of its feet (for swimming etc.) (1)
b) 0.4 N per cm^2 (2)
c) 0.02 N per cm^2 (2)
d) The weight is spread over a larger area. (1)

2 a) 250 cm^2 (1)
b) 1000 cm^2 (1)
c) larger (1) by a factor of four (1) (2)
d) They spread the weight over a larger area. (1)

3 a) It increases. (1)
b) It stays the same. (1)
c) One mark for each point plotted. One mark for each scale on each axis. One mark for a straight line joining the points. (7)

❑ 12.1.1 All done with mirrors ❑

1 a) virtual (1)
b) 2 metres in all (1)
c) angle of incidence = A (1)
mirror reflects the light (1)
angle of reflection = B (1)
angle A = angle B (1)

2 Concave mirrors make images upside down (1)
if the object is beyond focal point. (1)
Convex mirrors produce images correct way up. (1)

Answers

3 Light from the Sun (1)
is reflected by mirrors (1)
and focused on to water (1)
which is boiled to make steam to
generate electricity. (1)

4 The light reflects (1) off the inside of the optical fibres (1) (total internal reflection). (2)

5 a) two (1)
b) to see above the sea surface when still underwater (1)
c) e.g. in a crowd (1)
d) Cut holes in the box on opposite sides and at opposite ends. (1)
Angle mirrors at 45° to the holes and facing each other. (1)

❑ 12.1.2 All done with mirrors ❑

1 a) The image is not real. (1)
b) 2 metres in all. (1)

2 a) A (1)
b) B (1)
c) A = B (1)

3 a) distorted (1) but right way up. (1) (2)
b) upside down (1)
c) the inside of a spoon (1)
d) convex (1)

4 a) the Sun (1)
b) They focus the Sun's heat into one place. (1)
c) It boils water into steam to drive generators. (1)

5 a) two (1)
b) to see above the sea surface when still underwater (1)
c) e.g. in a crowd (1)

6 a) The mirrors produce repeating patterns (1)
because they are angled. (1)
b) The mirror directs the light (1) because it is curved. (1) (2)

❑ 12.2.1 Light and shadow ❑

1 a) not see through (1)
b) see through (1)
c) see through but the light is diffused (1)

2 a) About the same size as yourself and with sharp edges to the shadow. (2)
b) The shadow will be larger and the edges will be blurred. (2)
c) Light waves spread out after passing edges. (1)

3 A silhouette is the outline of an object which is blacked out by having a light source behind it. (2)

4 a) A pinhole is made in the front of a box. (1)
A translucent screen is stuck over a hole at the back of the box. (1)
Light enters the pinhole which is pointed towards a bright object. (1)
The image of the object appears on the screen. (1)
b) The image gets brighter and more blurred. (2)
c) Move the camera nearer to the object. (1)
d) There would be 2 images. (1)
e) Changing the size of the hole and changing the time of exposure. (2)

❑ 12.2.2 Light and shadow ❑

1 opaque = not see through
absorb = soak up
diffused = spread out
transparent = see through (4)

2 a) smaller shadow which has sharp edges (2)
b) larger shadow which has blurred edges (2)

3 A silhouette is the outline of an object which is blacked out. (1)

Answers

4
a) front (1)
b) to make the image sharp (1)
c) On the screen at the back of the camera (2)
d) The image is upside down (1)
e) Make the pinhole larger (1)
f) The image is more blurred (1)
g) The image gets larger (1)
h) Make two pinholes (1)
i) lenses (1)
j) Changing the size of the hole or changing the time of exposure (2)

❑ 12.3.1 Light ❑

1 It has patterns on the petals which can only be seen under ultraviolet light. (1) This is to attract bees for pollination (1) because bees can see ultraviolet patterns. (1) (3)

2
a) red, orange, yellow, green, blue, indigo, violet (7)
b) Light is refracted (1) and different coloured light is refracted at different angles. (1) (2)
c) violet (1)
d) red (1)
e) They have the same speed. (1)

3 Examples include:
a) sterilising surgical equipment (1)
b) TV remote controls (1)
c) cooking food (1)
d) TV/radio signals (1)
e) Sun beds (1)
f) photographing bones (1)
g) gamma, X rays, UV radiation, IR radiation, microwaves, radio waves (1)

4 Light is bent (1) or refracted (1) when it passes from one medium (water) to another (air). (1) (3)

❑ 12.3.2 Light ❑

1
a) Ultraviolet (1)
b) They have patterns visible in UV light (1)
c) To help pollination (1)

2
a) prism (1)
b) left to right (1)
c) red, orange, yellow, green, blue, indigo, violet (7)
d) spectrum (1)

3
a) gamma = sterilising surgical instruments
infra red radiation = TV remote control
microwave radiation = heating food
radio waves = radio signals
ultra radiation = Sun beds
X rays = photographing bones (6)
b) Gamma radiation has the shortest wavelength and radio waves are the longest (2)
c) UV can cause skin cancer (1)

4
a) People give off heat, or infra-red radiation (1)
b) Stones are cold (1)
c) eg. checking heat loss from a house (1)

❑ 12.4.1 Recording pictures ❑

1
a) Nicéphore Niepce (1)
b) Buildings next to his home (1)
c) 8 hours (1)
d) Louis Daguerre (1)
e) French (1)
f) Copper plate coated with silver (2)
g) mercury vapour (1)
h) Fixed with salt (1)

2 They enabled an instant picture to be developed from the film inside the camera in a few minutes. (2)

3
a) VHS (1)
b) the centre (1)
c) the edge (1)

4
a) mirror (1)
b) shutter (2)
c) night (1)
d) Less light means longer exposure time to allow the chemicals to change on the film. (2)

Answers

❑ 12.4.2 Recording pictures ❑

1 a) French (1)
b) Niepce (1)
c) Buildings next to his home (1)
d) 8 hours (1)
e) It needed a much shorter time to work so it was faster to work. (1)
f) Plates of copper covered in silver (1)
g) mercury vapour (1)
h) salt (1)
i) The image can be exposed to light and remain unchanged. (1)

2 They enabled an instant picture to be developed from the film inside the camera in a few minutes. (2)

3 a) VHS (1)
b) the centre (1)
c) the edge (1)

4 a) mirror (1)
b) shutter (2)
c) night (1)
d) Less light means longer exposure time to allow the chemicals to change on the film. (2)

❑ 12.5.1 Seeing and hearing ❑

1 a)

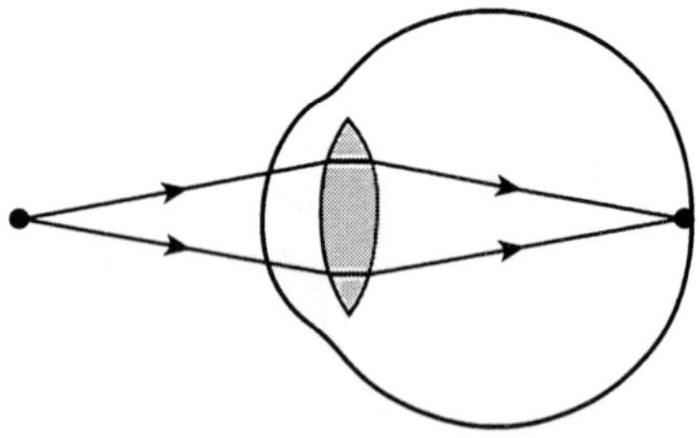

They need light rays to be spreading out when they reach their lens so the lens can focus them. (2)
b) Light from a distant object is parallel and is focused in front of the retina. (2)
c) concave (1)
d) It bends the ray outwards. (1)

2 a) retina (1)
b) behind the retina (1)
c) convex (1)
d) It bends the light inwards. (1)

3

OUTER	MIDDLE	INNER
pinna	ossicles	semicircular canals
eardrum	oval window eustachian tube	cochlea auditory nerve

(8)

4 a) Carry sound vibrations to the oval window (1)
b) Converts vibrations to nerve messages (1)
c) Sends messages to the brain (1)
d) Connects ear to nasal cavity and keeps air pressure of eardrums the same as that outside. (2)

5 TWO of ossicles, cochlea, auditory nerve. (2)

❑ 12.5.2 Seeing and hearing ❑

1

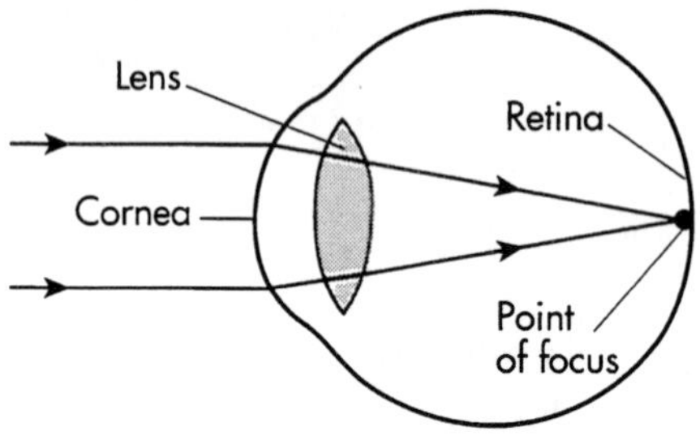

(8)

2 a) in front of the retina (2)

3 concave (2)

4 a) pinna (2)
b) eardrum (2)
c) cochlea (2)
d) semicircular canals (2)
e) auditory nerve (2)

5 ossicles, cochlea, auditory nerve (3)

❑ 12.6.1 How sound is made ❑

1 Movement caused when an object travels backwards and forwards very quickly. (1)

Answers

2 When the hammer hits the bell it makes the bell vibrate, which makes air particles next to the bell vibrate, then glass, then air again to our ears. There are no particles in a vacuum so vibrations cannot be passed on, so no sound. (4)

3 Space is a vacuum, no particles so no sound should be transferred. (2)

4 Longitudinal (1)

5 a) Particles come closer together. (1)
b) Particles move further apart. (1)

6 Normally the sound is spread out but a megaphone concentrates it and points it in one direction so it seems louder. (2)

7 Water is a liquid made up of particles. These particles can vibrate and so carry the sound. (1)

8 a) glass (1)
b) Particles carry sound and particles are closer together in a solid than a gas so it is quicker to pass on vibrations. (1)

❏ 12.6.2 How sound is made ❏

1 a) air (1)
b) glass and air (2)
c) Space with no air in. (1)
d) Particles are needed to carry sound, no particles so no sound. (2)
e) Yes. There are particles to carry the sound. (2)
f) i) particles closer together (1)
ii) further apart (1)

2 Rings of waves spread out in all directions from where the stone hit the water. (2)

3 a) all around (1)
b) All the sound is now pointed in that one direction so it is concentrated. (1)
c) The sound will be less. (1)

❏ 12.7.1 Frequency, pitch and volume ❏

1 Electrical signals (1)

2 a) Oscilloscope (1)
b) Transverse (1)

3 a) i) Distance between the same points on two adjacent waves, or the tops of two adjacent waves. (1)
ii) The number of wavelengths that go by in a second. (1)
iii) The higher the frequency, the higher the pitch and the shorter the wavelength. (1)
iv) One complete wave is produced in one second, the frequency is one hertz. (1)
b) Higher frequency (1)
Higher pitch (1)
Shorter wavelength (1)

4 a) They lose their full range of frequencies. (1)
b) Dogs can hear much higher frequencies. (1)

5 The skin would vibrate less for a quiet note than for a loud one. (1)

6 'a' is the amplitude of the wave and this is bigger the more energy the sound has. See diagram to 12.7.2 Q4. (2)

❏ 12.7.2 Frequency, pitch and volume ❏

1 a) Electrical signals (1)
b) Transverse wave (1)

2 a) wavelength (1)
b) frequency (1)
c) pitch (1)
d) hertz (1)

3 a) left – quiet (1)
right – loud (1)

Answers

b) Hitting the note hard moves the hammer up quickly to hit the string firmly making a loud note. (2)
Play gently, the hammer rises slowly, hits the string gently and makes a quiet note. (2)

4 a) left – quiet (1)
right – loud (1)
b) The greater the volume the greater the amplitude. (1)

❏ 12.8.1 The speed of sound ❏

1 Sound travels by particle vibration, passing the wave on to the next particle. Particles are touching in solids with a regular arrangement, touching in liquids but with irregular arrangement, and a long way apart in gases. So, vibration is easiest to pass on in solids, then liquids, then gases, so fastest in solids, then liquids, then gases. (4)

2 a) 3 km
b) $1\frac{2}{3}$ km
c) at the storm centre (3)

3 a) 4 times (1)
b) The body of a violin is 4 times longer than its narrowest width. (1)

4 a) subsonic (1)
b) supersonic (1)

5 a) sound barrier (1)
b) They began to shake. (1)
c) The shape pushed air in front of it causing shock waves on its wings and body. (1)
d) Planes made more pointed, swept back their wings, and used more powerful engines. (3)
e) Bell X-1. (1)
f) Captain 'Chuck' Yeager. (1)
g) 14 October 1947. (1)

❏ 12.8.2 The speed of sound ❏

1 a) 4 km
b) 3 km
c) 1 km
d) 0 km
e) $1\frac{2}{3}$ km (5)

2 a) Concorde (1)
b) The sonic boom is very sudden and has a startling effect on people and can damage buildings. (1)
c) 50 miles (1)
d) fly supersonic and break the sound barrier. (1)
e) 1947 (1)
f) Bell X-1 (1)
g) USA (2)
h) subsonic (2)

3 The violin is about 4 times longer than the width at its narrowest point. (2)

4 density, elasticity. (2)

❏ 12.9.1 Noise in the environment ❏

1 a) The amount of energy in the sound waves. (1)
b) The apparent strength of the sound as it reaches the eardrum and is transmitted to the brain. (1)

2 a) The unit used to measure the intensity of sound. (1)
b) A normal ear can just detect the sound. (1)
c) 0 dB (1)
d) This intensity can hurt the ears. (1)
e) 120 dB (1)

3 The same conditions can be ensured and so the figures may be compared fairly. (2)

4 The engine gives off a lot of noise and the silencer helps reduce this and so lowers noise pollution in the environment. (2)

Answers

5   Fences at the side. Sink the motorway into a cutting. Both methods deflect noise upwards away from nearby houses. (2)

6   Double glazing has a gap between the two layers of glass and this helps reduce noise. (1)

7   Speeds them up. (1)

❑ 12.9.2 Noise in the environment ❑

1   a) The same conditions can be ensured and so the figures may be compared fairly. (2)
    b) Clarinet, Trumpet, Trombone, Bass drum. (1)
    c) A normal ear can just detect the sound. (1)

2   Both methods deflect noise upwards away from nearby houses. (2)

3   The engine gives off a lot of noise and the silencer helps reduce this and so lowers noise pollution in the environment. (1)

4   a) intensity (1)
    b) loudness (1)
    c) soundproofing (1)

5   Thick loft insulation. Double/triple glazing. Any other sensible answers. (2)

6   Car engine working hard. 'Wind' rushing by. Tyres on the road surface. Any other sensible answers. (3)

❑ 12.10.1 A brief history of sound recording ❑

1   a) Thomas Edison – American (2)
    b) Tin foil on a cylinder (1)
    c) Wax cylinder replaced the tin foil (1)
    d) Gramophone (1)
    e) Steel piano wire (1)
    f) Paris Exhibition, 1900 (2)
    g) Instead of steel wire, they became plastic coated with a magnetic coating (2)
    h) 1980 (1)
    i) plastic, aluminium, lacquer (3)

2   Filtered out unwanted noise, mainly hiss, on cassettes and improved sound quality. (2)

❑ 12.10.2 A brief history of sound recording ❑

1   a) 1877 (1)
    b) Thomas Edison – American (2)
    c) Tin foil (1)
    d) Tin foil was replaced by wax. (1)
    e) Emile Berliner in 1880 (2)
    f) Gramophone (1)
    g) Aluminium (1)

2   a) steel piano wire (1)
    b) Paris Exhibition, 1900 (2)
    c) 1929 (1)
    d) It was much lighter. (1)
    e) Sony of Japan. (1)

❑ 13.1.1 Voyage to the Moon ❑

1   a) Yuri Alekseyevich Gagarin orbited the Earth once on the **12th April 1961** in Vostok 1. The journey lasted 1 hour and 48 minutes. (1)
    b) The Moon varies between **221,460 miles and 252,700 miles** from the Earth. (1)
    c) The astronauts in Apollo 11 stepped on to the surface of the Moon on **21st July 1969**. (1)
    d) Neil Armstrong was followed by **Edwin "Buzz" Aldrin**.
    Alan Shephard was the third astronaut who remained orbiting the Moon. (1)
    Neil Armstrong's first words were **"This is one small step for man, one giant leap for mankind"**. (1)
    e) Apollo 13 developed a fault on the way to the Moon. **A panel on the spacecraft was lost along with the crew's oxygen supply.** (1)
    The crew survived in the **small lander module** as they gathered speed around the Moon to land back on Earth. (1)

Answers

2 a) The Moon takes **28.53** days to orbit
the Earth. (1)
b) The Earth, Moon and Sun are all in
alignment. (1)
The Moon's **shadow** is cast on the
Earth and anyone in the shadow sees
the Moon blocking the Sun and
all goes dark. (1)
c) i) A total eclipse is where the whole
of the Sun is covered by the
Moon. (1)
ii) A partial eclipse is where only a
section of the Sun is covered by the
Moon. (1)
d) A lunar eclipse is the Earth's
shadow being cast on to the Moon's
surface (1) and a solar eclipse is the
Moon's shadow cast on to the Earth's
surface. (1) (2)

❑ 13.1.2 Voyage to the moon ❑

1 October 1957   Sputnik
March 1961   Laika
April 1961   Gagarin
February 1962   Glenn
June 1963   Tereshkova
July 1969   Apollo 11
April 1981   Colombia
September 1989   Voyager (8)

2 a) The light will darken (1) as the Moon
passes in front of the Sun. (1) (2)
b) Sun > Earth > Moon (3)
c) The Earth, Moon and Sun are all in
alignment. (1) The Moon's **shadow** is
cast on the Earth and anyone in the
shadow sees the Moon blocking the
Sun and all goes dark. (1) (2)

❑ 13.2.1 The Planets ❑

1 a) Mercury, Venus, Earth, Mars, Jupiter,
Saturn, Uranus,
Neptune, Pluto. (1)
b) Jupiter, Saturn, Neptune, Uranus,
(Earth, Mars), Pluto, Mercury, Venus. (1)
(Mars is slightly longer than Earth)
c) Pluto, Mercury, Mars, Venus, Earth,
Uranus, Neptune, Saturn, Jupiter. (1)

2 The further the planet is from the
Sun, the longer it takes for the
planet to go round the Sun. (2)
*(allow one mark for a simpler
answer – eg planets take longer)*

3 128.963 (1)
Less than half the mass of Jupiter (1)

4 Jupiter is much larger than Earth. (1)
This large volume means that its
density is reduced. (1)

5 Mercury should be the hottest (1) and
Pluto the coldest *(it is)* (1) (3)

6 It is close (second) to the Sun (1) and
carbon dioxide traps the heat
(greenhouse effect). (1) (2)

7 Mars (1)

8 Jupiter, Saturn, Uranus and Neptune (4)

9 Planets do not go round the Sun in
perfect circles (the path is eliptical)
so their distance varies. (1)

❑ 13.2.2 The Planets ❑

1 Mercury (1)

2 Pluto (1)

3 Uranus (1)

4 Saturn (1)

5 Mercury and Venus (2)

6 Jupiter, Saturn, Neptune and Uranus (4)

7 Venus (1)

8 Mercury (1)

9 Venus (1)

10 Jupiter (1)

11 four (1)

Answers

12 Jupiter, Saturn, Uranus, Neptune, Earth, Venus, Mars, Mercury Pluto. (1)

13 It is a small planet. The other four are gas giants. (1)

14 The drawing should be 144 mm in diameter. (1)

15 The drawing should be 12 mm in diameter. (1)

16 Neptune (1)

❑ 13.3.1 The Sun as a star ❑

1 a) hydrogen (1)
b) Atoms fuse together in nuclear reactions (2)
c) The Sun is so bright it obscures other stars (1)

2 a) Areas of the Sun which is cooler (1)
b) We follow the rate at which Sunspots move across the surface (1)

3 a) 500 seconds (1)
b) 8 minutes (1)

4 a) gravity (1)
b) temperature rises (1)
c) Planets formed from the gas and dust swirling round the Sun (1)
d) The solar system (1)

5 a) It will run out of fuel (1)
b) White dwarfs (1)
c) They explode in supernovas (1)

❑ 13.3.2 The Sun as a star ❑

1 a) They fuse together (1)
b) fusion (1)
c) It is too far away, the Sun is closer. (1)
d) There is not enough hydrogen to cause the temperature to rise enough to start nuclear reactions. (1)

2 Sunspots (1)

3 The brightness of the Sun will damage your eyes (1)

4 Sun – the others are planets (2)
Galaxies – the others are part of the solar system (2)
Moon – the others are planets (2)

5 a) Gravity made the cloud collapse (1)
b) The temperature rose (1)
c) Solar system (1)
d) It will run out of fuel (1)

❑ 13.4.1 Our views of the universe ❑

1 In the morning and afternoon Sun is low in the sky. (1)
At midday the Sun is high in the sky. (1)

2 a) The Sun takes longer to travel across the sky (the Earth's axis is tilted towards the Sun). (2)
b) The rays of the Sun are concentrated over a smaller area of the Earth. (1)

3 a) Planets appear to loop back on themselves (1)
b) Copernicus (1)
c) Keplar (1)
d) Jupiter (1)
e) The moons were going round Jupiter, not the Earth (1)
f) The telescope (1)
g) Newton (1)

4 a) Stars (1)
b) The Earth is rotating (1)
c) The pole star (1)

❑ 13.4.2 Our views of the universe ❑

1 a) east (1)
b) west (1)
c) north (1)
d) longer (1)
e) The Sun is lower in the sky (1)

Answers

2 Nicolaus Copernicus
Johannes Keplar
Galileo Galilei
Isaac Newton (4)

3 a) The telescope (1)
b) Jupiter's moons (1)
c) The church said the Earth was the centre of the universe and this showed that there were objects in the sky which did not go around the Earth. (2)

4 a) constellations (1)
b) Orion (1)

1.1.3 What do living things do?

1 Look at the diagrams below and answer the questions. Then take the letters from the shaded boxes and arrange them into a characteristic of living things.

Radio

Plant cells

Train

Leaves
Roots

Seeds

Eyes
Nose
Ears

Fuel + oxygen ⟶ carbon dioxide + water + energy
Respiration

a) Some plants reproduce by producing **S** E E D S

b) A R A D **I** O makes noise but does not move and is not living.

c) The E A R **S** are the organs of hearing.

d) The L E A **V** E S of a plant make food.

e) A **T** R A I N has wheels but cannot move by itself.

f) Making energy using fuel plus oxygen is R E S P **I** R A T **I** O N

g) P L A **N** T cells have walls made of cellulose.

h) The N O S **E** is the organ of smell.

i) R O O **T** S grow downwards.

j) The E **Y** E S are the organs of sight. (10)

The characteristic is S E N S I T I V I T Y (3)

2 Find 27 words to do with living things in the wordsearch.

N	B	M	F	E	E	D	I	N	G	G	H	E	A	R	I	N	G	T	A
O	A	O	U	A	E	I	D	F	R	E	S	P	I	R	A	T	I	O	N
I	C	V	E	T	R	E	P	R	O	D	U	C	T	I	O	N	X	U	I
S	M	E	L	L	O	P	R	Q	W	R	N	N	T	A	S	T	E	C	M
E	X	C	R	E	T	I	O	N	T	P	L	A	N	T	S	L	M	H	A
C	L	E	A	V	E	S	O	W	H	M	I	N	E	R	A	L	S	K	L
E	S	T	S	I	G	H	T	V	L	I	G	H	T	A	D	U	L	T	S
L	U	P	H	O	T	O	S	Y	N	T	H	E	S	I	S	J	I	H	G
L	M	E	M	B	R	A	N	E	X	Y	T	Z	A	B	W	A	T	E	R

(27)

1.2.3 Organs and organ systems

1 Below are three organs cut up into pieces. Choose diagrams from the boxes to draw the complete organs.

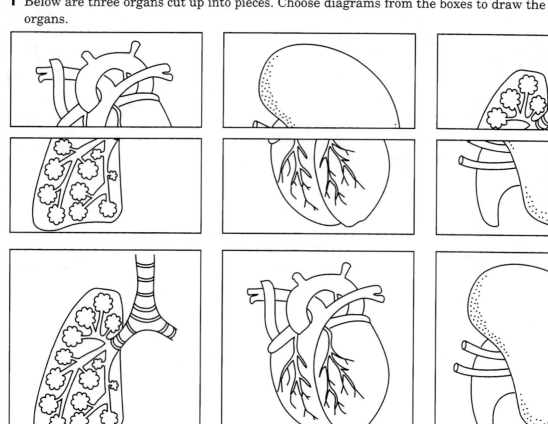

Lungs　　　　　Heart　　　　　Kidney　　　(30)

2 Add labels and arrows to the diagrams below to show how the digestive system is made up.

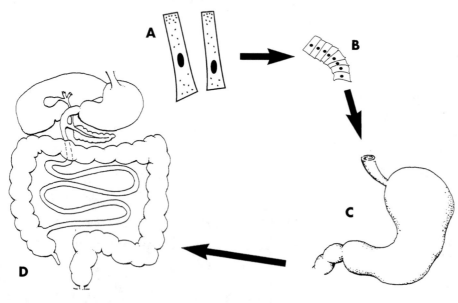

Organisms are made up of cells, organs and tissues.

Labels to use:

D Organs make up organ systems e.g. digestive system

A Individual cells

C Tissues make up organs e.g. stomach

B Layers of cells make tissues e.g. stomach lining

(20)

Science Companion © A Porter, M Wood, T Wood and Stanley Thornes (Publishers) Ltd 1995

1.3.3 Human reproduction

1 Label the male and female reproductive organs. (5)

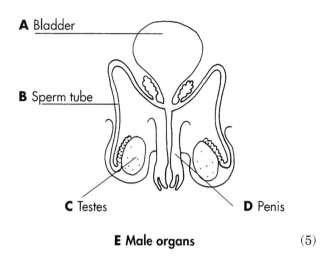

A Bladder
B Sperm tube
C Testes
D Penis
E **Male organs** (5)

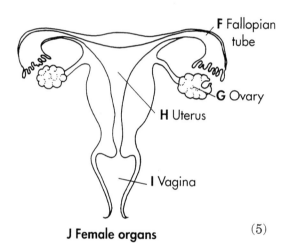

F Fallopian tube
G Ovary
H Uterus
I Vagina
J **Female organs** (5)

2 Fill in the missing words in the passage below.

Fertilisation occurs when an _egg_ meets a _sperm_. An egg is released from the ovaries once every _28_ days. The sperms are released into the _vagina_ during sexual _intercourse_. The sperms swim in the fluid of the _uterus_ to the fallopian tubes. If an egg is present, only _one_ sperm is needed to fertilise it. The rest die. The egg travels down to the lining of the _uterus_ where it is _implanted_. The egg is now called an _embryo_ and takes nine months to develop. (10)

3 Use the clues to fill in the words on the grid.

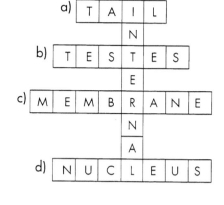

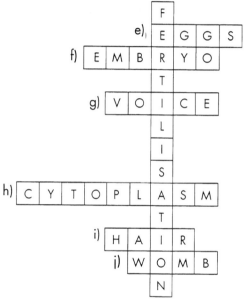

(10)

CLUES

a) This makes the male sperm move.
b) Sperms are produced by these in the male.
c) The outside layer of the female egg cell.
d) The centre of the female egg cell.
e) These are produced by the ovaries in the female.
f) A baby is called this in its early development.
g) This deepens as boys develop into adults.
h) This contains the supply of food in the female egg.
i) As boys develop into adults this grows on their bodies.
j) A baby develops in this part of the female's body.

1.4.3 Blood and circulation

1 The diagram shows a cross section of the heart.

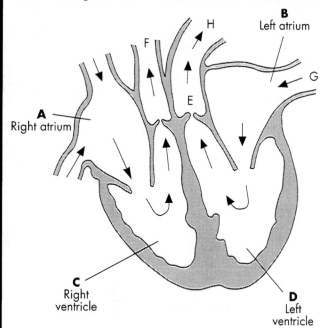

a) Put in **arrows** to show the direction in which the blood flows through the heart (4)

b) Label chambers **A–D**. (4)

c) Label one of the valves with the letter **E**. (2)

d) Label the following places with these letters:
F Blood leaves the heart to pick up oxygen
G Oxygenated blood comes from the lungs
H Blood is pumped to the rest of the body. (3)

e) Colour the oxygenated blood **red** and de-oxygenated blood **blue**. (2)

2 Fill in the spaces in the following passage:

The heart is an **ORGAN** which pumps **BLOOD**. It is made of a special type of muscle called **CARDIAC** muscle. The heart beats **70** times a minute. It has **FOUR** chambers, two **ATRIA** and two **VENTRICLES**. The **RIGHT** side of the heart pumps blood to the **LUNGS** to pick up **OXYGEN**. The left side of the heart pumps blood to the **REST** of the **BODY**. (12)

3 The table shows the four blood groups:

Group	% British population	Can receive blood group
A	41%	A & O
B	9%	B & O
AB	3%	A, B, AB, O
O	47%	O

a) How many blood groups are there?   4

b) Which blood group is the most common?   O

c) Which blood group is the least common?   AB

d) What percentage of the population is blood group B?   9%   (4)

e) A person with group AB is called a universal recipient, what does this mean?   HE CAN RECEIVE BLOOD FROM ANY GROUP   (2)

f) A person with group O is a universal donor. What does this mean?   HE CAN GIVE BLOOD TO ANY GROUP   (2)

Science Companion © A Porter, M Wood, T Wood and Stanley Thornes (Publishers) Ltd 1995

1.5.3 Respiration

1 Complete the passage below about baking bread. Use the following words to help you.

**carbon dioxide   rest   cooked   dough
kneaded   trapped   evaporates
alcohol   ferment   salt   yeast
rise   water**

Bread is made by mixing flour, SALT, WATER and YEAST. It is KNEADED into a DOUGH and then left in a WARM place to RISE. The dough rises because the yeast cells FERMENT producing ALCOHOL and CARBON DIOXIDE. The dough is elastic and carbon dioxide bubbles are TRAPPED making it rise. The dough is then COOKED in an oven, the alcohol EVAPORATES during baking. (13)

2 The experiment below shows the effect of sugar on yeast. Study it and answer the questions.

a) Why was the water in tube **A** boiled? TO REMOVE THE OXYGEN (1)
b) What colour would the limewater be at the start of the experiment? CLEAR (1)
c) Name and describe the reaction between yeast and sugar. YEAST BREAKS DOWN SUGAR INTO ALCOHOL AND CARBON DIOXIDE. IT IS CALLED ALCOHOLIC FERMENTATION (4)
d) What would the colour of the limewater be at the end of the experiment? CLOUDY OR MILKY (1)
e) Account for this colour CARBON DIOXIDE TURNS LIMEWATER CLOUDY OR MILKY (1)
f) What might you smell in tube **A** at the end of the experiment? ALCOHOL (1)
g) Account for this smell IT IS GIVEN OFF IN THE REACTION (1)

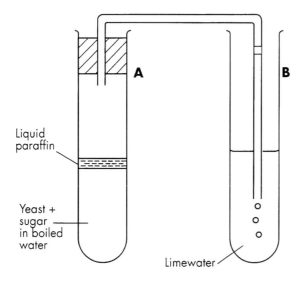

1.6.3 Breathing

1 The diagram below shows the organs of breathing.

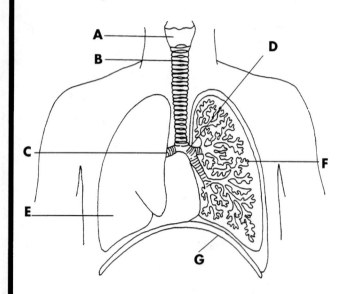

a) Label the parts of the diagram with the following letters:
- **A** Larynx
- **B** Trachea
- **C** Bronchi
- **D** Bronchioles
- **E** Lung
- **F** Alveoli
- **G** Diaphragm (7)

2 Look at the part of the lungs shown below then answer the questions.

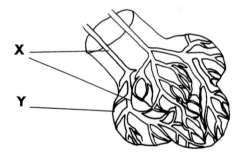

a) Name this part of the lungs   ALVEOLI   (1)

b) What are parts **X**?
   BLOOD CAPILLARIES   (1)

c) What passes in and out of the blood between **X** and **Y**?   CARBON DIOXIDE AND OXYGEN   (2)

d) Which cells transport oxygen?   RED   (1)

e) Name the process by which gas is exchanged.   DIFFUSION   (1)

3 The diagram below represents model lungs:

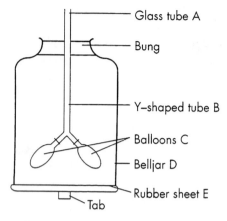

a) Name the parts represented by the letters **A–E**.
- **A** TRACHEA
- **B** BRONCHI
- **C** LUNGS
- **D** CHEST/THORAX
- **E** DIAPHRAGM   (5)

b) Explain what happens when the tab is pulled down.
THE AREA IN THE BELL JAR INCREASES AND THE PRESSURE DECREASES. AIR ENTERS TUBE A AND THE BALLOONS INFLATE   (2)

4 Transfer the figures below about lung capacity, on to the measuring cylinder.

a) Lungs always hold 1.5 l of air. Colour this part yellow.   (1)

b) Lung capacity for normal breathing 3.0 l. Draw a blue line.   (1)

c) Lung capacity for faster breathing 5.0 l. (e.g. exercising) Draw a red line.   (1)

d) Why is there always air left in the lungs?

SO THAT THE LUNGS DO NOT COLLAPSE   (1)

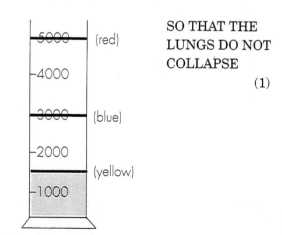

1.7.3 Good health

1 Examine the table below which shows the changes in nutrient intake over a twenty year period, then answer the questions.

Energy per person per day	1970	1975	1980	1985	1992
Total energy (kcal)	2560	2290	2230	2020	1960
Total fat (g)	119	107	106	96	87
Saturated fat (g)	52.0	51.7	46.8	40.6	34.3

a) What was the total calorie intake per day in 1970 and 1992?   2560 AND 1960 kJ   (2)

b) Suggest two reasons for the reduced intake.
  i) PEOPLE EAT LESS FAT
  ii) PEOPLE EAT LESS SUGAR   (2)

c) What is saturated fat?
   FAT FROM ANIMAL TISSUES   (2)

d) What can happen to your body if too much saturated fat is eaten?
   IT CONTAINS CHOLESTEROL WHICH BLOCKS UP ARTERIES   (2)

e) Suggest ways in which people have reduced their intake of saturated fat
   SUITABLE ANSWERS:
   BY EATING LESS RED MEAT
   BY EATING LESS PROCESSED FOODS   (2)

f) Identify the organ below:
   THE HEART   (1)

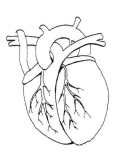

g) How does eating less fat help to keep this organ healthy?
   THE ARTERIES SURROUNDING THE HEART HAVE LESS CHANCE OF GETTING BLOCKED. THE HEART MUSCLE IS KEPT SUPPLIED WITH OXYGEN AND WORKS CORRECTLY. ARTERIES IN THE HEART ARE NOT BLOCKED AND BLOOD CAN FLOW. LESS RISK OF HEART ATTACK.   (3)

2 a) From these foods, underline those which keep you healthy in blue, those bad for you in red.   [———— blue   - - - - - - - red]

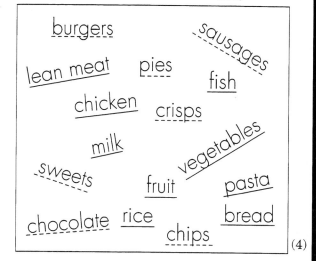

(4)

b) Look at the foods below and say why they should be eaten only occasionally.

THESE FOODS CONTAIN A LOT OF HIDDEN SATURATED FAT WHICH IS HIGH IN CALORIES AND CHOLESTEROL. THEY SHOULD ONLY BE EATEN SOMETIMES. EATEN IN EXCESS CAN LEAD TO HEART DISEASE AND BEING OVERWEIGHT.   (2)

(10)

1.8.3 When things go wrong

1 Read the newspaper article and answer the questions which follow.

> **Have a Heart**
> Heart disease is still responsible for most deaths in Britain today. One cause of heart disease is eating too much saturated fat. This is the fat from animal tissues which includes beef, pork, lard, dripping butter, cheese and cream. Saturated fat is also found as 'hidden fat' in foods like pies, sausages, cakes and biscuits. This type of fat contains the substance cholesterol. This blocks arteries which causes heart attacks. The Minister for Health is today launching a huge campaign to make the public aware of the importance of eating less fat.
> Dr. P. Vein 24/11/94

a) What is saturated fat? FAT FROM ANIMAL TISSUES LIKE BEEF, PORK AND BUTTER (1)

b) Give two examples of foods high in saturated fat. PORK, BEEF, BUTTER (2)

c) What is 'hidden fat'? FAT IN PROCESSED FOODS LIKE BISCUITS (1)

d) What is cholesterol? A SUBSTANCE IN SATURATED FATS (1)

e) How does cholesterol cause heart disease?

IT BLOCKS ARTERIES (2)

f) Suggest ways of cutting down on the amount of saturated fat we eat. EAT MORE WHITE MEAT AND FISH. CUT DOWN ON FOODS WHICH CONTAIN HIDDEN FAT (2)

g) Which other factors increase the risk of getting heart disease? SMOKING, STRESS AND LACK OF EXERCISE (2)

2 Study the table showing damage caused to the body by smoking.

Organ	Disease caused
Throat, mouth, nose, bladder and kidneys	Cancer
Lungs and trachea	Cancer. Bronchitis. Emphysema.
Oesophagus and stomach	Ulcers. Cancer.
Female reproductive organs	Cancer of the cervix. Premature babies. Low birth weights in babies.
Heart	Blocked arteries. Narrowing of the blood vessels. Build up of fat. Heart attack.
Legs	Buerger's disease. (blocked blood vessels, leads to amputations.)

Now label the diagram to show how smoking damages the body.

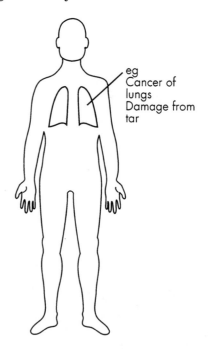

eg Cancer of lungs Damage from tar

(12)

1.9.3 Micro-organisms

1 Complete the passage on microbes.

There are four types of microbe, BACTERIA VIRUSES, FUNGI and PROTOZOA. Bacteria are microscopic organisms which live EVERYWHERE. There are three types: ROUND, rod-SHAPED and SPHERICAL. They multiply by DIVIDING into two. Not all bacteria are harmful; those which cause diseases are called PATHOGENS. Fungi are made of tiny MICROSCOPIC threads called HYPHAE. Examples include YEAST and MUSHROOMS. Fungi contain no chlorophyll and feed on the DEAD remains of plants and ANIMALS. Viruses are smaller than bacteria but are not LIVING. They can only live on living tissues and cause diseases. Protists are SINGLE-celled organisms which live in MOIST places. (19)

2 Label and complete the diagram below to show how minerals are recycled in nature.

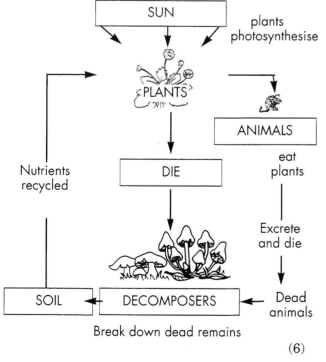

(6)

3 Read the article below and answer the questions

> **Robert Koch 1843–1910**
> Robert Koch was a German bacteriologist. He became interested in the disease anthrax which was killing thousands of cows. He grew the anthrax bacteria in his laboratory on flat dishes which he named after his assistant J.R. Petri. The dishes were filled with a gel called agar which fed the bacteria. Koch found a way of staining the bacteria so that they could be seen easily under the microscope. He also found that steam rather than dry heat was better at killing bacteria, this was a great help to the medical profession. Robert was also responsible for identifying the T.B. bacillus and worked on many other diseases including cholera, malaria and sleeping sickness.

a) What is anthrax? A DISEASE IN COWS (1)

b) How did Koch grow bacteria?
ON AGAR PLATES (1)

c) How did he make bacteria easier to see?
HE STAINED THEM (1)

d) How did he help the medical profession?
BY FINDING THAT STEAM KILLED
BACTERIA (1)

e) What is a T.B. bacillus? THE BACTERIUM
WHICH CAUSES TUBERCULOSIS (1)

f) Which other diseases did Koch work on?
MALARIA, CHOLERA AND
SLEEPING SICKNESS (1)

1.10.3 Soil, decay and compost

1 Use the following words to complete the passage below:

recycles   mix   oxygen   heat   warm   invertebrates   fertile   decays   bacteria   nutrients   respire   humus   moist

When plant material from the garden is put into a compost bin it DECAYS to make HUMUS. This is rich in NUTRIENTS and makes soil FERTILE. Plants decay best in a MOIST and WARM environment. A compost bin provides such a place. Plants in a compost bin have BACTERIA and fungi on them. A lid on the bin keeps the HEAT in. The bacteria and fungi break down plants while feeding on them. A compost bin has no bottom so that OXYGEN and INVERTEBRATES can get inside. Microbes need oxygen to RESPIRE and the invertebrates MIX up the compost. Putting compost back on to the garden RECYCLES nutrients. (12)

2 A pupil carried out an experiment to find out which soil drains the fastest, clay or sand.

d) How was the experiment kept fair? Give two ways. THE SAME AMOUNT OF WATER WAS USED. THE SAME AMOUNT OF SOIL WAS USED. OR ANY OTHER SUITABLE ANSWER (2)

e) Explain how you could improve the drainage in a heavy wet soil. MIX IN ORGANIC MATTER AND SAND AND GRIT (2)

f) Draw the particles in a clay and sandy soil.

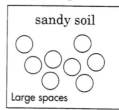
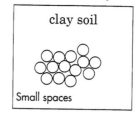

(4)

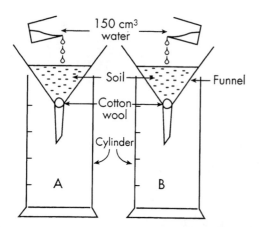

Results Table	A	B
Time for water to drain into cylinder (mins)	1.30	5.00
Volume of water collected after 30 minutes	145 cm³	20 cm³

a) Which sample drained the fastest? A (1)

b) How long did it take sample B to drain? FIVE MINUTES (2)

c) Suggest which sample was clay and why. B WAS CLAY BECAUSE IT TOOK LONGER TO DRAIN (2)

1.11.3 Reproduction in flowering plants

1 a) Label the diagram of a broad bean seed. Use the words given.

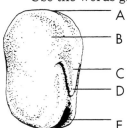

A Testa
B Cotyledon
E Scar
C Plumule
D Radicle
(5)

b) Give the functions of each part next to the letters below.
A TO PROTECT THE SEED (1)
B THE STORE OF FOOD (1)
C THE SHOOT (1)
D THE ROOT (1)
E WHERE ATTACHED TO THE SEED POD (1)

2 The diagram below shows an experiment to find the conditions needed for germination of seeds.

a) Draw what the seeds might look like after a few days in the test tubes.

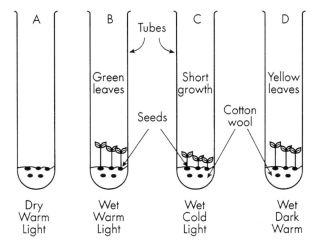

Conditions

b) Complete the table to show how you expect the seeds to look, giving reasons for your answers.

TUBE	RESULTS	REASON
A	NO GROWTH	NO WATER
B	SEEDS GROWN AND HAVE GREEN LEAVES	ALL CONDITIONS CORRECT
C	SEEDS GERMINATE BUT GROWTH SMALL	GROWTH SLOW IN COLD
D	GERMINATION. SEEDS GROWN BUT LEAVES YELLOW	SEEDS GERMINATE IN THE DARK BUT NEED LIGHT FOR PHOTOSYNTHESIS

3 Seeds are dispersed in different ways. Look at the diagrams below and explain how each is dispersed.

a)
Blackberry fruit

ANIMALS EAT THE FRUIT AND THE SEEDS ARE LEFT IN ANIMALS EXCRETA WHERE THEY CAN GROW

b)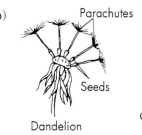
Dandelion

EACH SEED IS ATTACHED TO A PARACHUTE. THE PARACHUTES ARE CARRIED BY THE WIND

c)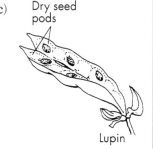
Lupin

DRY SEED PODS EXPLODE AND SEEDS ARE SCATTERED OVER A WIDE AREA

(6)

1.12.3 Leaves and photosynthesis

1 Complete the passage about photosynthesis. Photosynthesis is the process by which plants make FOOD. Most photosynthesis occurs in the LEAVES of a plant. The raw materials CARBON DIOXIDE and WATER diffuse into the CHLOROPLASTS where light ENERGY is absorbed. They react to form the products GLUCOSE and OXYGEN. Glucose MOLECULES cannot be stored in plants because they are small and SOLUBLE in water, so they are changed into STARCH molecules which are LARGE and insoluble in water. OXYGEN gas is a WASTE product of photosynthesis and is removed from the plant.

(14)

2 Label the boxes on the diagram to show the raw materials and products of photosynthesis.

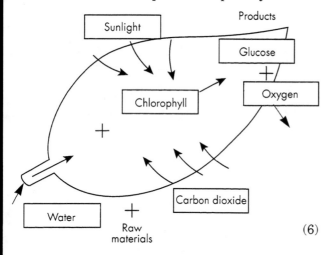

(6)

3 a) The diagrams below show how to test a leaf for starch. Complete the labels.

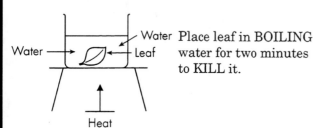

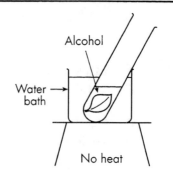

Place leaf in a tube of ALCOHOL and place in a WATER bath. This DECOLOURISES the leaf.

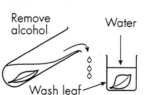

Wash the leaf to soften it. Alcohol makes leaves BRITTLE.

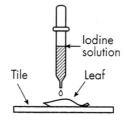

Test with IODINE solution. A BLUE BLACK result indicates STARCH is present.

(10)

b) Which parts of these leaves would turn blue black after a starch test? Colour those areas blue.

(2)

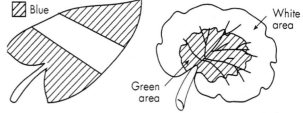

2.1.3 Living organisms

1 Place the letter of the animals below next to their descriptions.
A bat   B robin   C octopus   D seal
E spider   F trout   G jellyfish

a) A nocturnal mammal.   ___A___
b) A bird   ___B___
c) An arachnid with eight legs   ___E___
d) A mollusc with eight legs   ___C___
e) A freshwater fish   ___F___
f) A sea mammal   ___D___
g) A sea animal with a transparent body   ___G___
(7)

2 Examine the plant in the following diagram.

a) What type of plant is it? TREE   (1)

b) This plant is evergreen. What does this mean? IT DOES NOT SHED ITS LEAVES IN WINTER   (1)

c) Describe the leaves of this plant.
THIN NEEDLES   (2)

d) Where are the seeds produced in this plant?
IN CONES   (1)

3 The diagram below shows an alligator. Answer the questions about it.

a) To which group of animals does the alligator belong?   REPTILES   (1)

b) What is its habitat?   WATER, SWAMPS   (1)

c) Describe its skin.
TOUGH HORNY PLATES   (2)

d) Explain how the following features of the alligator help it to survive.
 i) Powerful tail   PROPELS IT THROUGH WATER   (1)
 ii) Webbed feet   FOR SWIMMING   (1)
 iii) Sense organs remain above the water when it is swimming along and its body is submerged.   IT CAN SEE AND SENSE PREY WITHOUT BEING DETECTED   (2)

4 Locusts belong to the largest group of invertebrates, the arthropods. Their features include:

compound eyes   segmented body parts   antennae   body in three parts: head, thorax and abdomen   jointed limbs   exoskeleton

OWN ANSWER ON DIAGRAM

a) Label those parts on the locust.

b) The following insects are also from the same group.

butterfly   beetle   wasp

Match them to their descriptions below:

i) Narrow waist between abdomen and thorax.   WASP   (1)

ii) Body covered with velvety fur and wings with scales.   BUTTERFLY   (1)

iii) Wings protected by wing cases.
BEETLE   (1)

2.2.3 Grouping animals and plants

1 Follow the key to identify the reef fish. Write the names under the diagrams.

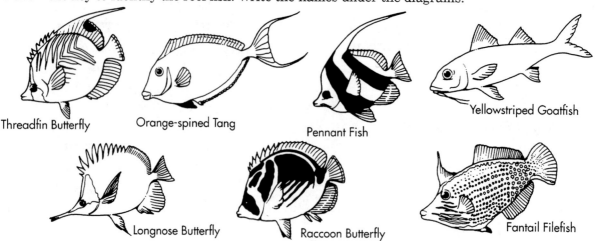

Key

1. Has it a striped body? .... go to question 2.
   Has it no stripes on its body? .... go to question 4.
2. Black and white markings over eye and upper body. .... **Raccoon Butterfly**.
   Dorsal fin is elongated to a point. .... go to question 3.
3. Body has two broad dark stripes diagonally across it. .... **Pennant Fish**.
   Narrow stripes on body and spot on dorsal fin. .... **Threadfin Butterfly**.
4. Tail fins divided into two points. .... go to question 5.
   Single tail fin. .... go to question 6.
5. Large flat body. Tail fin tapers into two long points. .... **Orange-spined Tang**.
   Streamlined body. Two dorsal fins. Two barbels on mouth. ... **Yellowstriped Goatfish**.
6. Elongated snout. Serrated dorsal fins. .... **Longnose Butterfly**.
   Spotted body. First dorsal fin curved. Tail fanshaped. .... **Fantail Filefish**.

(14)

2 The following are all fungi. Draw the correct fungus in the box which has its description.

 C

 D   A    B

(16)

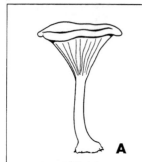 A

 B

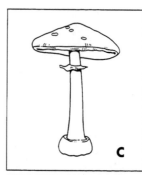 C

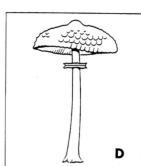 D

FALSE CHANTERELLE
Funnel-shaped cap curled at edges.

ANTABUSE INK CAP
Ribbed bell shaped cap. Dark gills.

FALSE DEATH CAP
Convex cap. White spots on cap. Thick stem with bulb at base.

PARASOL MUSHROOM
Umbrella shaped cap. Scales on cap.

Science Companion © A Porter, M Wood, T Wood and Stanley Thornes (Publishers) Ltd 1995

2.3.3 Variation

1 The heights of pupils in a class was measured. The results are shown in the table.

Height group (cm)	134–40	141–2	143–4	145–6	147–8	149–50	151–2	153–4	155–6
Number of students	1	3	3	5	7	5	3	1	1

a) Draw a block graph to show the results.

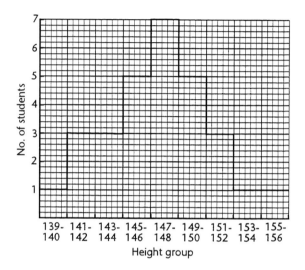

(8)

b) How many students were in the class?
   29 (1)

c) Which groups holds the most students?
   147–148 (1)

d) How many students are 142 cm in height or less?
   4 (1)

e) Do the results show continuous or discontinuous variation?
   CONTINUOUS (1)

f) What do we call a graph with this shape?
   A NORMAL DISTRIBUTION CURVE (1)

g) Which two of the following characteristics show the same type of variation as those in the graph?

   blood group   eye colour   <u>FOOT SIZE</u>
   <u>HANDSPAN</u>   sex (2)

2 Study the passage below and then answer the questions which follow:

Industrial melanism

Some insect species in Britain show **industrial melanism**. Melanism is the word used to describe a darker form of a naturally light-coloured species. Melanism occurs in natural conditions where organisms adapt to changing environments. Industrial melanism can happen where species adapt to live in sooty environments caused by industrialisation. The most common example of this is the peppered moth, *Biston betularia*. It has two forms, one black, the other speckled grey.

Before the Industrial Revolution most peppered moths in Britain were the grey form. This is because they rested on lichen-covered trees which were grey, and provided camouflage from predators. During the Industrial Revolution, soot killed the lichens on trees and the light moths could clearly be seen on the dark tree trunks. In 1848 the percentage of grey moths was 99%. In contrast, in 1894 the percentage of dark moths was 99%. The percentage of grey moths remained high in unpolluted areas of Britain. Other species showing industrial melanism are the oak eggar moth, the two-spot ladybird and the zebra spider. In each case the melanic form became more prevalent during the Industrial Revolution.

Dark peppered moth (the melanic form), and light peppered moth

a) What is melanism? THE DARKER FORM OF A LIGHTER COLOUR SPECIES (1)

b) Explain why the number of dark peppered moths increased during the industrial revolution. THE LIGHTER FORM WAS NO LONGER CAMOUFLAGED ON SOOT COVERED TREES. THE DARKER FORM SURVIVED (1)

c) What kind of predators might catch the moths? BIRDS (1)

2.4.3 Why do we look like our parents?

1 Read the following information about how sex is determined in a baby, then answer the questions which follow.

SEX OF A BABY
Sex chromosomes determine the sex of a baby. There are two types of sex chromosome called the Y chromosome which is short and the X chromosome which is long. Males contain an X and a Y chromosome and females contain two X chromosomes. Male sex cells, sperms have either an X or a Y chromosome, (half are X and half are Y) female sex cells all have X chromosomes. When fertilisation takes place an egg joins with a sperm which contains either an X or a Y chromosome. A combination of two X chromosomes will produce a girl and a combination of a Y and X chromosome will produce a boy.

a) How many types of sex chromosomes are there? TWO (1)

b) Which sex chromosome is a long one?
X (2)

c) Which sex chromosome do males have?
X AND Y (2)

d) What percentage of male chromosomes are X?
50% (2)

e) If a sperm has an X chromosome and an egg has an X chromosome, what sex will the baby be?
GIRL (2)

2 Solve the clues to find the words in the grid. Then find the word in column 6 which links them.

			I	N	**H**	E	R	I	T	E	D	
				G	**E**	N	E	T	I	C	S	
		C	H	**R**	O	M	O	S	O	M	E	S
				G	**E**	N	E	S				
B	R	E	E	**D**	I	N	G					
I	N	P	A	**I**	R	S						
		N	A	**T**	U	R	A	L				
		C	Y	**T**	O	P	L	A	S	M		

(8)

CLUES
1. Features passed on to offspring are this.
2. The study of heredity and variation.
3. Thread-like structures in the nucleus.
4. Parts of clue 3 which carry genetic information.
5. Creating new individuals.
6. Chromosomes are found like this.
7. The opposite of artificial selection.
8. Jelly-like substance found in a cell.

The word in column 6 is HEREDITY (1)

3 Quiz

1 Differences between individuals are called
VARIATION (1)

2 An example of discontinuous variation is
SEX/ BLOOD TYPE/ EYE COLOUR (1)

3 The process by which species which most suited to an environment will survive is called
NATURAL SELECTION (1)

4 Passing on features from parents to offspring is called HEREDITY (1)

5 Features like hair colour and ear shape are called INHERITED FEATURES (1)

6 Label parts **A**–**C** on the diagram below.

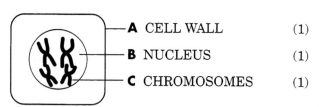

A CELL WALL (1)
B NUCLEUS (1)
C CHROMOSOMES (1)

7 Parts of chromosomes which contain genetic information are called
GENES (1)

8 Most human cells contain 46 chromosomes (1)

9 How many chromosomes do these cells contain?

 23 (1)   23 (1)

2.5.3 Extinction – the end of a species

1 From the list below, choose eight features which belong to *Tyrannosaurus rex* and write them on the diagram.

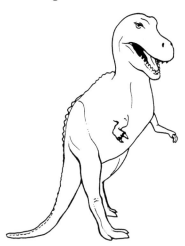

Meat-eater
Plant-eater
Walks on two legs
Walks on four legs
Huge head
Tiny head
Serrated teeth

Small teeth
Huge curved teeth
Tiny clawed hands
Long neck
Short powerful neck
Long whip-like tail
Short strong tail   (8)

2 Explain how *T. rex* was adapted to meat eating.   POWERFUL NECK FOR GRASPING PREY. SHARP CURVED TEETH AND CLAWS FOR TEARING FLESH   (2)

3 Underline the correct answers below:
  a) Dinosaurs belonged to the group of animals called:
     (birds) (amphibians) (<u>reptiles</u>) (mammals)
  b) Which of the following hitting the earth could have caused the extinction of dinosaurs?
     (comet) (<u>meteorite</u>) (star) (lightning)
  c) What does a paleontologist study?
     (planets) (comets) (<u>fossils</u>) (plates)
  d) A rare metal present in meteorites is called:
     (carbon) (copper) (<u>iridium</u>) (iranium)
  e) A species which is found in only one place on earth is called:
     (epidemic) (<u>endemic</u>) (igneous) (indigenous)   (7)

4 Read the passage below and answer the questions which follow.

Extinction in the Hawaiian islands

Hawaiian Goose

It took over 70 million years for plants and animals to colonise the Hawaiian islands. Large animals like reptiles and mammals could not reach the islands because they were unable to cross the Pacific ocean. The living things which arrived first were those which could travel by sea currents, insects, seeds and snails. Birds arrived by air currents. Life in the islands evolved slowly but flourished because there were no predators or competition for food. For this reason many species evolved without protective mechanisms like stingless nettles and flightless birds. The arrival of people to the island brought extinction to many species and many more are now in danger like the Hawaiian Goose and the Happyface Spider. People destroyed the forests to plant food crops and brought predators like rats, cats, and pigs. Because 90% of Hawaii's flora and fauna is endemic, that means that it is not found anywhere else on Earth, species which die are therefore made extinct in the world.

Happyface Spider

  a) Why were large animals unable to reach the Hawaiian islands when they were being colonised.   COULD NOT CROSS THE PACIFIC OCEAN   (2)
  b) How did life get to the islands?   BY AIR AND SEA CURRENTS   (2)
  c) Why did some species evolve without protective mechanisms?   THERE WERE NO PREDATORS AND NO COMPETITION FOR FOOD   (2)
  d) Name one of those species.   STINGLESS NETTLE   (1)
  e) Name two predators brought by man to the islands.   CATS, RATS, PIGS   (2)
  f) Why did man destroy forests?   TO GROW FOOD CROPS   (2)
  g) What does the word 'endemic' mean?   ONLY FOUND IN THAT PART OF THE WORLD   (2)

3.1.3 People – a threat to the Earth

1 Place the resources below next to their uses. Use each one only once.

**carbon dioxide   coal   flour
hardwood   heating homes   oil
limestone   oxygen   water   wood**

a) Respiration   OXYGEN

b) Fire extinguishers   CARBON DIOXIDE

c) Petrol and diesel fuels OIL

d) Window frames   HARDWOOD

e) Paper products   WOOD

f) Bread and baking   FLOUR

g) Cement   LIMESTONE

h) Drinking and washing   WATER

i) Producing electricity COAL

j) Heating homes   GAS

(10)

2 Look at the illustration below and answer the questions which follow.

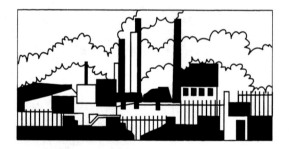

a) How is this industry affecting the environment?
Give at least two ways

CAUSING AIR POLLUTION

UGLY WORKING

KILLS ANIMALS   (1)

b) How might the health of people be damaged?

POISONOUS FUMES / NOISE

POLLUTION / CAUSES DEPRESSION   (1)

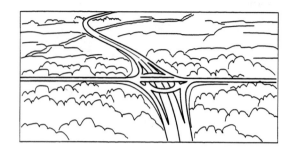

c) Give three ways in which a motorway can damage the environment.

1. DESTROYS LAND   (1)

2. CAUSES POLLUTION   (1)

3. KILLS PLANTS AND ANIMALS   (1)

3 Solve the clues to complete the word puzzle. Then find the resources in the boxes.

	A	B	C	D	E	F	G	H	I
1	R	E	S	O	U	R	C	E	S
2	S	H	E	L	T	E	R		
3	M	I	N	I	N	G			
4	E	X	H	A	U	S	T	E	D
5	P	O	L	L	U	T	I	O	N
6	M	A	T	E	R	I	A	L	S
7	W	A	S	H	I	N	G		
8	E	A	R	T	H				
9	G	A	R	D	E	N	I	N	G
10	O	X	Y	G	E	N			

(10)

CLUES
1. Things we need to live
2. A house is a form of this.
3. Taking ores and minerals from the ground.
4. When resources run out the supply is this.
5. Damage to the earth.
6. Products are made using raw   .
7. We need water for this.
8. Our planet.
9. Peat is used mainly for this.
10. A gas we need for respiration.

G	A	S
3F	4D	2A

O	I	L
5B	5G	5C

A	I	R
6B	7E	8C

W	A	T	E	R
7A	8B	5F	2C	2G

W	A	R	M	T	H
7A	9B	9C	6A	5F	2B

(5)

Science Companion © A Porter, M Wood, T Wood and Stanley Thornes (Publishers) Ltd 1995

3.2.3 The problem of pollution

1 The Bloggs family have had a picnic and have rubbish to take home with them. Use a highlighter pen or coloured pencil to shade each item of rubbish to show how they should dispose of it.
Blue [B] = recycle   Green [G] = compost heap   Red [R] = dustbin

aluminium foil R	cola bottle B	plastic cup B
apple core G	crisp packet R	plastic carrier bag B
banana peel G	egg shell G	teabags G
burger tray B	fish and chips paper R	wine bottle B
beer can B		yogurt pot B

2 Look at the diagram about global warming and complete the passage which follows.

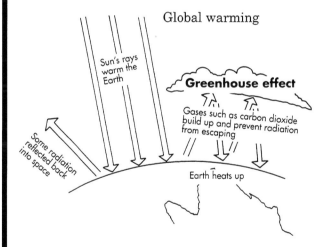

The SUN heats up the Earth by RADIATION. Normally some radiation is REFLECTED back into space. We produce CARBON DIOXIDE gas which builds up around the Earth and prevents RADIATION from escaping. This heats up the Earth and is called the GREENHOUSE EFFECT. This has increased the temperature of the Earth's ATMOSPHERE.   (7)

3 Look at the label from the product below and answer the questions.

WHIZ WASHING POWDER
Active ingredients
Anionic surfactant   Phosphates   Enzymes

a) Which ingredient causes pollution in rivers and lakes?

PHOSPHATES   (2)

b) These ingredients cause algae to grow in lakes and rivers. How does this kill fish and plants?

THEY ENCOURAGE THE GROWTH OF ALGAE. THIS BLOCKS LIGHT WHICH KILLS PLANTS. FISH DO NOT GET OXYGEN THEY DIE   (2)

c) How can we help to stop this substance getting into the water cycle?

USE PHOSPHATE FREE DETERGENTS   (2)

4 Use arrows to match the words to their meanings below.

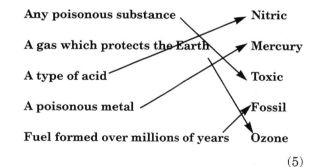

(5)

3.3.3 Are we doing enough?

1 Read the information about animals caught in oil spills and answer the questions which follow.

Animals swallow oil and lungs, kidneys and liver are damaged.

Mammals are blinded by oil.

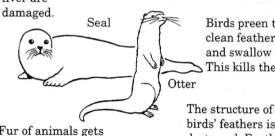
Seal — Otter — Seabird

Birds preen to clean feathers and swallow oil. This kills them.

The structure of birds' feathers is destroyed. Feathers are no longer waterproof and no longer provide insulation against cold.

Fur of animals gets matted. Insulating properties of fur are lost. Animals die of cold.

Oil is sticky and does not dissolve in water.

Oil destroys red blood cells in birds. They become anaemic.

a) What is crude oil?
   UNREFINED OIL (1)

b) Which properties of oil make it difficult to clean up? IT IS STICKY / DOES NOT DISSOLVE IN WATER / FLOATS (2)

c) Name three animal organs damaged by oil. LUNGS / KIDNEYS / LIVER (3)

d) How do the birds swallow the oil?
   WHEN CLEANING THEIR FEATHERS (1)

e) How does oil damage fur and feathers?
   FUR GETS MATTED, INSULATION LOST. FEATHERS STRUCTURE IS DESTROYED WATERPROOFING LOST AND INSULATION (1)

f) Which cells in birds are damaged by oil?
   RED BLOOD (1)

g) Suggest ways in which oil spills at sea could be reduced.
   OWN ANSWERS (2)

2 Complete the table about poisonous gases.

GAS	SYMBOL	FORMED	POLLUTION CAUSED
Sulphur dioxide	SO_2	Burning compounds containing sulphur	Acid rain Damages buildings and kills plants
Carbon dioxide	CO_2	Burning carbon compounds with oxygen	Greenhouse effect
Carbon monoxide	CO	Burning carbon compounds in a restricted area	Toxic exhaust fumes

(6)

3 Underline the correct definitions of the words below.

a) TOXIC — (edible) (<u>poisonous</u>) (pleasant)

b) COMBUSTION — (<u>burning</u>) (mixing up) (pollution)

c) RAW — (cooked) (<u>untreated</u>) (unnatural)

d) LEACHED — (<u>washed away</u>) (bleached) (blocked)

e) CONVERTED — (separated) (<u>changed</u>) (purified)

f) ALTERNATIVE — (<u>different</u>) (useful) (changing)

g) LEVELS — (heights) (<u>amounts</u>) (stages)

h) REMOVE — (<u>take out</u>) (replace) (travel)

i) EMITTED — (<u>given out</u>) (shouting) (burnt)

j) FILTERED — (smoothed) (<u>cleaned</u>) (untreated)

(10)

3.4.3 Living things and their environment

1 Fill in the missing words to complete the passage below.

Living things can DETECT changes in their environments. They need to do this in order to FEED and protect themselves from PREDATORS. Organisms react to changes in the environment like the seasons and the WEATHER. The seasons affect the temperature and the length of DAY and NIGHT. In winter animals and plants find it difficult to find FOOD. Plants cannot GROW well because the ground is FROZEN and there is less light. Many plants spend the winter in a DORMANT state. Animals HIBERNATE, MIGRATE or store up food for the winter. (12)

2 The list below shows how these animals spend the winter. Write the correct letter in the box.

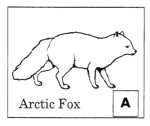
Arctic Fox A

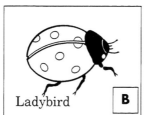
Ladybird B

Penguin F

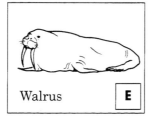
Walrus E

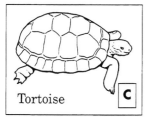
Tortoise C

Arctic Tern D

(6)

A Grows a dense woolly coat.

B Huddle together in a sleeplike torpor.

C Migrate to a warmer place.

D The cold makes their bodies slow down so they sleep most of the winter.

E Has an extra thick layer of blubber.

F Migrates by swimming to warmer places.

You may colour the diagrams if you wish.

3 Read the information about the hedgehog, then answer the questions which follow.

The Hedgehog
Hedgehogs are animals which hibernate during the winter. Food is scarce during the winter and they would probably starve if they did not hibernate. They feed themselves to make extra fat before the winter. Their bodies change in a number of ways during hibernation. Their body temperature and heart rate slow down so that they conserve energy. Curling up into a tight ball reduces the surface area of the hedgehog and also conserves energy. When the warm weather returns the animal's body returns to normal and it wakes up. If you have a pile of twigs or leaves in your garden which is left undisturbed during the winter you might get a hedgehog hibernating in your garden.

a) Why does the hedgehog hibernate in winter?
FOOD IS SCARCE (1)
b) How does the hedgehog prepare for winter?
IT FEEDS TO PUT ON FAT (1)
c) How does the heart rate change during hibernation?
IT SLOWS DOWN (1)
d) How does curling up into a tight ball conserve energy? IT REDUCES SURFACE AREA OF THE BODY (1)
e) Write down the names of four other hibernating animals.
ANY FOUR, E.G. BEAR, MOUSE (1)

3.5.3 Why do animals and plants live in different places

1 Complete the passage about penguins below.

> WORDS: Antarctica birds blubber feathers fly freezing warm predators
>
> ### Penguins
>
> Penguins are BIRDS but they do not FLY. Penguins like the emperor penguin live only in ANTARCTICA. The temperature is nearly always below FREEZING. It can fall to minus forty degrees celsius. Penguins skin is protected by dense FEATHERS and a thick layer of BLUBBER under their skin. They stand in groups to keep WARM, taking it in turn to stand on the inside of the group. Keeping together also protects them from PREDATORS. (8)

2 The table below shows the different layers in a tropical rainforest. Answer the questions about the information given.

LAYER	LIFE
Emergent layer	Giant trees grow above the canopy. Large predators like Harpo eagle live here and spy on their prey below.
Canopy	Trees form a dense roof of leaves. Most animals are found here because there is protection and food. Animals found include toucans, spider monkey, Lar gibbon and flying frog.
Understorey layer	This layer has shade-loving plants only. Creepers grow here and reach to the forest floor.
Shrub layer	No light here so sparse vegetation like ferns. Many insects live here.

a) Which layer contains the least vegetation?
   SHRUB LAYER (1)
   Why? THERE IS VERY LITTLE LIGHT (1)

b) Why do most animals live in the canopy layer?
   THERE IS FOOD AND PROTECTION (1)

c) How are animals adapted to living in this layer? ANIMALS LIKE MONKEYS HAVE SPECIALISED HANDS AND FEET FOR CLIMBING. ANIMALS CAN LEAP FROM BRANCH TO BRANCH LIKE FLYING FROG. (1)

d) What type of plants live in the understorey layer? THOSE WHICH CAN TOLERATE SHADE (1)

e) Write a sentence about what it might be like on the forest floor.
   OWN ANSWER (1)

3 Unscramble the names below of some animals which inhabit rain forests.

a) **FLYING FROG**   b) **TAPIR**
c) **TARANTULA**   d) **OCELOT**
e) **TOUCAN**   f) **IGUANA**
g) **HUMMINGBIRD**
h) **WOOLLY MONKEY** (8)

3.6.3 Populations within ecosystems

1 Look at the pictures below and answer the questions which follow.

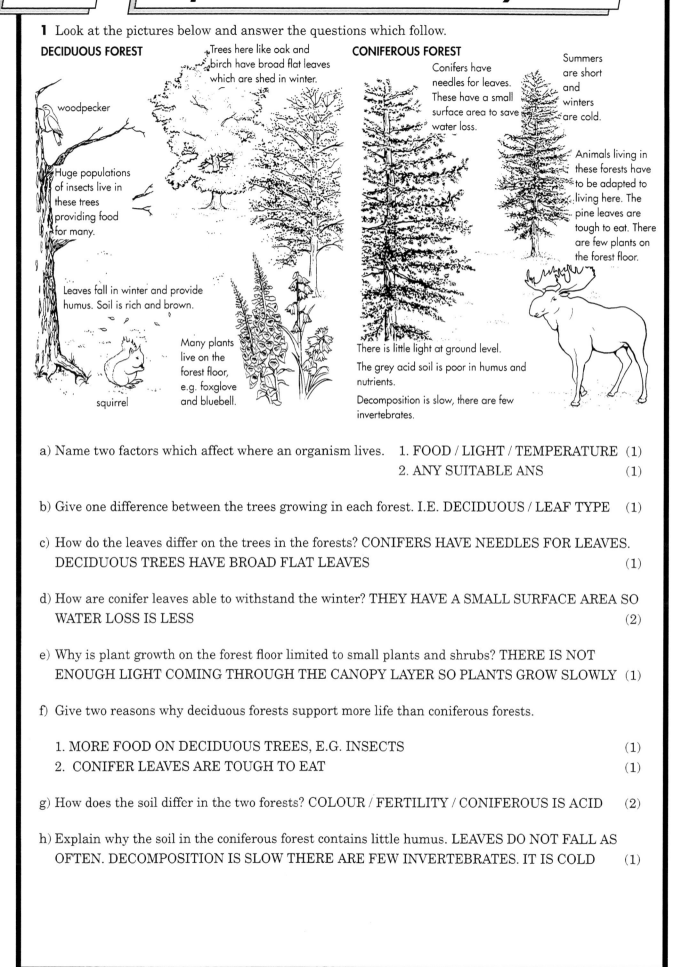

DECIDUOUS FOREST — Trees here like oak and birch have broad flat leaves which are shed in winter. Huge populations of insects live in these trees providing food for many. Leaves fall in winter and provide humus. Soil is rich and brown. Many plants live on the forest floor, e.g. foxglove and bluebell.

CONIFEROUS FOREST — Conifers have needles for leaves. These have a small surface area to save water loss. Summers are short and winters are cold. Animals living in these forests have to be adapted to living here. The pine leaves are tough to eat. There are few plants on the forest floor. There is little light at ground level. The grey acid soil is poor in humus and nutrients. Decomposition is slow, there are few invertebrates.

a) Name two factors which affect where an organism lives.   1. FOOD / LIGHT / TEMPERATURE (1)
                                                                                                                  2. ANY SUITABLE ANS (1)

b) Give one difference between the trees growing in each forest. I.E. DECIDUOUS / LEAF TYPE (1)

c) How do the leaves differ on the trees in the forests? CONIFERS HAVE NEEDLES FOR LEAVES. DECIDUOUS TREES HAVE BROAD FLAT LEAVES (1)

d) How are conifer leaves able to withstand the winter? THEY HAVE A SMALL SURFACE AREA SO WATER LOSS IS LESS (2)

e) Why is plant growth on the forest floor limited to small plants and shrubs? THERE IS NOT ENOUGH LIGHT COMING THROUGH THE CANOPY LAYER SO PLANTS GROW SLOWLY (1)

f) Give two reasons why deciduous forests support more life than coniferous forests.

   1. MORE FOOD ON DECIDUOUS TREES, E.G. INSECTS (1)
   2. CONIFER LEAVES ARE TOUGH TO EAT (1)

g) How does the soil differ in the two forests? COLOUR / FERTILITY / CONIFEROUS IS ACID (2)

h) Explain why the soil in the coniferous forest contains little humus. LEAVES DO NOT FALL AS OFTEN. DECOMPOSITION IS SLOW THERE ARE FEW INVERTEBRATES. IT IS COLD (1)

4.1.3 Recycle it

1 Complete the passage below about recycling aluminium. Use each word once.

**household   environment   ore   litter
stick   recycle   magnetic
aluminium   energy   resources**

Millions of drinks cans are sold in Britain each year. Half of these are made of ALUMINIUM. You can find out if a can is aluminium by using a magnet. Aluminium is a non-ferrous metal so it is not MAGNETIC and a magnet will not STICK to its side. It is cheaper to RECYCLE aluminium metal because it takes 95% less ENERGY than producing the metal from its ORE. Recycling aluminium saves RESOURCES and helps the ENVIRONMENT because it reduces the amount of LITTER buried in the ground. Drinks cans make up about 10% of HOUSEHOLD waste. (10)

2 Read the passage about plastic and answer the questions which follow.

> Many bottles are now made of plastic instead of glass. Plastic bottles are lighter in weight than glass or metal which makes them easier to carry. Plastic is made from oil and is non-biodegradable. It is not easily recycled because there are many different types. Some manufacturers are now labelling their plastic containers with the symbols opposite which identify the type of plastic used. They can then be sorted into the different types and recycled.

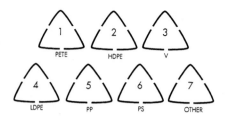

a) Name one advantage of using plastic instead of glass to make bottles

LIGHTER IN WEIGHT   (1)

b) From which material is plastic made?

OIL   (2)

c) Why does plastic cause pollution?

IT DOES NOT BREAK DOWN /
NON BIODEGRADABLE   (2)

d) What do the symbols below the passage mean?

THEY INDICATE THE TYPE OF PLASTIC
FROM WHICH IT IS MADE   (2)

e) Find four products with these symbols on them.

OWN ANSWER   (2)

Science Companion © A Porter, M Wood, T Wood and Stanley Thornes (Publishers) Ltd 1995

4.2.3 Passing on energy

1 Look at the food web and answer the questions.

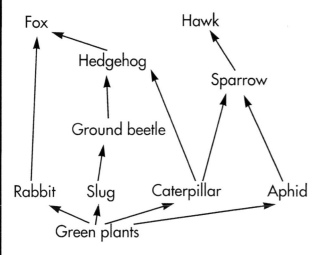

a) Draw three food chains from the food web.
ANY THREE. MUST START WITH A PRODUCER. E.G.
GREEN PLANTS → SLUG → GROUND BEETLE   (3)

b) Name an organism eaten by foxes   RABBIT

c) What eats caterpillars?   SPARROW / HEDGEHOG

d) Name a prey of sparrows   APHID

e) Name a predator of hedgehogs   FOX

f) Name a carnivore   FOX / HAWK

g) Name the producers   GREEN PLANTS   (6)

2 Fill in the gaps to complete the definitions:
A type of energy stored in green plants

cHEmiCAl eNERgY

Organisms which make food

pRODuCErS

Animals which eat only animal flesh

CaRNIvORES

Animals which eat only plants

hERbIVOrEs

Many food chains drawn together

fOOd wEB

How plants make food

PHotOsYNthESIs   (6)

3 A new predatory mite has been found feeding on cabbages. Read the information and draw it in the box.

New mite! Description
Round body
Four eyes
Two long antennae
Pointed tail
Ten jointed hairy legs
Zig-zag markings on its back

Give this new species a name.

ALL PARTS MUST BE DRAWN

Name: ANY. *Cabagus scrumptus*   (10)

4.3.3 The carbon and nitrogen cycles

1 Complete the passage below by filling in the missing words.

**atmosphere   combine   elements
gaseous   fixing   plants   proteins
reactive**

Nitrogen is a GASEOUS element. It makes up 78% of the Earth's ATMOSPHERE. Living things need nitrogen to make PROTEIN in their cells. Most PLANTS are unable to take nitrogen directly from the air because it is an unreactive gas. It will not COMBINE easily with other ELEMENTS. Nitrogen is changed into a more REACTIVE form by processes called nitrogen FIXING. (8)

2 The diagram shows the nitrogen cycle.

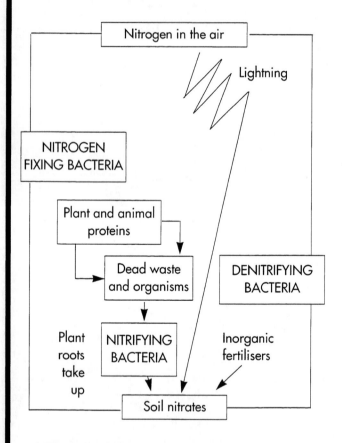

a) Write the following on the diagram:
**Nitrogen fixing bacteria
Denitrifying bacteria
Nitrifying bacteria** (3)

b) Finish the following sentences:
  i) Nitrogen fixing bacteria are found in root NODULES and turn nitrogen into NITRATES. (2)

  ii) Nitrifying bacteria act on AMMONIA from decaying organisms and animal WASTE and turn NITROGEN into nitrates. (3)

  iii) Denitrifying bacteria turn nitrates from the soil back into nitrogen GAS and release it into the air. (1)

3 Look at the diagram below.

a) Why are fertilisers added to soil?
SO THAT PLANTS CAN TAKE UP NUTRIENTS (2)

b) Which substance in fertilizers is converted into nitrates in soil?
AMMONIA (2)

c) Name an advantage of farmers using inorganic fertilisers.
IT WORKS FASTER / EASY TO APPLY (1)

d) What are organic fertilisers?
THOSE FROM ANIMAL AND PLANT WASTE (2)

e) Excess fertilisers leach into rivers. What does the word 'leaching' mean?
WASHING MINERALS OUT OF THE SOIL (2)

f) Explain how leached minerals affect the plants and animals in rivers.
MINERALS ENCOURAGE THE GROWTH OF ALGAE ON THE WATER'S SURFACE. THIS BLOCKS LIGHT. PLANTS CANNOT PHOTOSYNTHESISE AND DIE. OXYGEN IS NOT RELEASED INTO THE WATER. ANIMALS DIE. (4)

5.1.3 Types of materials

1 Use the pictures below to help you complete the name of the product, and the raw material from which it is made. Then place an **N** in the box if the raw material is natural or an **M** if it is manufactured.

a)
PRODUCT   RAW MATERIAL

PETROL   **CRUDE OIL**   (3)

b)
PRODUCT   RAW MATERIAL

PAPER   **WOOD**   (3)

c)
PRODUCT   RAW MATERIAL

TEE SHIRT   **COTTON**   (3)

2 Sort out the following mixed up words. They are all to do with materials.

a) RWA   RAW   (1)
b) CRIBAF   FABRIC   (1)
c) ROE   ORE   (1)
d) YORTACF   FACTORY   (1)
e) RREEOUSC   RESOURCE (1)
f) CLEERCY   RECYCLE   (1)
g) LEALB   LABEL   (1)

3 Aluminium is found in the ground as an ore.

a) The ore of aluminium is   BAUXITE   (1)

b) Aluminium is often used to make drink

   CANS   (1)

c) Recycling aluminium can save WASTE

   and the ORE   (2)

4 Use the clues below to complete the maze. A new word starts in each numbered square.

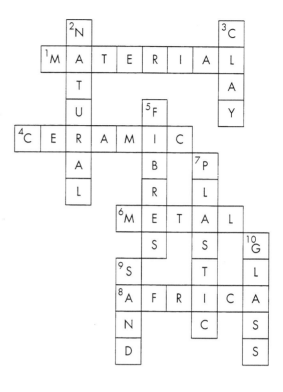

CLUES

1 *across*
This is the stuff which things are made up of.
2 *down*
Found in the world and unchanged by people.
3 *down*
The raw material for pottery.
4 *across*
This word comes from the Greek word for pottery.
5 *down*
These are hair-like strands.
6 *across*
Aluminium is an example of this.
7 *down*
A modern material used to make buckets.
8 *across*
Most gold is found in the south of this continent.
9 *down*
Used with limestone and sodium carbonate to make 10 down.
10 *down*
You see through this material out of the classroom.
(10)

A copy of The Periodic Table will be needed

5.2.3 Properties, uses and development

Below are 19 clues to help you find 19 words. Most of the words are linked with 'properties, uses and developments of materials'. The number of letters in each word is shown in brackets after the space for you to write your answer.

CONCRETE	(8)	Sand, cement, pebbles and water when mixed, dry into this material.
OPEN	(4)	Opposite of closed.
MOHS	(4)	The last name of the man who invented the scale of hardness.
PILKINGTON	(10)	These 'brothers' invented the float glass process.
RED	(3)	First colour of the rainbow.
EARTH	(5)	Our planet.
SCRATCH	(7)	Hard minerals will do this to soft minerals.
SUN	(3)	The star of our solar system.
INVENTION	(9)	You make this when you invent something.
VOWELS	(5)	There are five of these in our alphabet.
EXPERIMENT	(10)	Hopefully, you do lots of these in science.
SILK	(4)	This fibre is made by silkworms.
TENSILE	(7)	Strength which resists pulling.
RAYON	(5)	An early manufactured fibre for use in ladies' stockings.
ENERGY	(6)	This is saved by recycling things.
NYLON	(5)	Fibre invented by Wallace Carothers.
GLASS	(5)	Window panes are made of this.
TALC	(4)	The softest mineral.
HARDNESS	(8)	The last word in the third clue up above (19)

Use the first letter of each answer to name one property of materials.

COMPRESSIVE STRENGTH (1)

Science Companion © A Porter, M Wood, T Wood and Stanley Thornes (Publishers) Ltd 1995

5.3.3 Separating and purifying

Chromatography

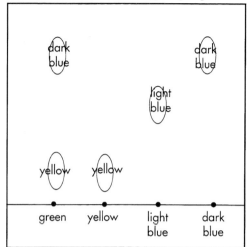

The diagram above shows how the ink from a green pen compares with three other pens (yellow, light blue and dark blue). All four pens were tested using the technique called chromatography.

1. Use coloured pens or crayons to show where the colours are on the diagram. (5)
2. Complete the following paragraph which describes how the chromatogram above was made. (8)

A square of paper was cut out. The paper must be able to ABSORB water. A line was drawn two centimetres from the bottom of the paper using a PENCIL. Four marks were made along this line. A dot of ink from the GREEN pen was put on the first mark. The yellow on the second mark and so on. The paper was stood in a tank with WATER used as the liquid in the tank. It was important that the liquid level was BELOW the line. This was to stop the colours from DISSOLVING in the liquid. The liquid travels up the paper and takes the colours with it. The colours travel at different SPEEDS which means they separate out. The green pen is made up of TWO different colours.

Separating a sand and salt mixture

This diagram shows the steps needed in separating a mixture of sand and salt. There are four steps. The mixture is added to water. Then it is stirred and then passed through a filter paper. Finally it is heated to evaporate the water.

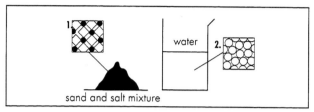

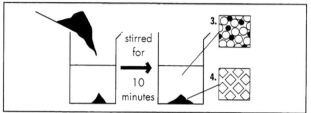

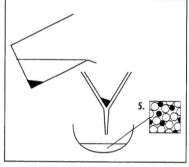
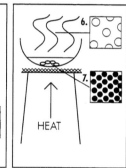

There are empty squares on the diagram numbered 1 to 7. You must fill in the squares using drawings to represent the particles at that point. Use the diagrams below which represent particles of water ◯ sand ◇ and salt ●.

You will need to use one of the diagrams more than once. (7)

5.4.3 Acids, alkalis and indicators

1 Complete the nine horizontal words using clues (a) to (i). Then answer part (j).

	1	2	3	4	5	6	7	8	9	10
a)			A	C	I	D				
b)			V	I	N	E	G	A	R	
c)	B	L	O	O	D					
d)				M	I	L	K			
e)					C	I	T	R	I	C
f)		A	L	K	A	L	I			
g)		N	E	U	T	R	A	L		
h)					O	V	E	N		
i)	W	A	T	E	R					

CLUES

a) Sulphuric is a type of this substance. (1)

b) This liquid is often added to fish and chips. (1)

c) A liquid which is pumped around our bodies. (1)

d) A white liquid often delivered to our doors. (1)

e) This acid is found in lemons. (1)

f) The opposite of an acid. (1)

g) These liquids all have a pH of seven. (1)

h) Cooking equipment which needs a strong alkali to clean it. (1)

i) When pure, this common liquid has a pH of 7. (1)

j) Now find the word in column 5. (1)

2 In the letters below are hidden the names of seven acids and hydroxides. The names are printed left to right or read vertically from the top downwards. Circle the names of the seven chemicals. (7)

T	F	S	O	D	I	U	M	Y	A
U	H	U	R	S	G	K	X	J	M
C	A	L	C	I	U	M	O	E	M
R	I	P	C	S	W	Z	A	G	O
E	T	H	A	N	O	I	C	D	N
R	H	U	P	F	C	A	V	E	I
E	H	R	F	J	I	U	M	M	U
A	C	I	T	R	I	C	A	B	M
G	Y	C	A	R	B	O	N	I	C

3 Use the words in the list below to complete the passage. You may use the words once, more than once or not at all.

alkalis   caustic   citric   hydrochloric   more   neutral   seven   sodium   rubber

All acids have a pH of less than SEVEN. All alkalis have a pH of MORE than seven. Acids can react with exactly equal amounts of ALKALIS to make a NEUTRAL solution of pH 7. Strong acids include HYDROCHLORIC acid. Weak acids include CITRIC acid. The strongest common alkali in the home is SODIUM hydroxide which is used as oven cleaner. The common name for this alkali in oven cleaner is CAUSTIC soda. The word caustic means burning. When cleaning an oven it is wise to use RUBBER gloves and eye protection. (9)

4 Complete the pH number maze using the following clues. All the answers go left to right or downwards. In the maze, the pH number is written as a word. So, if the answer is 6, in the maze this is written as SIX.

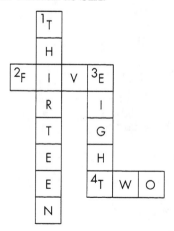

CLUES

1 *down* pH of soap flakes
2 *across* pH of soda water
3 *down* pH of toothpaste
4 *across* pH of lemon juice

(4)

5.5.3 The Periodic Table

A copy of The Periodic Table will be needed to answer these questions. It will need names, symbols and the atomic numbers of the elements.

1 Below is a word square. The letter M is in the square already at square (a1).

	1	2	3	4	5	6	7	8	9
a)	M	A	N	G	A	N	E	S	E
b)	N	I	C	K	E	L			
c)	M	A	G	N	E	S	I	U	M
d)			M	E	R	C	U	R	Y
e)		I	O	D	I	N	E		
f)		C	A	L	C	I	U	M	
g)	A	R	G	O	N				
h)				S	I	L	V	E	R
i)	C	A	E	S	I	U	M		

PART 1
Find the nine words to fill the rows (a) to (i). The clues are all atomic numbers. The answers are the names of the elements.

CLUES
a) 25, b) 28, c)12, d) 80, e) 53, f) 20, g) 18, h) 47, i) 55.   (9)

PART 2
From the word square, find the letters which fit in these spaces.

G	E	R	M	A	N	I	U	M
a4	i3	h9	c1	g1	b1	e5	f7	d3

PART 3
a) What is the atomic number of the element you have discovered in part 2?   32   (1)
b) What country is it named after?
   GERMANY   (1)

2 Below are the melting points of elements 10 to 18 of the Periodic Table.

Atomic number	Element	Melting point at atmospheric pressure (°C)
10	Neon	−249
11	Sodium	98
12	Magnesium	650
13	Aluminium	660
14	Silicon	1410
15	Phosphorus	44
16	Sulphur	113
17	Chlorine	−101
18	Argon	−190

a) Plot a graph, on the graph paper below, of melting point against atomic number.
(3 for points)
(1 for line)

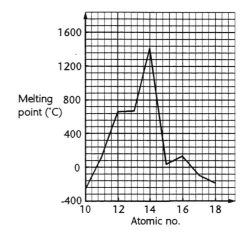

b) In which group of the Periodic Table is the element with the highest melting point?
FOUR   (1)

3 Use the following words, once, more than once, or not at all, to complete the passage below.

**atom   decreasing   element
elements   groups   increasing
neutrons   periods   protons
rows   similar**

In the Periodic Table of the ELEMENTS, the elements are arranged in order of INCREASING atomic number. The atomic number of an ELEMENT is the number of electrons or PROTONS in one atom of the element. Vertical columns are called GROUPS and contain elements which react in a SIMILAR way. Horizontal ROWS of elements are called PERIODS.   (8)

5.6.3 Reactivity of metals

Displacement reactions

1 When a metal is more reactive than another metal in a compound, there will be a reaction. The positions of the metals will swap over. We say one metal has displaced another.

e.g. IRON + LEAD OXIDE → LEAD + IRON OXIDE

a) Each of the four reactions shown below are displacement reactions which will work. Use the idea above to predict what the names of the metals, solids and solutions are.

IRON OXIDE + ALUMINIUM → METAL **A** + SOLID **B**

METAL **C** + SOLUTION **D** → BATLIUM + IRON SULPHATE

METAL **E** + SOLUTION **F** → BATLIUM + ALUMINIUM BROMIDE

BATLIUM + SILVER OXIDE → METAL **G** + SOLID **H**

Metal **A** is   IRON

Solid **B** is   ALUMINIUM OXIDE

Metal **C** is   IRON

Solution **D** is   BATLIUM SULPHATE

Metal **E** is   ALUMINIUM

Solution **F** is   BATLIUM BROMIDE

Metal **G** is   SILVER

Solid **H** is   BATLIUM OXIDE   (8)

b) Now place the four metals – silver, iron, batlium and aluminium in the correct order of reactivity by completing the table below. (4)

REACTIVITY	NAME OF METAL
Most reactive	ALUMINIUM
2nd most reactive	IRON
3rd most reactive	BATLIUM
Least reactive	SILVER

2 Use your knowledge of metals to judge whether the following statements are true or false. Tick the box which you think is right. (15)

Statements	True?	False?
1. Gold is classed as a reactive metal		√
2. Silver is a good conductor of electricity	√	
3. Metals are good conductors of heat	√	
4. Gold corrodes more easily than silver		√
5. Iron is commonly found naturally as the metal		√
6. Ores are rocks which are rich in metals	√	
7. Ores are turned into oxides before extracting the metal	√	
8. Carbon is used to turn iron oxide into iron in a furnace	√	
9. Iron is more reactive than carbon		√
10. Copper reacts vigorously with water		√

3 Complete the labels on the diagram of the blast furnace. The words to use are:
CARBON DIOXIDE, COKE, HOT AIR, IRON METAL, IRON ORE, LIMESTONE, SLAG (7)

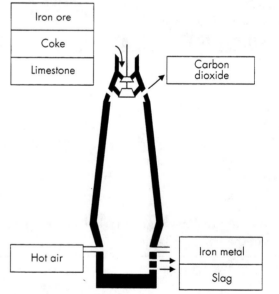

b) Use the seven labels to complete the table below in order to show which are raw materials and which are products of the blast furnace. Iron ore has been done for you. (6)

Raw Materials	Products
iron ore	CARBON DIOXIDE
COKE	SLAG
LIMESTONE	IRON METAL
HOT AIR	

6.1.3 Solids, liquids and gases

1 Complete this word triangle. The clues are for words going across. The number tells you how many letters the word has.

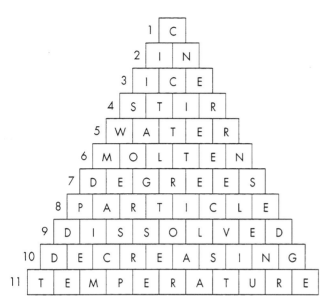

CLUES
1. The symbol for the Celsius scale.
2. To drink coffee, the coffee is put _____ a cup or mug.
3. The solid state of steam.
4. To mix a solid and liquid using a glass rod.
5. The liquid state of steam.
6. A word to describe a solid when it is a liquid.
7. The units of temperature.
8. Solids are regular arrangements of these.
9. A solute did this to make a solution.
10. Lowering
11. A scale used to measure this in °C. (11)

2 Use the words in the list below to complete the following passage. Each word may be used once, more than once, or not at all.

**dissolves   insoluble   solid
solids   soluble   solute   solution
solvent   suspension**

An insoluble solid is one which stays as a SOLID when it is added to a liquid. If the solid dissolves then we say it is SOLUBLE in the liquid. A liquid which dissolves SOLIDS is called a SOLVENT. A solid which dissolves may be called a solute. When a SOLID dissolves in a SOLVENT it forms a SOLUTION. (7)

3 Study the following information and then answer the questions which follow.

Gore-tex fabric is the world's most advanced weather-proof fabric. Its success lies in the special membrane (2) made from an expanded form of Teflon – PTFE. This membrane contains 14 million holes per square milli-metre. Each hole is 20 000 times smaller than a drop of water, but 700 times larger than a particle of water as a gas. This means that rainwater cannot get through so the wearer stays dry. But perspiration can evaporate and escape, so the material 'breathes' and is pleasant to wear. The outer fabric (1) and the liner (3) are selected depending on the use of the final garment, and are per-manently bonded either side of the Gore-tex membrane.

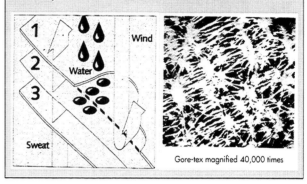

a) From what is Gore-tex fabric made?

   AN EXPANDED FORM OF TEFLON   (1)

b) How much smaller is a hole in Gore-text than a drop of water?

   20,000 TIMES   (1)

c) Why is Gore-tex described as breathable?

   IT LETS SWEAT OUT AND STOPS THE WEARER GETTING UNCOMFORTABLE   (1)

d) Why are an outer fabric and liner used?

   THE MEMBRANE IS DELICATE AND MORE DURABLE MATERIALS SURROUND IT   (2)

e) Why is a plastic garment not as comfortable to wear as Gore-tex?

   PLASTIC TRAPS THE SWEAT AND GETS STICKY   (2)

Science Companion © A Porter, M Wood, T Wood and Stanley Thornes (Publishers) Ltd 1995

6.2.3 Atoms and temperature scales

1 Study the following information on temperature scales and then complete the table which follows.

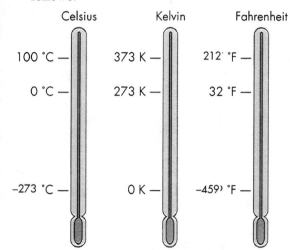

To convert temperatures from Fahrenheit to Celsius, take 32 from the Fahrenheit temperature and multiply by $\frac{5}{9}$.

To convert temperatures from Celsius to kelvin simply add 273 on to the Celsius figure.

Complete the table below giving your answers to the nearest whole number.

	Celsius	Kelvin	Fahrenheit
a)	−18	255	0
b)	16	289	61
c)	66	339	151
d)	239	512	462
e)	120	393	248

(10)

2 Fill in the following horizontal words using the clues that follow. Then find the two words in sentence (i) using the letters from the numbered squares.

	1	2	3	4	5	6	7	8	9	10	11
a)	C	E	N	T	I	G	R	A	D	E	
b)	F	A	H	R	E	N	H	E	I	T	
c)	D	E	G	R	E	E	S				
d)	C	E	L	S	I	U	S				
e)	T	E	M	P	E	R	A	T	U	R	E
f)	T	H	E	R	M	O	M	E	T	E	R
g)	A	L	C	O	H	O	L				
h)	M	E	R	C	U	R	Y				

CLUES
a) One hundredth of a degree. Also commonly used as the name for the Celsius temperature scale.
b) 70°F. Complete this F scale word.
c) Units of temperature.
d) The scale with the freezing point of water at 0.
e) We measure this {answer(e)} using the apparatus for answer (f).
f) See clue (e).
g) A red or blue dye is added to this liquid in these pieces of apparatus to measure temperature.
h) The only liquid metal at room temperature. (8)

Use letters from the squares above to complete this sentence:
i) Gabriel Fahrenheit was a

G	E	R	M	A	N
c3	d2	h6	f5	e7	b6

scientist who worked in

H	O	L	L	A	N	D
g5	f6	g7	d3	b2	a3	c1

for a time as a glassblower. (2)

3 Below is a bar chart showing the melting points of some common metals. The temperature units are kelvin. Study it carefully before answering the questions which follow.

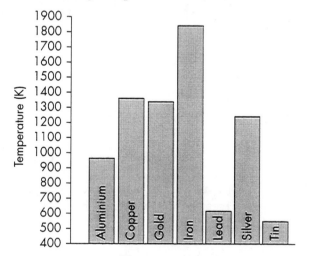

a) Place the seven metals in order of decreasing melting point with the highest first. IRON COPPER GOLD SILVER ALUMINIUM LEAD TIN (2)

b) Name the metals which melt between 927°C and 1127°C.

SILVER GOLD COPPER (3)

Science Companion © A Porter, M Wood, T Wood and Stanley Thornes (Publishers) Ltd 1995

6.3.3 Particles and changes of state

1 Wordsearch. Hidden among the 120 letters below are words all to do with 'the effect of heating a substance'. Circle the words as you find them, and tick them off on the list. Words go from left to right, top to bottom or diagonally top left to bottom right.

The words are:
boiling
condensing
cooling
freezing
gas
heat
liquid
melting
particles
solid
sublimation (11)

Which word appears twice? COOLING (1)

2 The triangle below links together the three states of matter. Five words are represented by letters. Write the five words in the spaces provided.

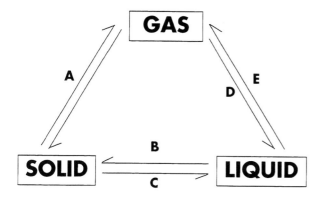

I think the missing words are:

A SUBLIMATION (1)

B FREEZING (1)

C MELTING (1)

D CONDENSING (1)

E BOILING (1)

3 The sets of apparatus shown at the top of the next column have been set up to study the diffusion of gases.
Apparatus **A** is for the diffusion of hydrogen and air.

Apparatus **B** is for the diffusion of carbon dioxide and air.

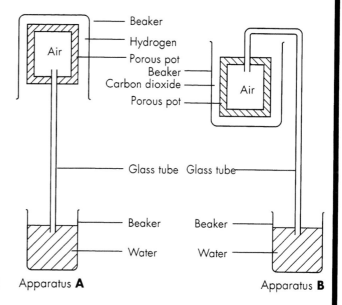

Apparatus **A**   Apparatus **B**

a) Why is it necessary to use two different designs of apparatus?

HYDROGEN IS LESS DENSE THAN AIR
CARBON DIOXIDE IS MORE DENSE
THAN AIR (2)

b) State, and explain, what will be seen in apparatus **A**.

BUBBLES OF AIR WILL COME OUT OF
THE TUBE IN THE WATER. HYDROGEN
DIFFUSES INTO THE POROUS POT
FASTER THAN AIR CAN DIFFUSE OUT,
PRESSURE RISES, SO AIR IS BLOWN
OUT (3)

c) State, and explain, what will be seen in apparatus **B**.

WATER WILL BE SUCKED UP THE
TUBE IN THE WATER. AIR DIFFUSES
OUT OF THE POT, FASTER THAN
CARBON DIOXIDE CAN DIFFUSE IN,
PRESSURE FALLS, SO WATER IS
SUCKED UP. (3)

6.4.3 — Atoms, ions and molecules

1 In this puzzle there are seven clues. Write the answers to the clues into the grid on the right. When you have done this, transfer the letters to the grid below. If you get the right letters in the correct spaces, a sentence will appear.

CLUES

a) Practicals in a science lesson.

E	X	P	E	R	I	M	E	N	T	S
1a	1b	1c	1d	1e	1f	1g	1h	1i	1j	1k

b) The amount of liquid in cm^3 is its _____.

V	O	L	U	M	E
2a	2b	2c	2d	2e	2f

c) Short for one thousand grams.

K	I	L	O
3a	3b	3c	3d

d) Sand can be separated from salt water by this method.

F	I	L	T	E	R	I	N	G
4a	4b	4c	4d	4e	4f	4g	4h	4i

e) To warm something up we add this.

H	E	A	T
5a	5b	5c	5d

f) Element number six in the Periodic Table which makes diamond and graphite.

C	A	R	B	O	N
6a	6b	6c	6d	6e	6f

g) The state of matter of which ice is an example.

S	O	L	I	D
7a	7b	7c	7d	7e

(7)

A	N		A	T	O	M		I	S		T	H	E
5c	6f		6b	4d	2b	1g		7d	1k		1j	5a	1a

S	M	A	L	L	E	S	T		P	A	R	T		O	F
7a	2e	5c	7c	4c	1d	7a	5d		1c	6b	4f	1j		3d	4a

A	N		E	L	E	M	E	N	T		T	H	A	T
6b	1i		1h	2c	2f	1g	4e	4h	5d		4d	5a	6b	1j

C	A	N		E	X	I	S	T		A	N	D
6a	5c	1i		5b	1b	4g	1k	5d		6b	4h	7e

S	T	I	L	L		B	E	H	A	V	E		L	I	K	E
1k	5d	4b	2c	7c		6b	1a	5a	5c	2a	1d		4c	3b	3a	1h

T	H	E		E	L	E	M	E	N	T
4d	5a	2f		4e	2c	5b	1g	2f	6f	1j

(18)
(1 mark per word)

Science Companion © A Porter, M Wood, T Wood and Stanley Thornes (Publishers) Ltd 1995

6.5.3 Radioactivity

1 Find the missing words in rows (a) to (i) below using the clues. Then find the name of the scientist in column 4.

	1	2	3	4	5	6	7	8
a)				B	E	T	A	
b)		P	I	E	R	R	E	
c)	C	O	N	C	R	E	T	E
d)		R	E	Q	U	I	R	E
e)			C	U	R	I	E	
f)			D	E	C	A	Y	
g)		M	A	R	I	E		
h)		E	N	E	R	G	Y	
i)			A	L	P	H	A	

CLUES
a) Radiation which needs a thin sheet of aluminium to stop it.
b) The first name of the husband of a famous husband and wife team involved in the discovery of radioactivity. Her first name is your clue for (g). Their surname is your clue for (e).
c) This material is needed to stop the most deadly radiation.
d) A long word meaning need. It starts and ends with RE.
e) See (b) above.
f) Half-life is a measure of this feature of a radioactive substance.
g) See (b) above.
h) A nuclear reactor produces this.
i) Radiation which can be stopped by a thin sheet of paper. (9)

The name of the famous scientist in column 4 is BECQUEREL. (1)

2 Below is a list of words.
**alpha   bacteria   beta   boiling
damage   gamma   germs
hospitals patients   sterilisation**

Read the passage which follows very carefully, and use words from the list to complete the gaps. Each word may be used once, more than once, or not at all.

Gamma radiation can be used to kill cancer cells. It can also be used to kill BACTERIA or germs. When used to kill bacteria or GERMS the method is called STERILISATION. This method is used for dressings and instruments in HOSPITALS which must be free from bacteria and germs before they are used on PATIENTS. Boiling in water is a common form of sterilisation. (8)

3 The following statements are about smoke detectors. Fill in the gaps.

a) The type of radiation used in smoke detectors is ALPHA RADIATION. (2)

b) The radiation carries the ELECTRICITY across an air gap. (1)

c) In a fire SMOKE PARTICLES stop the radiation. (2)

d) An ALARM sounds when the circuit is broken in a fire. (1)

e) Smoke detectors are fairly cheap and can save a lot of LIVES. (1)

4 The following questions are about alpha, beta and gamma radiation.

a) Which of the three types of radiation are deflected by a magnetic field?

ALPHA and BETA (2)

b) Why is gamma radiation the most dangerous to people?

GAMMA RAYS CAN PENETRATE THE HUMAN BODY AND CAN CAUSE GENETIC MUTATION WHICH COULD LEAD TO CANCER. (3)

6.6.3 Solvents, glues and hazards

1 Alongside is a label for Domestos multi surface kitchen and bathroom cleaner. Study it carefully before answering the questions below.

a) Which parts of the body are liable to be irritated by this product?
EYES and SKIN (2)

b) What should you do if you get any in your eyes?
RINSE IMMEDIATELY WITH PLENTY OF WATER AND SEEK MEDICAL ADVICE (2)

c) Which gas may be produced if this product is mixed with an acid?
CHLORINE (1)

d) What does the product contain which should not be spilt on clothes, carpets and upholstery?
BLEACH (1)

e) From the list of ingredients give the chemical name of a bleach.
SODIUM HYPOCHLORITE (1)

f) In use, you are given capfuls of the product to add to water. How many cm^3 of product does each capful hold?

(note 1 ml = 1 cm^3). 30cm^3 (1)

g) How much of the product is needed to disinfect a dishcloth? 90cm^3 (2)

h) From the label suggest the approximate volume of water which the manufacturers expect to find in 1 full bucket 5 LITRES (2)

2 Alongside is the label from a well-known make of liquid paper. Study the label carefully and then answer the questions below.

a) What is the hazard label warning about?
THE PRODUCT IS HARMFUL (1)

b) Which part of the body should you 'avoid contact with'?
THE EYES (1)

c) What chemical is in the product which causes concern?
1,1,1 -TRICHLOROETHANE (1)

Science Companion © A Porter, M Wood, T Wood and Stanley Thornes (Publishers) Ltd 1995

7.1.3 Physical and chemical changes

Physical changes

Write the answers to the questions in the correct boxes in the grid. Rearrange the letters in the shaded boxes to complete the sentence below. (6)

Physical changes are TEMPORARY

1. The process which happens to most solids when they are heated. (7 letters)
2. What steam will become when it condenses. (5)
3. Roads are this when the surface water freezes in winter. (3)
4. What liquids will do when they are heated. (4)
5. A word for a gas which evaporates from a liquid. (6)

1.	M	E	L	T	I	N	G
2.	W	A	T	E	R		
3.	I	C	Y				
4.	B	O	I	L			
5.	V	A	P	O	U	R	

Chemical changes

Write the answers to the questions in the correct boxes in the grid. Rearrange the letters in the shaded boxes to complete the sentence below. (6)

Chemical changes are PERMANENT

6. This sulphate changes from blue to white when you heat it. (6 letters)
7. This is a raw material used to make glass. (4)
8. What you often need to do to a chemical to change it. (4)
9. A mixture of these materials is called an alloy. (6)
10. How to chemically change cake mixture. (6)

6.	C	O	P	P	E	R
7.	S	A	N	D		
8.	H	E	A	T		
9.	M	E	T	A	L	S
10.	B	A	K	I	N	G

Supergreen

10 PHOTODEGRADABLE SWING BIN LINERS

RIM: 46 ins/1.17 m
DEPTH: 30 ins/76 cm

There are two types of plastic called thermo-softening and thermosetting plastics. One way in which they are different is in their reaction to heat. Complete the sentences below.

Thermosoftening plastics when heated will turn from a solid into a LIQUID. This transformation is called MELTING. It is an example of a PHYSICAL change. These plastics are often collected from waste because they can be RECYCLED (used again). This is done with the plastic used to make BOTTLES. Thermosetting plastics are different. They undergo a CHEMICAL change when they are heated and become new MATERIALS. The plastic used in the bin liners shown above also undergoes a chemical change. In this case it changes when exposed to LIGHT (8)

In the grid below try to find six physical changes and four ways to make a chemical change.

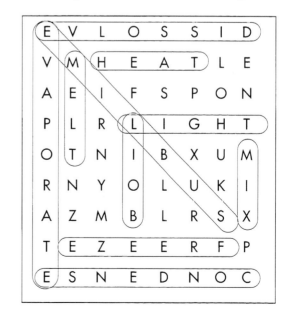

Science Companion © A Porter, M Wood, T Wood and Stanley Thornes (Publishers) Ltd 1995

7.2.3 Changes for the better from land and air

1 Uses of Oxygen

The table below shows the uses to which the gas oxygen is put. Use the figures in the table to complete the pie chart below. Colour in and label the sections. Each section equals 5%. (5)

Use of oxygen	%
Steel making	55
Making chemicals	25
Other uses (eg. medical)	10
Cutting metal	5
Rockets and explosives	5

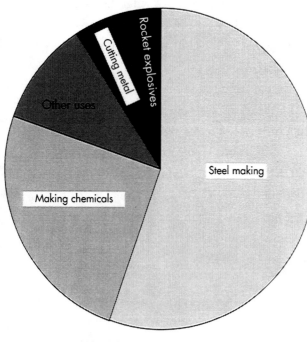

a) In steel making oxygen is used to get rid of unwanted carbon in the iron. What gas will be made when carbon reacts with oxygen?

CARBON DIOXIDE

b) What use would oxygen have in a hospital?

TO HELP BREATHING PROBLEMS

c) Acetylene (or ethyne) is used with oxygen in torches which can cut through metal. Why is the oxygen used instead of air?

IT MAKES THE FLAME HOTTER

d) In the space shuttle oxygen is used with hydrogen to help launch it into space. What is the product of hydrogen burning in oxygen?
WATER (STEAM) (4)

2 Uses of Limestone

Limestone is a rock which was formed millions of years ago from sea shells and coral. The main chemical in limestone is calcium carbonate. This can be changed into calcium oxide by heating the limestone in a kiln. To get the high temperatures needed, the limestone is mixed with a fuel called coke (mainly carbon). The mixture is poured into the top of the kiln and air is blasted in from the bottom. The coke burns in the air, raising the temperature. The limestone is turned into quicklime (calcium oxide) and carbon dioxide. The blast of air removes the carbon dioxide as soon as it forms. Quicklime will turn into slaked lime when water is added to it. Slaked lime is called calcium hydroxide.

Use the words below to complete the flowchart of the production of slaked lime from limestone. You may need to use some words more than once. (6)

calcium carbonate, calcium hydroxide, calcium oxide, carbon, carbon dioxide, quicklime

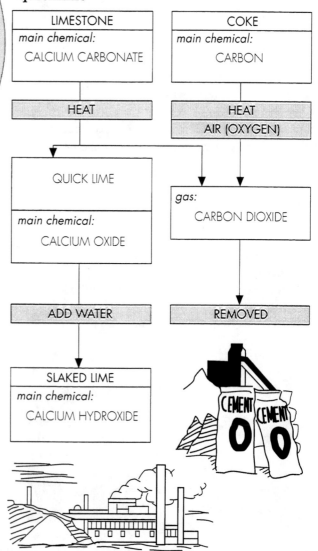

Science Companion © A Porter, M Wood, T Wood and Stanley Thornes (Publishers) Ltd 1995

7.3.3 Changes for the better from water and living things

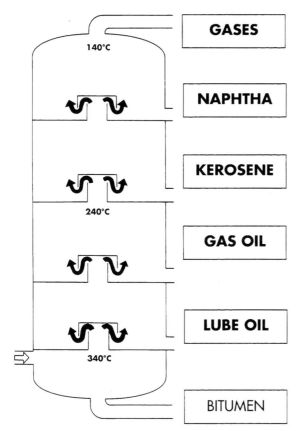

1 Look at the simplified diagram of the fractionating tower used at an oil refinery. Use the information in the table below to complete the labels on the diagram. (5)

FRACTION	BOILING RANGE
Bitumen	solid
Gases	below 25°C
Gas oil	200–400°C
Kerosene	174–275°C
Lube oil	over 300°C
Naphtha	20–200°C

2 Match the name of the fractions to their main uses in the diagram below. One has been done for you. (5)

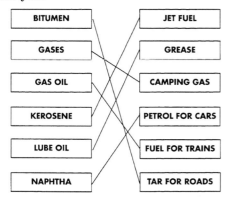

3 Write the answer to each question in the space provided. The shaded column should reveal the name of a heart drug which is found in foxgloves.

a.		S	O	D	A				
b.	P	E	N	I	C	I	L	L	I	N
c.	D	R	U	G	S				
d.	A	S	P	I	R	I	N		
e.	C	O	T	T	O	N			
f.		S	O	A	P				
g.	C	E	L	L	O	P	H	A	N	E
h.	M	E	D	I	C	I	N	E	
i.		L	I	N	S	E	E	D	

a) What **S** comes after caustic and is used to turn fats into soap?
b) What **P** is a mould which can stop the growth of bacteria?
c) What **D** are extracted from many plants and can have an effect on the workings of the body?
d) What **A** is a headache pill and can be made from willow trees?
e) What **C** comes from seed heads and can be woven into fabrics?
f) What **S** can be made from natural plant materials and is used to wash our skin?
g) What **C** is a wrapping material made from extracted cellulose in plants?
h) What **M** is often made from plants and is used to treat or prevent diseases?
i) What **L** is an oil made from the flax plant? (10)

4 Look up the following words in a dictionary to try to find out what living thing they come from.

SILK – FIBRE PRODUCED FROM THE LARVA OF A SILK MOTH

LATEX – THE MILKY JUICE OF RUBBER PLANTS

COCHINEAL – SCARLET DYE EXTRACTED FROM COCCUS INSECT

LITMUS – INDICATOR EXTRACTED FROM CERTAIN LICHENS

CAMPHOR – OIL FROM CINNAMON TREE (5)

7.4.3 Chemicals from oil

1 Match the name of the monomer in the first column to the name of its polymer in the second column. Then match the name of the polymer to its use in the third column. One example has been done for you. (6)

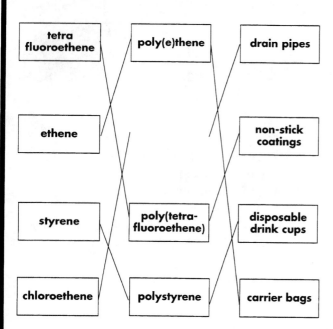

2 The table below shows the names of three groups of hydrocarbons (alkanes, alkenes and alkynes). The naming of the hydrocarbons follows a logical pattern based on the number of carbon atoms and the type of bond. Use this pattern to complete the table. (7)

3 Enter the answers to the following seven questions into the spiral grid below. The letter which ends one word, begins the next word and should be written in the shaded squares. (7)

1. Another word for crude oil (9 letters)
2. A small molecule which forms a repeating unit in a polymer (7 letters)
3. To make a plastic less flexible, makes it more _____ (5 letters)
4. A type of engine which uses fuel which is thicker than petrol (6 letters)
5. This means to move together smoothly, as when there is oil between two surfaces (11 letters)
6. Some frying pans have Teflon surfaces to make them _____ (two words 3-5)
7. The fraction of crude oil from which jet fuel is made (8 letters)

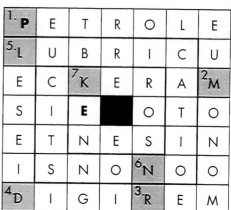

	ALKANES all single bonds	**ALKENES** one double bond	**ALKYNES** one triple bond
ONE CARBON	methane		
TWO CARBONS	ETHANE	ethene	ethyne
THREE CARBONS	propane	PROPENE	PROPYNE
FOUR CARBONS	BUTANE	BUTENE	butyne
FIVE CARBONS	PENTANE	pentene	PENTYNE

Science Companion © A Porter, M Wood, T Wood and Stanley Thornes (Publishers) Ltd 1995

7.5.3 Changing our food

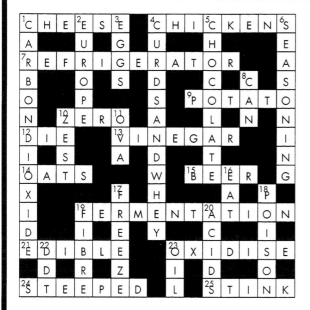

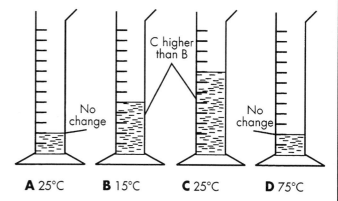

A 25°C   **B** 15°C   **C** 25°C   **D** 75°C

The diagram above shows the apparatus at the start of the following experiment:
A dough mixture was made from flour and water. A small amount of this dough was placed in cylinder **A**. Some yeast was mixed into the rest of the dough and this was added to cylinders **B**, **C** and **D**. The same amount of dough was added to each cylinder. Cylinders **A** and **C** were placed in a warm room. Cylinder **B** was placed in a cold room and cylinder **D** was heated to 75°C. All the cylinders were left for the one hour exactly.

1 What was the temperature of the warm room?

25°C (1)

2 What was the temperature of the cold room?

15°C (1)

3 Give two things done in the experiment that made it a fair test.

a) THE VOLUME OF DOUGH WAS THE SAME IN EACH CYLINDER (1)

b) EACH CYLINDER WAS LEFT FOR THE SAME LENGTH OF TIME (1)

4 Which cylinder is the control experiment?

D (1)

5 On the diagram above mark clearly the positions of the dough after one hour. If there has been no change in the amount of dough then write 'no change' in the cylinder. (4)

6 What has made the dough in some of the cylinders rise up?

CARBON DIOXIDE (1)

Across
1 A solid food made from milk
4 These come from 3 down
7 A cold place to preserve food
9 This vegetable is the raw material for crisps
10 The temperature below which most food freezes
12 What could happen to you with food poisoning
13 This is made when beer or wine goes off
14 The ingredients for porridge
15 This is made by fermenting malted barley
19 The process of changing sugar into alcohol
21 This means 'can be eaten'
23 Exposure to the air can cause food to do this
24 Soaked in water like dried peas
25 The bad smell of rotten food

Down
1 A gas made when yeast mixes with sugar (2 words 6,7)
2 The place where many of our food laws are drawn up
3 These can sometimes carry salmonella
4 What milk separates into (what Miss Muffett ate) (3 words 5,3,4)
5 A sweet made from cocoa beans
6 Like salt, pepper or spices added to food for flavour
8 Putting food in one of these will keep it for a long time
10 A word used for the rind of a lemon or orange
11 The scientific word for 3 down
16 What we do with food
17 A cold way to preserve food
18 A toxic substance made by some bacteria
19 Plant tissue, healthy in a diet
20 These liquids (like lemon juice and vinegar) can preserve food
22 An insecticide which can end up poisoning food
23 Thick liquid extracted from olives (30)

7.6.3 Oxygen – the gas of life

John Mayow (1640–79) was a lawyer by profession but enjoyed doing scientific experiments. He did experiments with air long before anyone knew what air really was. In his first experiment he stood a candle in a bowl of water and covered it with a bell jar. The candle continued to burn for a little while and then went out, but long before all the air in the jar was used up. Mayow could tell how much air in the jar was used up because the water level rose inside the jar. He noticed that about a fifth of the air had been used up.

Next he placed a mouse inside another jar. Rather cruelly he watched as the mouse slowly died inside the bell jar. The mouse like the candle flame, died long before all the air had been used up.

Mayow decided that the candle was only using the part of the air that was useful for burning. The mouse used only the part of the air that was useful for living. He began to wonder if they were both using the same part of the air.

1 a) The diagram below shows the first experiment which Mayow set up. Complete the second diagram to show what happened when the candle went out. (2)

 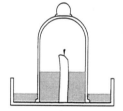

b) The paragraph opposite describes what would happen if the mouse and the candle were placed in the jar together. Complete the missing words.

Both the mouse and the candle use the gas OXYGEN. When this has been used the mouse will DIE and the candle will GO OUT. The water will rise ONE-FIFTH of the way up the jar. The water will rise FASTER than it did with just the mouse in. (5)

2 Enter the answers to questions 1 to 10 in the grid below and use the numbered letters to complete the sentence at the foot of the page. (18)

	a	b	c	d	e	f	g	h	i	j	k
1	N	I	T	R	O	G	E	N			
2	W	A	S	T	E						
3	E	X	P	E	L	L	E	D			
4	C	H	L	O	R	O	P	H	Y	L	L
5	L	U	N	G	S						
6	H	E	A	R	T						
7	I	N	T	E	S	T	I	N	E	S	
8	G	I	L	L	S						
9	F	O	U	R							
10	S	T	O	M	A	C	H				
11	B	L	O	O	D						

1 The major gas in the air
2 Unwanted (like carbon dioxide in our blood)
3 Got rid of (like carbon dioxide from our lungs)
4 The green pigment in plants
5 The organs we use to take in air
6 The organ we use to pump blood
7 Where food is digested
8 What fish use to take in oxygen
9 The percentage of carbon dioxide in exhaled air
10 Where food goes once swallowed
11 This carries gases round the body

6d	1g	8e	4g	7a	4e	2b	1c	7g	9b	5c					
R	E	S	P	I	R	A	T	I	O	N	C	H	A	N	G	E	S

9a	10c	1e	3h			4d	3b	4i	8a	7i	1a
F	O	O	D	A	N	D	O	X	Y	G	E	N

				10f	6c	9d	11a	11c	7b		11e	8b	4f	3b	1b	3h	2e
I	N	T	O	C	A	R	B	O	N		D	I	O	X	I	D	E

2a	10e	6e	3a	1d			3d	7h	3g	9d	5d	4i
W	A	T	E	R	A	N	D	E	N	E	R	G	Y

Science Companion © A Porter, M Wood, T Wood and Stanley Thornes (Publishers) Ltd 1995

7.7.3 Oxidation

1. Say what happens to the following foods when they are oxidised by oxygen from the air.

   a) wine GOES SOUR / TURNS TO VINEGAR

   b) a peeled apple TURNS BROWN (2)

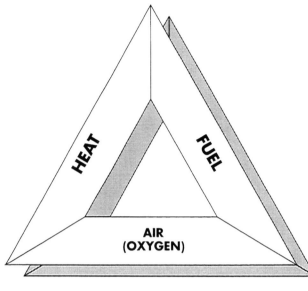

2. Complete the fire triangle above by writing in the three sides, the three things which are needed to make a fire burn. (3)

3. Match the metals and non metals to the properties in the middle with lines. Two have been done for you. (4)

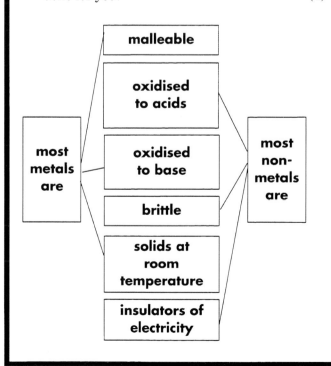

4. The picture at the top of the page shows the kind of chemicals which can be found in the exhaust fumes of a car. They are not in any order.

   a) The fuel in a car is made from hydrocarbons. What word tells you that some of these are not oxidised?

      UNBURNT (1)

   b) Underline the two chemicals below which are found in hydrocarbons.

      nitrogen   **hydrogen**   **carbon**
      sulphur   lead   (2)

   c) Name the three chemicals in the picture which are made when hydrocarbons burn.

      WATER

      CARBON MONOXIDE

      CARBON DIOXIDE (3)

   d) Which of the oxides comes from a gas in the air (besides oxygen)?

      OXIDES OF NITROGEN (1)

   e) Which of the oxides comes from impurities found naturally in the fuel?

      SULPHUR DIOXIDE (1)

   f) What environmental damage can these last two oxides cause?

      ACIDIFIES RAIN WATER (1)

   g) What test would show that carbon dioxide was present in the exhaust?

      ADD LIMEWATER (1)

   h) What would be the result?

      LIMEWATER GOES CLOUDY (1)

7.8.3 Chemical energy

1 A student investigated the temperature change when dissolving chemicals in water. Complete the last column in the table of results below. (4)

Temperature of water	Chemical added	Temperature of solution	Exothermic or endothermic?
18°C	A	10°C	ENDOTHERMIC
17°C	B	19°C	EXOTHERMIC
18°C	C	−2°C	ENDOTHERMIC
19°C	D	36°C	EXOTHERMIC

2 Complete the following paragraph about chemical energy changes. (7)

Some reactions involve an INCREASE in the temperature. This can be measured with a THERMOMETER. In these reactions HEAT is given out to the surroundings. These type of reactions are called EXOTHERMIC reactions. The opposite to these reactions are ENDOTHERMIC reactions. Here the surroundings have to supply ENERGY to the reaction. So that this can happen, the TEMPERATURE of the reaction falls.

3 Match the reactions in the left hand column to the energy changes in the right hand column. Use one of the changes twice. Choose the most important change for each reaction. (5)

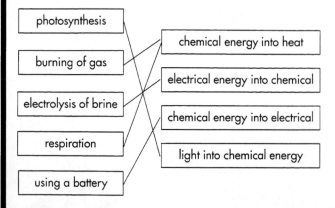

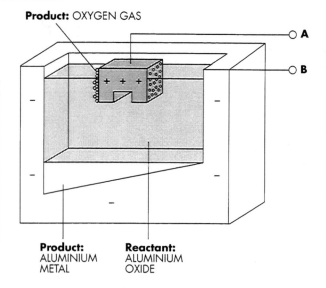

4 The diagram above shows how aluminium is made by passing electricity through liquefied aluminium oxide. This is the word equation for the reaction.

aluminium oxide → aluminium + oxygen

Add the words in the box below to the labels on the diagram. (3)

```
ALUMINIUM METAL
OXYGEN GAS
ALUMINIUM OXIDE
```

Delete the incorrect word in each of the following sentences.

a) The positive electrode is **A** (1)

b) Energy is **supplied** (1)

c) The reaction in the word equation is **endothermic** (1)

d) The aluminium oxide has **less** energy than the products (1)

e) ΔH for the reaction in the word equation is **positive** (1)

f) When aluminium and oxygen react the temperature will **increase** (1)

7.9.3 — Rates of reaction

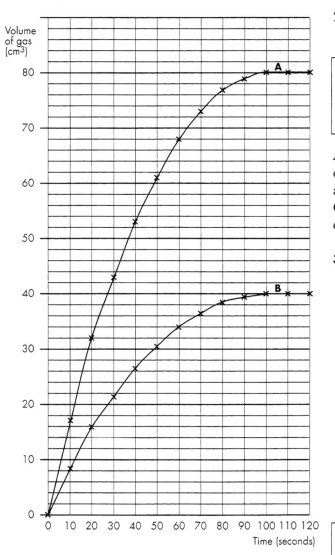

1 A student measured the volume of gas produced every 10 seconds when marble chips (calcium carbonate) reacted with dilute hydrochloric acid.
The marble chips were about 1 mm in diameter and the temperature was 20°C. There was more acid than marble chips.

a) Plot the results shown at the bottom of the page on to the above graph. Label it **A**. (4)

b) Sketch another curve on the graph which shows another experiment with the same dilute acid and the same sized marble chips, but half the amount of chips were used. Label it **B**. (2)

2 Use some of the words in the box below to complete the paragraph about catalysts. (6)

> chemically, decrease, increase, large, physically, small

A catalyst is usually used to INCREASE the rate of a chemical reaction. It is used in SMALL amounts. Catalysts are unchanged CHEMICALLY during the reaction but may be changed PHYSICALLY.

3 Add the missing words from these five sentences to the grid below. (5)

R – Diluting solutions will _____ the rate of a reaction.
A – Changing the temperature _____ the rate of a reaction.
T – The rate can be measured by _____ a reaction.
E – Increasing the temperature gives particles more _____.
S – Having a smaller surface area will make a reaction _____.

R	E	D	U	C	E
A	L	T	E	R	S
T	I	M	I	N	G
E	N	E	R	G	Y
S	L	O	W	E	R

Time (in seconds)	0	10	20	30	40	50	60	70	80	90	100	110	120
Volume of gas (in cm³)	0	17	32	43	53	61	68	73	77	79	80	80	80

8.1.3 The weather and weather forecasting

1 Study carefully these weather figures for Edinburgh and Jersey in March 1990. Then answer the questions which follow.

Edinburgh

Date	4	5	6	7	8	9	10
Rain (mm)	0.25	1.50	1.25	0.05	0.75	0.40	0.75
Maximum temperature (°C)	5	8	10	11	12	9	6
Sunshine (hours)	2.1	4.5	1.8	0.2	0.5	0.0	2.3

Jersey

Date	4	5	6	7	8	9	10
Rain (mm)	1.50	0.00	0.00	0.00	0.00	0.00	0.00
Maximum temperature (°C)	9	10	11	12	11	14	13
Sunshine (hours)	6.6	4.9	9.4	1.3	0.3	9.0	5.0

a) What was the total number of hours of sunshine for the week in Edinburgh? 11.4 hrs (2)

b) What was the total number of hours of sunshine for the week in Jersey? 36.5 hrs (2)

c) Use the graph paper below to draw a bar chart of the sunshine in Edinburgh (one colour) and Jersey (second colour). (14)

d) On which date did Edinburgh receive more sunshine than Jersey? 8 MARCH (1)

e) What was the average maximum temperature in Edinburgh during the week? 8.7 °C (2)

f) What was the average maximum temperature in Jersey during the week? 11.4 °C (2)

g) What was the total rainfall for the week in Edinburgh? (1) 5.40 mm (2)

h) What was the total rainfall for the week in Jersey? 1.50 mm (1)

2 Find the seven horizontal words using the clues (**a**) to (**g**) below and then find the name of the type of cloud in column 5.

	1	2	3	4	5	6	7	8	9	10	11	12	13
a)	P	R	E	S	S	U	R	E					
b)		A	N	T	I	C	Y	C	L	O	N	E	
c)		D	E	P	R	E	S	S	I	O	N		
d)	P	R	E	V	A	I	L	I	N	G			
e)	F	R	O	N	T	S							
f)	C	I	R	R	U	S							
g)			I	S	O	B	A	R					

CLUES

a) Air _____ is the name given to the weight of air above us.

b) This is an area of high pressure.

c) This is an area of low pressure.

d) These winds blow from the south-west in the UK.

e) There are three types of weather _____.

f) The first high cloud which shows that a warm front is approaching.

g) Line on a map joining places with the same air pressure. (7)

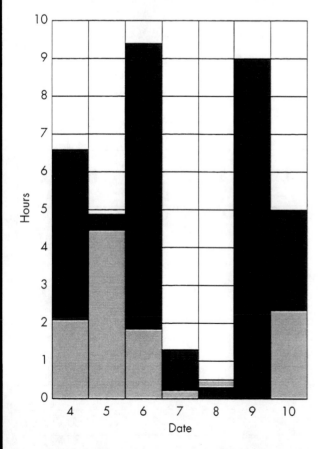

The cloud in column 5 is STRATUS (2)

Science Companion © A Porter, M Wood, T Wood and Stanley Thornes (Publishers) Ltd 1995

8.2.3 Weather science

1 Read the following information and then answer the questions which follow.

> **Reflection of radiated heat**
> When heat energy is absorbed by an object, its molecules vibrate more and it gets hotter. Black surfaces absorb radiation the best. White or silvery surfaces are very poor at absorbing radiation and reflect most radiated heat energy which reaches them. The polar icecaps reflect most of the heat radiation they receive because they are white. The best absorbers of radiation are also the best emitters.

a) Why does a black teeshirt feel hotter in sunshine than a white one?

BLACK ABSORBS HEAT RADIATION BETTER THAN WHITE. WHITE REFLECTS HEAT AND STAYS COOLER (2)

b) Which part of the passage helps to explain why substances expand when they get hotter?

AS HEAT ENERGY IS ABSORBED, THE VIBRATION OF THE PARTICLES INCREASES, THEY TAKE UP MORE SPACE AND SO IT EXPANDS (2)

c) A radiator not only heats a room by radiating heat. It also uses conduction and convection. Explain how it heats by these last two methods.

IT TRANSFERS HEAT ENERGY BY CONDUCTION TO THE AIR. THEN THE HOT AIR RISES BRINGING IN COOLER AIR TO BE HEATED, AND THIS IS CONVECTION (2)

2

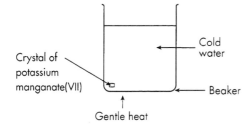

a) Explain what you would see happening in the beaker. Draw some arrows on the diagram to help your answer. (1)

THE CRYSTAL WOULD DISSOLVE MAKING THE WATER PINK. AS IT (1) DISSOLVES THE HEAT WOULD CARRY THE COLOUR UPWARDS (2) SHOWING CONVECTION CURRENTS. AS IT COOLED, THE PINK WATER WOULD FALL TO COMPLETE THE CIRCLE. (2)

b) In a second beaker, the crystal has been placed on the bottom in the middle. Draw arrows on the diagram below to show any convection current which would be seen. (2)

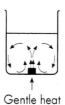
Gentle heat

3 Why do birds sometimes fluff up their feathers in winter?

THIS TRAPS AIR BETWEEN THE FEATHERS. AIR IS A POOR CONDUCTOR OF HEAT AND SO THE BIRD KEEPS ITS BODY HEAT IN. (2)

8.3.3 Climate, farming and catastrophies

1 Study the Beaufort Scale for wind which is shown below.

Beaufort scale number	Wind	Effects
0	Calm	Smoke rises vertically
1	Light air	Smoke drifts in wind
2	Light breeze	Wind felt on face, leaves rustle, weather vanes move
3	Gentle breeze	Light flag blows, leaves and small twigs move
4	Moderate breeze	Small branches move, dust and loose paper blow
5	Fresh breeze	Small trees sway, waves seen on ponds, etc
6	Strong breeze	Large branches move, telephone wires whistle
7	Moderate gale	Hard to walk into the wind, trees start to sway
8	Fresh gale	Very hard to walk into wind, twigs break off trees
9	Strong gale	Structural damage probable, slates and chimney pots lost
10	Whole gale	Trees uprooted, serious damage to buildings
11	Storm	Very rare inland, causes widespread damage
12	Hurricane	Major disaster

Now use the two squares which follow to draw simple pictures to illustrate the effects of the winds at two different values on the Scale. (8)

2 The west of the UK is warmer in the winter than the east.

a) Which current keeps the west warm?

THE NORTH ATLANTIC DRIFT (1)

b) Where does this current come from?

THE CARIBBEAN (1)

c) Why doesn't it affect the east?

THE CURRENT FLOWS ALONGSIDE THE WEST COAST ONLY, NOT THE EAST (1)

3 When 15 mm of rain falls in 3 hours, weather experts describe the conditions as a flash flood. Why does rain like this cause a flood?

THE WATER COMES DOWN TOO QUICKLY TO BE DRAINED AWAY. IT FLOWS QUICKLY ACROSS THE LAND CAUSING FLOODS (2)

4 Use the clues below to find the 13 horizontal words in the grid below. Then find the flood prevention device in column 5.

	1	2	3	4	5	6	7	8	9
a)		W	E	A	T	H	E	R	
b)			S	H	A	D	O	W	
c)			G	A	L	E			
d)	C	L	I	M	A	T	E		
e)		B	R	E	E	Z	E		
f)				S	U	N			
g)			B	O	R	E			
h)	A	T	L	A	N	T	I	C	
i)		D	R	I	F	T			
j)	N	O	R	T	H				
k)	H	U	R	R	I	C	A	N	E
l)	T	I	D	E					
m)	T	O	R	N	A	D	O		

(13)

CLUES

a) The day by day changes in the air.
b) Eastern England is in a rain —————.
c) Wind of 7–10 on the Beaufort Scale.
d) 30 years of (a) in a pattern.
e) Wind of 2–6 on the Beaufort Scale.
f) Provides energy for the Earth.
g) A small tidal wave or a dull person.
h) (i) and (j) in the order (j) (h) (i) name a current which influences the UK.
k) Catastrophic wind developed from a cyclone.
l) Caused mainly by the pull of the moon on the sea.
m) A strong wind sometimes called a twister.

The flood protection device in column 5 is
THAMES BARRIER (1)

8.4.3 Air masses and Charles' law

1 Study the information abut the main air masses that affect the British Isles, and then answer the questions which follow:

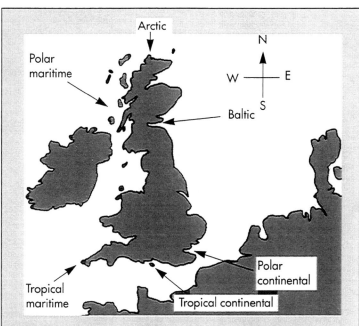

All our weather except thunderstorms is 'imported', because we are a small land area. Air which has come over the large land area of the continent has lost most of its moisture and so is usually dry when it reaches us. Maritime air has travelled over the sea for large distances and so is usually wet and brings rain. Air masses from the north (the polar region) are very cold, whilst in those from the south (towards the equator where it is hot) the air is warm.

The air masses are named after the places from which they come. The most common air mass to affect us is the polar maritime, which starts in the north-west and then swings round to the west or south-west.

a) What is an air mass?
A LARGE BODY OF AIR (1)

b) Here are four weather conditions:
- **A** hot, sunny, dry, dusty
- **B** cool, sunny periods, showers
- **C** cold, dry, bright, sunny
- **D** warm, humid, cloudy, light rain

From the four air masses given below, pick a letter, for the weather conditions you would expect, from the list above.
- i) polar continental — C
- ii) tropical continental — A
- iii) tropical maritime — D
- iv) polar maritime — B (4)

2 Below is a graph of some results from an experiment to demonstrate Charles' law.

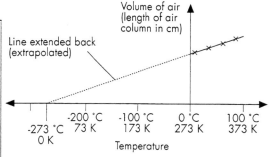

a) In this experiment the graph is extrapolated. What does this mean?

EXTENDING THE GRAPH BEYOND THE RESULTS THAT WERE OBTAINED IN THE EXPERIMENT (2)

b) The graph is a straight line. What does this show?

THE VOLUME CHANGES AT A CONSTANT RATE AS THE TEMPERATURE CHANGES (2)

3 In experiments about the gas laws there are four variables to consider. They are mass, pressure, temperature and volume. In experiments about Charles' law, which of these four variables must be kept constant?

MASS PRESSURE (2)

4 Use Charles' law to calculate the missing figures in the following table.

	Initial volume (cm³)	Initial temperature (K)	Final volume (cm³)	Final temperature (K)
a)	**100**	500	60	300
b)	140	**100**	88.2	63
c)	75	84	**28.6**	32
d)	600	475	320	**253.3**
e)	**187**	275	85	125
f)	12	**112.8**	60	564
g)	520	220	**567.3**	240
h)	147	1489	8	**81**
i)	**30**	90	50	150

(9)

8.5.3 Weathering and landscaping

1. Study the following information about the River Nile, and then answer the questions.

   > The River Nile in Egypt used to carry a lot of mud, called silt, with it. In 1970 the Aswan High Dam was completed and the quantity of mud in the river fell.
   > Before entering Egypt, the Nile flows through the Ethiopian Highlands. At Egypt's southern boundary it is held behind the Aswan High Dam, and forms Lake Nasser which is almost 500 km long.
   > Below are the silt readings (in parts per million) taken 100 km downstream of the Aswan High Dam site in 1889 and 1989.

	Jan	Feb	Mar	Ap	May	Ju	Jul	Aug	Sept	Oct	Nov	Dec
1889	60	50	40	45	45	90	700	2700	2400	900	130	75
1989	45	46	45	50	50	48	47	45	45	46	47	46

   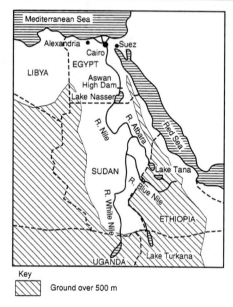

   a) Why did Egypt get a lot of flooding along the banks of the River Nile before 1970?

   HEAVY RAINS OVER THE ETHIOPIAN HIGHLANDS BETWEEN JUNE AND OCTOBER CAUSED A LOT OF WATER TO FLOW. THE RIVER BROKE ITS BANKS AND FLOODED THE COUNTRYSIDE (2)

   b) What was the advantage of this flooding?

   THE SILT IN THE WATER ACTED AS FERTILISER AND IT WAS FREE (2)

   c) Why do Egyptian farmers have to buy fertiliser today?

   NO FLOODING NOW MEANS NO SILT, SO NO NATURAL FERTILISER (1)

   d) The River Nile has a very large delta, built up over millions of years. What caused the delta to form?

   SILT WAS DEPOSITED DURING THE TIMES OF FLOODING (2)

2. Use clues (a) to (j) to complete the horizontal words in the grid below and then find the word in column 6.

	1	2	3	4	5	6	7	8	9	10	11	12	13
a)						W	I	N	D				
b)	M	E	D	I	T	E	R	R	A	N	E	A	N
c)						A	C	I	D				
d)	L	I	M	E	S	T	O	N	E				
e)						H	A	R	D				
f)		S	C	R	E	E							
g)					E	R	O	S	I	O	N		
h)				A	R	I	Z	O	N	A			
i)						N	I	L	E				
j)		N	I	A	G	A	R	A					

   CLUES
   a) Moving air which is said to blow.
   b) The sea into which the River Nile flows.
   c) This type of rain eats away our buildings.
   d) The most popular natural building stone in Britain.
   e) Rock like this stays. It is not eroded.
   f) A pile of rock at the bottom of a rock face.
   g) The process of being eroded.
   h) The Grand Canyon is in this state in America.
   i) The most famous river in Egypt.
   j) World famous waterfalls on the USA/Canada border. (10)

   The word in column 6 is WEATHERING (1)

Science Companion © A Porter, M Wood, T Wood and Stanley Thornes (Publishers) Ltd 1995

8.6.3 The water cycle

1. Study the following information and then answer the questions which follow.

WORLD NEWS 27.10.94

A new mineral water has been launched on to the world market today. It comes from a tiny spring near Stirling in Scotland. 'Glenwater' is formed from centuries of Scottish rains which have percolated through the rocks to form an underground source of natural mineral water with a taste and purity which is certain to become known worldwide. It is purified as it is formed by the rain filtering through the rocks. This removes solid impurities and dissolves minerals from the rocks which give 'Glenwater' its lovely taste. Years of heavy rain have produced a high water table leading to a forceful spring which should last for ever even with sales expected at 1 million litres per year. The slightly acid water (pH 7.3) contains many healthy minerals.

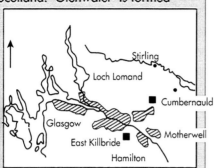

MINERAL ANALYSIS

Mineral	mg per litre
calcium	23
magnesium	6
potassium	0.2
sodium	6.7
hydrogen carbonate	81
sulphate	13
nitrate	1
chloride	9
fluoride	0.06
silicates	5.8
dry residue at 190°C	96
pH	7.3

a) What is the serious science error which the newspaper reporter has made?
pH 7.3 IS SLIGHTLY ALKALINE NOT SLIGHTLY ACID (1)

b) What is the main metallic element present? CALCIUM (1)

c) Which mineral is present in the smallest amount?
FLUORIDE (1)

d) What benefit is this mineral to humans?
TOUGHENING TEETH ENAMEL (1)

e) How are unwanted solids removed from the water? PERCOLATION THROUGH THE ROCKS IS NATURAL FILTRATION REMOVING SOLIDS (2)

2. Study the following information and then answer the questions which follow.

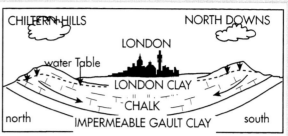

Clay is impermeable which means it does not let water through, and London sits on clay. Beneath the clay is a chalk syncline running from the Chiltern Hills to the North Downs. Rain on the Hills and Downs drains through the permeable chalk until it reaches the water table. The water table is higher under the Hills and Downs than below London. This means that the water below London is under pressure and up comes the water. This type of situation is known as an artesian well. The chalk syncline is an example of an artesian basin.

a) What is a water table?

THE LEVEL AT WHICH UNDERGROUND WATER SETTLES (1)

b) Why is the water under London at a high pressure?

THE WATER TABLE IS HIGHER UNDER THE HILLS NORTH AND SOUTH AND SO THIS APPLIES A PRESSURE TO THE WATER UNDER LONDON (1)

c) From the diagram, what do you think a syncline is?

A DIP IN A LAYER OF ROCK (1)

8.7.3 The rock cycle

1 Study the following three paragraphs on rock formation and then answer the questions which follow.

> **A** The hot sun beating down on warm, shallow seas means that lots of water evaporates. Calcium carbonate which has dissolved in the water is then forced to precipitate as crystals and settle on the sea bed. This process is continuous, and may last millions of years. So, along with deposits of shells from dead animals, limestone rocks many thousands of metres thick are created.

> **B** During volcanic eruptions, the magma pours out as lava and cools, releasing gases, turning into rocks such as basalt. This rock is often highly resistant to erosion and can result in the volcanic rock remaining long after the surrounding cone of ashes has eroded.

> **C** Marble is made from limestone which has been subjected to heat and pressure. The original sedimentary rock has been turned into a dense crystalline structure which can take lovely carvings and a high polish. Mudstone, another sedimentary rock, under similar conditions becomes slate which can be split into thin sheets.

a) What type of rock is being made in

   (A) SEDIMENTARY

   (B) IGNEOUS

   (C) METAMORHPIC   (3)

b) In (A), what causes the water to evaporate?

   THE HEAT FROM THE SUN   (2)

c) What mineral makes limestone?

   CALCIUM CARBONATE   (1)

d) How long may the process in (A) last?

   MILLIONS OF YEARS   (1)

e) In (B), is basalt formed inside the Earth or on the surface?

   THE SURFACE   (1)

f) Is basalt a hard or soft rock?   HARD   (1)

g) Explain why you think your answer to part (f) is correct.   IN (B) IT SAYS IT IS HIGHLY RESISTANT TO EROSION   (1)

h) Is an ash cone hard or soft?   SOFT   (1)

i) Explain why you think your answer to part (h) is correct.

   THE NECK IS LEFT LONG AFTER THE CONE OF ASHES HAS ERODED   (1)

j) How is marble made from limestone?

   GIVE IT HEAT AND PRESSURE   (2)

k) What sort of new structure is in marble which was not in limestone?

   DENSE CRYSTALINE STRUCTURE   (2)

l) From what named rock is slate made?

   MUDSTONE   (1)

2 Below are the names of eight rocks or minerals.
china clay clay coal copper granite iron ore limestone tungsten
Fit them into the following table in the place where each fits best. Each rock of mineral may only be used once.   (8)

Rocks and minerals	Uses
CHINA CLAY	pottery, paper
COPPER	plumbing, electrical cables
IRON ORE	iron and steel
LIMESTONE	cement, chemicals
TUNGSTEN	drill tips
CLAY	bricks, cement
GRANITE	building stone
COAL	fuels, chemicals

(8)

Science Companion © A Porter, M Wood, T Wood and Stanley Thornes (Publishers) Ltd 1995

8.8.3 — Earthquakes and volcanoes

1 Study the following information about the Pacific Ocean, and then answer the questions which follow.

> 10° north of the Equator, in the Pacific Ocean, south of Mexico, two plates are moving apart. Sea water seeps down into the gap, gets hot and rises in springs which are rich in minerals. These springs or vents support strange life forms 8,500 feet below sea level. Among the life forms are giant tube worms which love the highly acidic water which contains little salt, but a lot of hydrogen sulphide. Minerals including iron, zinc and copper are brought up from deep inside the Earth, by water which reaches a temperature of 380°C under the enormous pressure at this great depth. Tube worms die within weeks when earthquakes close the mineral-rich vents.

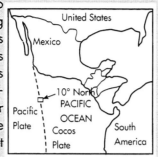

a) Name the two plates which are moving apart in this area of the Pacific Ocean.
   PACIFIC PLATE and COCOS PLATE   (2)

b) How can water reach 380°C and still be a liquid?   THE HIGH PRESSURE STOPS IT BOILING AT 100°C AND LETS IT BOIL HIGHER   (1)

c) Name three metals which are present in the water from the vents.
   IRON, ZINC and COPPER   (3)

d) What proof is suggested in the notes above, that tube worms live on the very hot mineral-rich water?
   THEY DIE WITHIN WEEKS OF VENTS BEING CLOSED   (1)

e) Suggest which chemical causes the water from the vents to be highly acidic.
   HYDROGEN SULPHIDE   (1)

2 Study the following information about Mount Pinatubo, and then answer the questions which follow.

> On 15 June 1991, 5,770 feet high Mount Pinatubo in the Philippines erupted. Sulphur dioxide was thrown 25 miles up into the atmosphere where along with moisture it created a thin cloud which circled the Earth in 21 days. Satelllite data showed that 2% of sunlight would be deflected from the Earth, lowering temperatures a little. Ash accompanied the sulphur dioxide. The ash contained bits of old volcano, pumice, sulphur-rich anhydrite and crystals of hornblende. The ash covered many square miles of the Philippines and poses a constant threat turning into an unstable slurry every time it rains.

a) Name the volcano which erupted in the Philippines on 15 June 1991.
   MOUNT PINATUBO   (1)

b) How long did it take for the high thin cloud from the volcanic eruption to circle the Earth?
   21 DAYS   (1)

c) How did scientists measure the effect of this cloud on our climate?
   THEY USED DATA FROM A SATELLITE   (1)

d) Name four things found in the ash.
   BITS OF OLD VOLCANO, PUMICE, SULPHUR-RICH ANHYDRITE, CRYSTALS OF HORNBLENDE   (4)

e) What is a slurry?
   AN UNSTABLE MIX OF WATER AND FINE POWDERED SOLID   (2)

f) Name the main gas produced in the eruption.
   HYDROGEN SULPHIDE   (1)

g) What type of rain is this gas known to produce?
   ACID RAIN   (1)

9.1.3 How things work

1 The symbols for a cell, a bulb and a switch are:

Use these symbols to complete the following circuits.

a) One bulb which will light when one switch is pressed. (3)

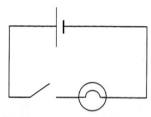

b) Two cells connected to three bulbs. (5)

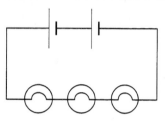

2 The bulb in the circuit below is said to glow with NORMAL brightness.

Say how bright the bulbs in the following circuits would be. Write one of the words, BRIGHT, NORMAL, DULL, OFF in the box below each circuit. (6)

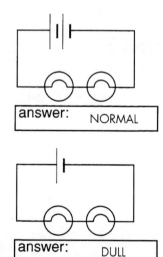

answer: NORMAL

answer: DULL

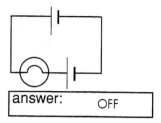

answer: OFF

answer: BRIGHT

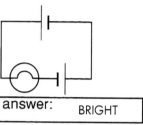

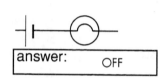

answer: OFF

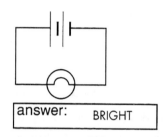

answer: BRIGHT

3 Put the following household objects into the correct side of the table below.

steel fork   jumper   key   magnet
newspaper   rubber   scissors
spanner   tap   teacup
glass window (11)

INSULATORS of electricity	CONDUCTORS of electricity
newspaper	steel fork
jumper	key
teacup	magnet
glass window	spanner
rubber	tap
	scissors

Science Companion © A Porter, M Wood, T Wood and Stanley Thornes (Publishers) Ltd 1995

9.2.3 Go with the flow

1 Put the missing words from the passage below into the crossword grid. (16)

1.C	U	2.R	R	3.E	N	4.T		5.F	I	R	6.E
H		E		D		E		I			L
7.A	M	P		8.I	N	S	U	L	A	T	E
R		E		S		T		A			M
G		9.L	O	O	S	E		10.M	O	V	E
E				N		11.D	I	E			N
12.S	E	13.A						N			T
		14.C	O	N	D	U	C	T	O	R	

If a material will let electricity flow through it, it is called a (*14 across*). Water from the (*12 across*) will conduct electricity. The flow of electricity is caused by electric (*1 down*) when they (*10 across*). In wires this produces an electric (*1 across*) which is measured in a unit called an (*7 across*). D.C. stands for direct current and (*13 down*) stands for alternating current. Electric charges which are the same will (*2 down*) each other.

The electric light bulb was invented by Thomas (*3 down*). He (*4 down*) different materials for the coil inside the bulb, called the (*5 down*). If the coil becomes (*9 across*), the bulb will not work. Electricity can be fatal. Heat from an electric (*5 across*) comes off the glowing (*6 down*). You have to (*8 across*) the casing to stop electricity going through you, otherwise you could (*11 across*) when you touch it.

2 The passage below contains words in bold. In each case, cross out the word which is incorrect.

Anything that **slows down** the current in a circuit is called a resistor. The current could be measured with **an ammeter**. The bigger the resistance in a circuit, the **smaller** the current would be. A resistor that can be altered is called a **variable** resistor. Resistance is measured in units called **ohms**. (5)

3 Match the symbol to the name of the electrical apparatus. (5)

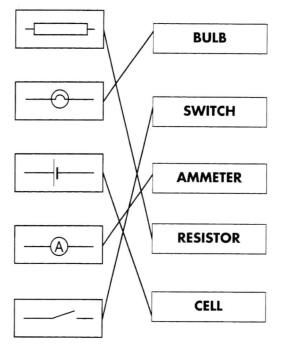

4 The bulb in the circuit below is said to glow with normal brightness.

Say how bright the bulbs in the circuits below would be. Write one of the words BRIGHT, NORMAL, DULL, OFF in the box below each circuit. (4)

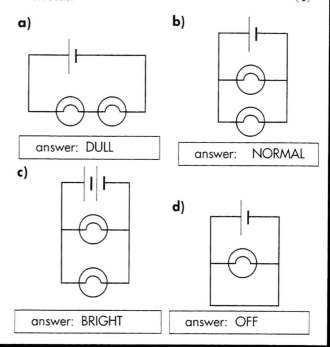

a) answer: DULL
b) answer: NORMAL
c) answer: BRIGHT
d) answer: OFF

9.3.3 Electricity in your home

1 The steps below are part of the story of the electricity to your home. The steps are in the wrong order. Put the steps into the correct order in the boxes **A** to **E**.

- ❏ Electricity passes through the National Grid at 440,000V
- ❏ Electricity enters homes through the meter
- ❏ Fuel is burned in a power station
- ❏ Electricity is transformed to 240V supply
- ❏ Water boils into steam to drive turbines (5)

A	FUEL IS BURNED IN A POWER STATION
B	WATER BOILS INTO STEAM TO DRIVE TURBINES
C	ELECTRICITY PASSES THROUGH THE NATIONAL GRID AT 440,000V
D	ELECTRICITY IS TRANSFORMED INTO 240V SUPPLY
E	ELECTRICITY ENTERS THE HOME THROUGH THE METER

2 Complete the sentences by writing the answers in the rows **a** to **d**. If the answers are correct, there should be a word reading down the first column.

a	P	L	A	S	T	I	C
b	L	I	V	E			
c	U	N	I	T	S		
d	G	R	E	E	N		

a) Electric wires are covered in PLASTIC to insulate them.
b) The brown wire is the LIVE wire.
c) Kilowatt hours are the UNITS of electricity on a meter.
d) The earth wire is yellow and GREEN. (4)

3 Below is a diagram of a standard electrical plug.

a) Colour in the three wires coming out of the flex. You will need a blue, brown, green and yellow pencil or felt-tip. (3)
b) Label the diagram with the words Earth wire, Live wire, Neutral wire, Fuse. (4)

4 The answers to the clues a to d will fit into the grid below. If they are correct, they will spell out an important part of a plug in the first column.

a	F	L	E	X		
b	U	N	P	L	U	G
c	S	O	C	K	E	T
d	E	A	R	T	H	

a) The electricity cable between the plug and the television.
b) To disconnect the power supply from a lamp.
c) The hole in the wall which supplies electricity.
d) The safety wire in the plug. (4)

9.4.3 The life of Michael Faraday

1 Here is an illustration of an electric vehicle.

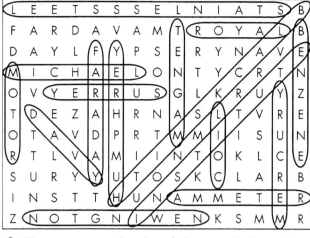

a) What is the vehicle used for?

DELIVERING MILK (1)

b) Where is the electrical energy stored?

BATTERIES (1)

c) Why do some people consider electric vehicles more environmentally friendly?

NO EXHAUST FUMES (1)

d) What advantages does an electric vehicle have over one powered by diesel in this case?

QUIETER (1)

e) In early mornings in winter, electric motors have a particular advantage over diesel engines. What is it?

DIESEL ENGINES DO NOT START EASILY IN THE COLD (1)

f) What two disadvantages do electric motors have over diesel engines?

LIMITED RANGE
LIMITED SPEED (2)

g) Name two other vehicles which use electric motors.

FORK LIFTS, TRAMS
TRAINS, ETC. (2)

2 The answers to the clues below can be found in the grid above. Circle each answer as you find it.

a) The main character in the title (2 words).
MICHAEL FARADAY

b) His tutor in London (2 words).
HUMPHREY DAVY

c) His town of birth.
NEWINGTON

d) His county of birth.
SURREY

e) His father's job.
BLACKSMITH

f) The place in London where he worked (2 words)
ROYAL INSTITUTE

g) A material he discovered – found in petrol.
BENZENE

h) Another material he discovered – used to make cutlery.
STAINLESS STEEL

i) An electric machine for driving vehicles.
MOTOR

j) An iron bar which attracts other iron objects.
MAGNET

k) Wire wound round in a circle.
COIL

l) This is used to measure electric current.
AMMETER

m) A poisonous metal used in early electricity experiments.
MERCURY (16)

9.5.3 Electromagnets

1 Put the answers to the clues in the grid. You should reveal two words in the shaded column. (18)

a) The opposite pole to the north.
b) The unit of frequency, symbol Hz.
c) An iron bar which attracts objects made from iron.
d) These move through metals when they conduct electricity.
e) The sort of current which a dry cell supplies.
f) Wire twisted round a cylinder.
g) Part of a radio which uses an electromagnet.
h) Something which does not conduct electricity.
i) A magnetic _____ forms between the north and south poles of a magnet.
j) This can become charged when you pull it through your hair.
k) A material which conducts electricity.
l) A metal used as the core of an electromagnet.
m) The kind of charge which is made when electrons are removed from a material.
n) These supply electricity for radios, torches, etc.
o) This turns electrical appliances on and off.
p) The opposite charge to m.
q) Coal and gas are _____ sources for generating electricity.

The words in the shaded column are STATIC ELECTRICITY

2 a) Put the following labels on the diagram below.
Car battery, Coil of wire, Electromagnet, Ignition switch, Soft iron core. (5)

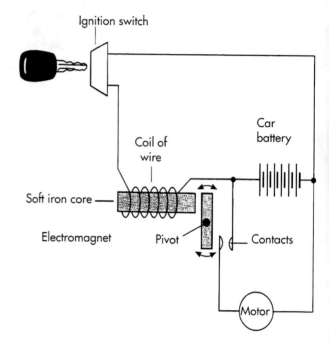

b) Fill in the missing words.

To start, the car motor needs energy supplied by

the BATTERY. The motor cannot start until the

CONTACTS in the circuit come together. The

motor circuit can be completed when the PIVOT

pushes the contacts. An electromagnet is used to

move the PIVOT. The electromagnet is made

from a COIL of wire wrapped around a core made

of IRON. When the IGNITION switch is turned,

it operates the electromagnet. (7)

9.6.3 Numbers and words

1 Each clue has a missing word. Put this word in the correct space in the grid below. (17)

	1.P		2.I				3.S	4.P	Y
5.D	I	A	G	R	A	M	6.M		
I		S		O		7.A	T	O	M
8.G	U	T	9.E	10.B	E	R	G		C
I			Y		A		N		E
11.T	Y	12.P	E	W	R	I	T	E	R	S
		E			C		T		S
13.P	I	N			O		I		O
T		14.C	R	E	D	I	T	C	A	R	D
15.O	N	E			E				S

Across

3 A _____ might use a code made up of letters and numbers.
5 A Chinese pictogram uses a simple _____ to represent a word.
7 Scientists have been able to move an individual _____ to make the smallest writing ever.
8 Johannes _____ invented the first movable type in the 1450s.
11 _____ are machines which print letters by pressing them against a ribbon soaked in ink.
13 To use a cash card you need to know your _____ number.
14 A _____ holds financial information on a magnetic strip set in its plastic.
15 In Roman numerals I represents the number _____.

Down

1 Pictures have been used in the _____ to represent words.
2 Magnetic tape is made from the metal we call _____.
4 The chips inside electronic devices are called _____.
5 Another word for a finger is a _____.
6 To store information on a plastic card you need to have _____ tape.
9 We can see words and numbers using our _____.
10 Most supermarkets use a laser to scan the _____ on an item of shopping.
12 Since decimalisation 100 _____ make one pound.
13 Three letters which tell you to look on the next page.

2 Each clue is a number. Change the number into Roman numerals and put it into the grid. (7)

1 down – 103
2 across – 14
2 down – 11
3 down – 8
4 across – 3
5 across – 54
6 across – 4

	1.C		2.X	I	3.V
4.I	I	I			I
I				5.L	I	V
6.I	V			I	

3 The table below shows the symbols which the ancient Mayan civilisation of South America used for their numbers.

a) Complete the table by drawing in the missing symbols. (5)

Number	Symbol
1	•
2	••
3	•••
4	••••
5	▬
6	•̱
7	••̱
8	•••̱
9	••••̱
10	▬̱
11	•̱̱

b) Write the answer to the following sums as Mayan symbols.

i) ••• × •• = •̱
ii) • + •••• = ▬
iii) ▬ − •• = •••
iv) •••• ÷ ••• = •••

(4)

4 By shading the right sections in the grids alongside show how a calculator displays the numbers 3 and 5.

9.7.3 Microelectronics

1. Binary information has been used to send coded messages into outer space. You can see the principle with this message in binary.

   0111001000011100101001110

   There are 25 digits and 25 squares in the grid alongside. Take the first five numbers of the message. Where there is a zero leave a square in the first line of the grid blank. Where there is a one, then shade in the square on the grid. Use the next five numbers to fill in the second line. Continue until you have completed the squares to reveal another number. (5)

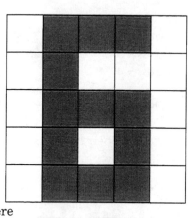

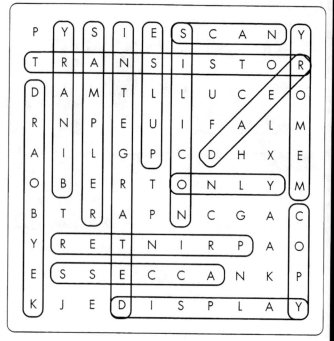

2. Complete the words in the passage below and then look for them in the wordsearch grid. (15)

   A microprocessor contains a miniaturised INTEGRATED circuit made from SILICON. The circuit contains small electronic switches. Each one, called a TRANSISTOR, lets a PULSE of electricity through the circuit. The pulses form a stream of on and offs called BINARY code. The memory chips come in two versions – RAM (stands for random ACCESS MEMORY) and ROM (READ ONLY memory).

   You can input words into a computer through the KEYBOARD and see the DISPLAY on a monitor screen. You can SCAN a photograph and use a SAMPLER to capture a sound. The computer can output its information to a PRINTER after which it is usually called a hard COPY.

3. Unscramble the following words. The clues given for each word should help.
   a) TERMCOUP   COMPUTER   (1)
      (a machine that handles words, pictures, numbers and sounds, electronically)
   b) LOLUTACCAR   CALCULATOR   (1)
      (This can help with difficult maths)
   c) KOWTERN   NETWORK   (1)
      (One way in which computers can 'talk' to each other)
   d) TEAFROWS   SOFTWARE   (1)
      (What computers need in order to work)
   e) MORYME   MEMORY   (1)
      (Where information is stored in a computer)

4. Underline one of the words in brackets which best matches the meaning of the word in capitals.
   a) INTEGRATED –
      (small) (**whole**) (separate)
   b) MICROPROCESSOR –
      (**chip**) (cell) (display)
   c) TRANSISTOR –
      (code) (**gate**) (tape)
   d) DATA –
      (robot) (pulse) (**information**)
   e) DIGITAL –
      (**numbered**) (continuous) (ten)   (5)

Science Companion © A Porter, M Wood, T Wood and Stanley Thornes (Publishers) Ltd 1995

9.8.3 Logic gates

1 There are three sensors shown below. For each one say how it can be switched on. (3)

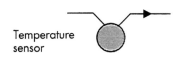
Temperature sensor

MAKE THE SENSOR WARM

Light sensor

SHINE LIGHT ON SENSOR

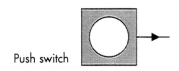
Push switch

PRESS SENSOR DOWN

2 The four circuits below show sensors connected to outputs (alarms, bells, lamps, relay switches). Most of the circuits use a logic gate (NOT, OR, AND). For each circuit say what must be done to switch on the output. Also give one practical use for each circuit. (8)

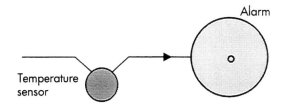

How to switch on alarm:

WARM SENSOR

Use for circuit:

FIRE ALARM

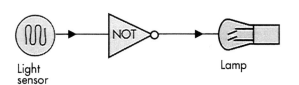

How to switch on lamp: COVER LIGHT SENSOR

Use for circuit: SWITCH LIGHT ON AUTOMATICALLY AT NIGHT

How to switch on bell:

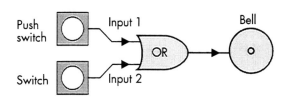

PRESS EITHER OF THE SWITCHES

Use for circuit:

DOORBELL FOR 2 DOORS

How to switch on relay:

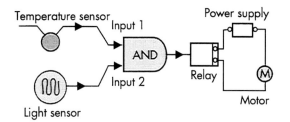

WARM SENSOR AND SHINE LIGHT

Use for circuit: RUNNING A MOTOR DURING THE DAY WHEN WARM (FAN)

3 Complete the tables below showing the outputs of NOT, OR and AND gates. (9)

a) NOT

Input 1	Output
ON	OFF
OFF	ON

b) OR

Input 1	Input 2	Output
ON	ON	ON
ON	OFF	ON
OFF	ON	ON
OFF	OFF	OFF

c) AND

Input 1	Input 2	Output
ON	ON	ON
ON	OFF	OFF
OFF	ON	OFF
OFF	OFF	OFF

Science Companion © A Porter, M Wood, T Wood and Stanley Thornes (Publishers) Ltd 1995

10.1.3 Fuels

1 Use the words below to complete the passage.

> animals, carbon, carbon dioxide, impermeable, layers, oxygen, rocks, sea

How oil and gas were formed

Dead plants and ANIMALS which were living in the SEA drifted down to the bottom. They became covered by LAYERS of sand and mud. This stopped the gas OXYGEN getting to them. As they decayed the CARBON in their bodies could not turn into CARBON DIOXIDE and they became oil and gas.

Over millions of years the ROCKS changed as the Earth's crust moved. The oil and gas rose to the surface. Where they met a layer of IMPERMEABLE rock, the oil and gas became trapped. (8)

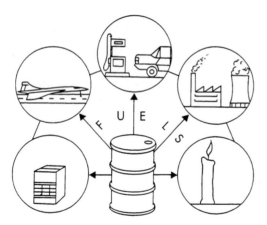

2 The diagram above shows five fuels which can be made from petroleum. Complete the table below to show what these fuels are and what they are used for. (10)

Name of fuel	What it is used for
CALOR GAS	portable fires
KEROSENE	jet fuel
PETROL	car engines
FUEL OIL	power stations
WAX	candles

3 Enter the missing words in the sentences below into the grid. This should reveal a hidden word in the shaded column. (6)

(a)	■		F	U	E	L	S
(b)	■	C	O	A	L	■	■
(c)	G	A	S	■	■	■	■
(d)	D	I	S	T	I	L	■
(e)	■		O	I	L	■	■
(f)	■		P	L	A	N	T	S

a) Substances used for heating, when they are burned, are called FUELS.
b) Many power stations use a solid fuel called COAL to generate electricity.
c) GAS is supplied to homes through pipelines.
d) If you want to separate a mixture of liquids, you must DISTIL them.
e) Petroleum is also known as crude OIL.
f) Many fuels were formed from dead PLANTS.

4 The pie chart below is divided into sections each worth 5%. The table shows approximate uses of fuels in the world. Use the figures in the table to complete the pie chart. (6)

Type of fuel	How much is used in the world
COAL	35
OIL	43
GAS	20
OTHERS	2

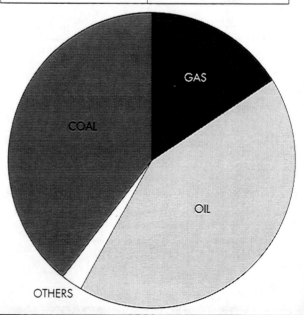

10.2.3 Making use of energy

1 Draw lines to match the type of energy to its use. (5)

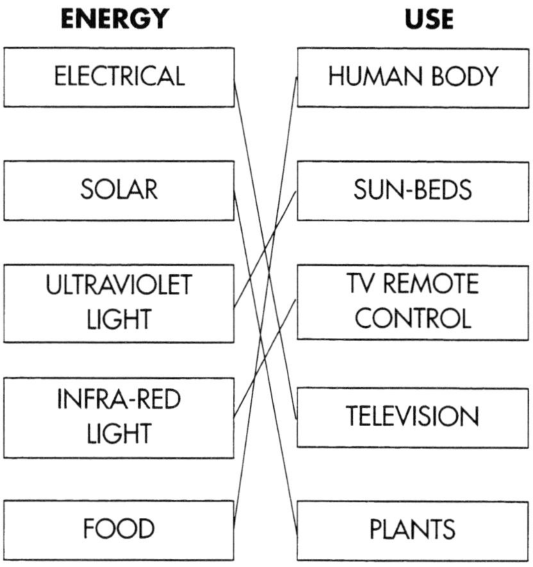

ENERGY	USE
ELECTRICAL	HUMAN BODY
SOLAR	SUN-BEDS
ULTRAVIOLET LIGHT	TV REMOTE CONTROL
INFRA-RED LIGHT	TELEVISION
FOOD	PLANTS

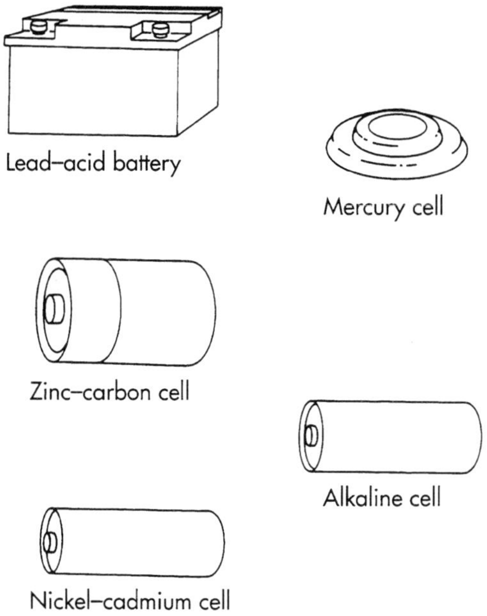

Lead–acid battery

Mercury cell

Zinc–carbon cell

Alkaline cell

Nickel–cadmium cell

2 Which of the cells in the diagram above,

a) is a single rechargeable cell?

   NICKEL CADMIUM CELL (1)

b) is used in watches?

   MERCURY CELL (1)

c) is used in a car?

   LEAD–ACID BATTERY (1)

d) lasts longer than normal dry cells?

   ALKALINE CELL (1)

3 The missing words in the nine sentences will fit into the word spiral. the last letter of one word is also the first letter of the next word. The start letter of each word is shown by the shaded square. (9)

1. Transistor _____ collect invisible signals through an aerial. (6 letters)
2. Energy from the Sun is called _____ energy. (5 letters)
3. You can _____ some batteries and use them again. (8 letters)
4. An _____ fire can heat a room without a flame. (8 letters)
5. A solid fuel made from plants is _____. (4 letters)
6. A form of energy our eyes can see is _____. (5 letters)
7. Burning gas _____ chemical energy into heat energy. (10 letters)
8. You can store energy when you _____ an elastic band. (7 letters)
9. Fuels burn to produce _____ energy. (4 letters)

¹R	A	D	I	O	²S	O
⁵C	O	A	⁶L	I	G	L
I	⁸S	T	R	E	H	A
R	M	A	T	T	⁷T	³R
T	R	E	⁹H	C	R	E
C	O	F	S	N	A	C
E	L	⁴E	G	R	A	H

4 Use the words chemical, electrical or heat to complete the sentences below.

e.g. a paraffin heat changes chemical energy into heat energy.

a) A digital thermometer changes HEAT energy into electrical energy. (1)

b) An electric fire changes ELECTRICAL energy into HEAT energy. (2)

c) A dry cell changes CHEMICAL energy into ELECTRICAL energy. (2)

d) Recharging a cell changes ELECTRICAL energy into CHEMICAL energy. (2)

10.3.3 On the move

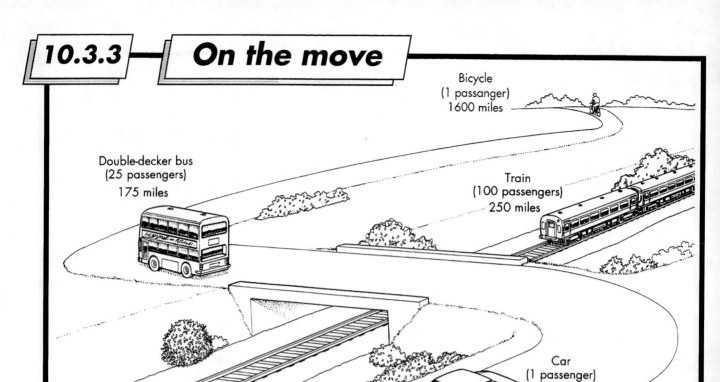

1. The drawing above shows you how far you could go for about £2 worth of fuel per person. For the car this is about one gallon (5 litres) of petrol. It also shows you the number of people you can transport this distance.
   a) Complete the table below. (4)

Transport	No. of passengers	Distance for £2 of fuel per person
bicycle	1	1600
bus	25	175
car	1	30
train	100	250

   b) List the four forms of transport from the most efficient to the least. (4)

   most efficient: BICYCLE

   TRAIN

   BUS

   least efficient: CAR

   c) What fuels do the forms of transport need?

   bicycle: FOOD   (1)

   bus/train: DIESEL   (1)

   car: PETROL   (1)

2. The prices below are for four energy sources – gas, coal, electricity and fuel oil. The cheapest is fuel oil and electricity is four times more expensive than gas.
   a) Complete the table by putting the energy source next to the correct price. (4)

Energy source	Price per unit
GAS	22.0p
FUEL OIL	13.5p
COAL	16.5p
ELECTRICITY	88.0p

   b) For each energy source say how it can be used to transport things or people.

   Coal: STEAM ENGINES   (1)

   Fuel oil: SHIPS   (1)

   Gas: CARS   (1)

   Electricity: MILK FLOATS   (1)

   c) Why do you think electricity is so much more expensive than the other sources?

   IT REQUIRES OTHER FUELS TO MAKE IT   (1)

Science Companion © A Porter, M Wood, T Wood and Stanley Thornes (Publishers) Ltd 1995

10.4.3 Energy conservation

1 The diagram below shows how double glazing keeps heat from being lost. The gap between the two panes of glass can be changed. Single glazing has a gap of zero mm.

a) On the grid below plot a line graph. Along the bottom put the gap between the panes. Up the side put the heat loss. Join the points with a smooth curve. (8)

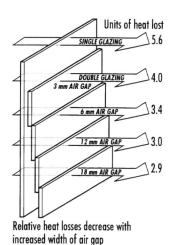

Relative heat losses decrease with increased width of air gap

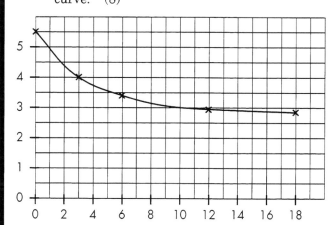

b) What is link between the air gap and the amount of heat lost?

AS THE GAP INCREASES THE WINDOW KEEPS IN MORE HEAT (1)

c) In what way can double glazing help the environment?

IT SAVES ON NATURAL RESOURCES (1)

d) Apart from heat insulation, what other advantage can double glazing offer?

NOISE REDUCTION (1)

e) What disadvantage is there in fitting double glazing?

IT IS EXPENSIVE (1)

2 The diagram below shows how energy can be extracted from the ground using hot rocks. Holes are drilled into the Earth's crust. Explosive charges are detonated at the bottom of each hole to connect them.

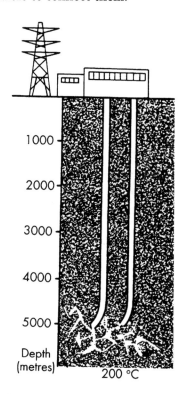

a) How deep are the holes?

5000 m (1)

b) How do the explosions connect the two holes

SMALL CRACKS IN THE ROCK (1)

c) Cold water is pumped down one hole. What happens to it at the bottom of the hole?

IT IS HEATED AND BECOMES STEAM (2)

d) How does this enable electricity to be generated?

STEAM TURNS TURBINES (1)

e) What name is given to this type of energy?

GEOTHERMAL (1)

f) Give one advantage of using this method to generate electricity.

CHEAP, DOES NOT USE RESOURCES (1)

g) Why would these power stations not last forever?

THE ROCKS COOL DOWN (1)

Science Companion © A Porter, M Wood, T Wood and Stanley Thornes (Publishers) Ltd 1995

10.5.3 Human life and activity – the energy we need

1 Read the table which shows the nutritional value of some foods, then answer the questions which follow.

FOOD	% PROTEIN	% FAT	% CARBOHYDRATE	ENERGY per 100 g
Brown bread	8.0	2.0	52.0	1042
Weetabix	11.0	2.0	77.0	1470
Rice	2.6	1.1	31.3	586
Peas	5.0	0	8.0	210
Yogurt	4.7	1.0	4.6	193
Chicken	30.0	7.0	0	770
Fish fingers	21.0	5.0	3.0	800
Baked beans	6.0	0	17.0	390
Chips	6.0	38.0	49.0	2940
Bananas	1.1	0.3	19.2	337
Cola	0	0	7.0	110
Sausages	12.0	25.0	13.0	1530

a) Write the percentage of fat, carbohydrate and protein in each food below.

Bread
- Protein 8.0 %
- Fat 2.0 %
- Carbohydrate 52.0 %

Fish fingers
- Protein 21 %
- Fat 5.0 %
- Carbohydrate 3.0 %

Yogurt
- Protein 4.7 %
- Fat 1.0 %
- Carbohydrate 4.6 %

Cola
- Protein 0 %
- Fat 0 %
- Carbohydrate 7.0 %

Chips
- Protein 6.0 %
- Fat 38.0 %
- Carbohydrate 49.0 %

(5)

b) Work out the energy values, in kJ of each meal below.

200 g sausages	200 g chicken
3060 kJ	1540 kJ
100 g chips	100 g rice
2940 kJ	586 kJ
100 g baked beans	100 g peas
390 kJ	210 kJ
Meal 1. 6390 kJ	Meal 2. 2336 kJ

(8)

c) Write a sentence about the nutritional value of each meal.

Meal 1. CONTAINS PROTEIN CARBOHYDRATE AND FAT AND IS HIGH IN FAT. IT GIVES LOTS OF ENERGY (1)

Meal 2. HAS CARBOHYDRATE, FAT AND PROTEIN. IT IS LOWER IN FAT THAN MEAL 1. (1)

d) Which meal provides the most energy?
MEAL 1 (1)

e) Which meal is healthier and why?
MEAL 2. IT HAS ALL THE NUTRIENTS BUT IS LOWER IN FAT. (1)

11.1.3 What is a force?

1 Jane's group collected a set of results after adding known masses to a hook on the bottom of a spring. The increase in the length of a spring from its unstretched length is called the extension.

a) Use the table of results alongside to plot a graph on the graph paper below which shows the extension of a spring as different masses are added.

Extension (cm)	Mass added (g)
0	0
5	10
10	20
15	30
20	40
25	50
30	60
35	70
40	80

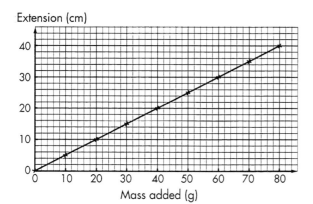

(4)

b) What will be the extension for the following masses?
   i) 0 g   0 cm³ (1)
   ii) 25 g   12.5 cm³ (1)
   iii) 45 g   22.5cm³ (1)
   iv) 75 g   37.5 cm³ (1)

2 Use the six clues (a) to (f) to find the horizontal words in the grid below and then find the word in column 5 with help from clue (g)

	1	2	3	4	5	6	7	8
a)	■	■	N	E	W	T	O	N
b)	■	■	S	P	E	E	D	■
c)	G	R	A	V	I	T	Y	■
d)	C	H	A	N	G	E	S	■
e)	■	■	P	U	S	H	I	N	G
f)	R	E	A	C	T	I	O	N

CLUES

a) Forces are measured in these units.
b) The limit of this on the motorway is 70 mph.
c) Masses are pulled downwards by this force.
d) An object can be moving and still have balanced forces acting on it. You only feel a force when the balance of the forces _____.
e) The opposite of pulling.
f) Forces act in pairs. This is the opposite force to action. (6)

Now the clue for the word in column 5.
g) The force with which gravity pulls down on an object is WEIGHT (1)

3 Forces are measured in units called newtons. If you hang a 100 g mass on a newtonmeter, then the meter will read 1 newton (1 N). A mass of 1 kilogram (1000 g) will give a reading of 10 newtons (10 N).

Use the above to help complete the following table.

Mass in grams	Newtonmeter reading in newtons
300	**3**
600	6
2000	**20**
4000	40
110	1.1
12	**0.12**
8	**0.08**
56	0.56
370	3.7
274	**2.74**

(10)

Science Companion © A Porter, M Wood, T Wood and Stanley Thornes (Publishers) Ltd 1995

11.2.3 — Going down and slowing down

1 A mass of 10 kg weighs 100 N on Earth. The table alongside shows the weight in newtons of 10 kg on each of the planets in the solar system.

Planet	Weight of 10 kg (N)
Mercury	40
Venus	70
Earth	100
Mars	40
Jupiter	230
Saturn	90
Uranus	80
Neptune	110
Pluto	40

a) Use the graph paper below to draw a bar chart to illustrate the figures in the table above.

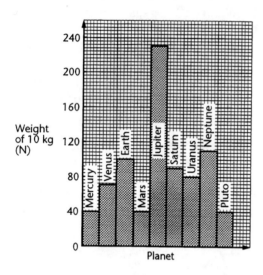

(9)

b) On which planets would an astronaut weigh more than on Earth?

JUPITER AND NEPTUNE (2)

c) On which planets would a 1 kg mass weight 4 N?

MERCURY, MARS AND PLUTO (3)

d) On which two planets is the gravity almost the same as on Earth?

SATURN AND NEPTUNE (2)

e) Explain your answer to part (d)

10 kg WEIGHTS 100 N ON EARTHS 90 N ON SATURN AND 110 N ON NEPTUNE (3)

f) On which planet would a mass of 1 kg have a weight of 7 N?

VENUS (1)

2 Study the newtonmeters **A – J**. Using the table to Question 1, give the names of **one** planet on which the newtonmeter is being used.

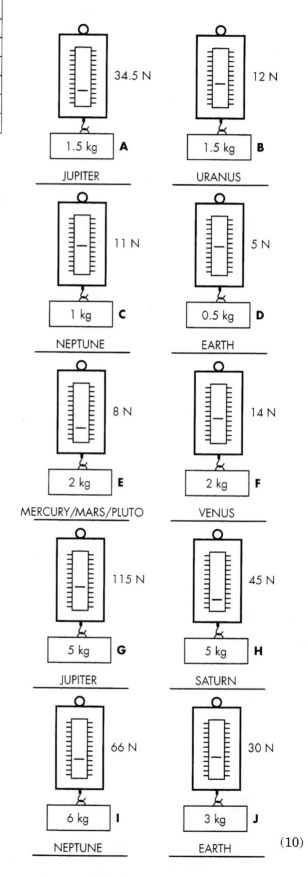

A 34.5 N, 1.5 kg — JUPITER
B 12 N, 1.5 kg — URANUS
C 11 N, 1 kg — NEPTUNE
D 5 N, 0.5 kg — EARTH
E 8 N, 2 kg — MERCURY/MARS/PLUTO
F 14 N, 2 kg — VENUS
G 115 N, 5 kg — JUPITER
H 45 N, 5 kg — SATURN
I 66 N, 6 kg — NEPTUNE
J 30 N, 3 kg — EARTH

(10)

11.3.3 — Starting and stopping

1 Use the words below to complete the following sentences about friction. You may use a word more than once.

doubled   further   lubricant   microscope   oil   rain   rough   surface

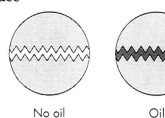

No oil   Oil

Friction can be reduced by using a LUBRICANT. One such substance is OIL. If you study the surfaces through a MICROSCOPE you can see they are ROUGH. Friction is reduced because OIL smooths out the sliding of one SURFACE over the other. RAIN reduces the friction on the roads. The distance it takes a car to stop is more than DOUBLED in wet weather. Cars need to keep FURTHER from one another in the wet to prevent accidents. (9)

2 Use clues (a) to (n) to find the words to complete the spaces below. As an additional clue, the word downwards in column 4 is the name of the painful disease of the joints suffered by many elderly people.

CLUES
a) A liquid used as a lubricant in car engines.
b) The top layer of a road.
c) The part of the leg just below the hip.
d) Short for polytetrafluoroethene the non-stick coating on pans.
e) There are many of these in our skeleton. Nickname of Dr. McCoy in Star Trek.
f) On a bicycle and car these help it to stop.
g) Tyres need a good _____ on the road to help a vehicle to stop.
h) An alloy of iron used to make car bodies.
i) Friction makes this. A bunsen burner supplies this.
j) A force which stops one surface sliding over another.
k) A car can do this when it loses its grip on a slippery road surface.
l) The covering on some joints which provides a slippery surface for movement.
m) A natural joint may be replaced by an _____ one.
n) The hip is this type of joint.

The word in column 4 is OSTEOARTHRITIS (15)

	1	2	3	4	5	6	7	8	9	10	11
a)				O	I	L					
b)				S	U	R	F	A	C	E	
c)				T	H	I	G	H			
d)	P	T	F	E							
e)				B	O	N	E	S			
f)			B	R	A	K	E	S			
g)				G	R	I	P				
h)				S	T	E	E	L			
i)				H	E	A	T				
j)			F	R	I	C	T	I	O	N	
k)		S	K	I	D						
l)	C	A	R	T	I	L	A	G	E		
m)	A	R	T	I	F	I	C	I	A	L	
n)				S	Y	N	O	V	I	A	L

3 There is a coded message below. Use the squares and their letters from question 2 above to place a letter in a blank space. There are no M and W letters in question 2 so these have been done for you.

I	C	E
5a	11	5i

I	S
6c	3h

A
8b

S	M	O	O	T	H
2k	—	7n	4e	4c	8c

S	U	R	F	A	C	E
4b	5b	3l	5m	4f	1l	6h

A	N	D
9m	10j	5k

S	O
7e	7n

F	R	I	C	T	I	O	N
5m	6b	5g	6j	4h	5i	9j	5e

I	S
8j	3h

L	O	W
11n	7n	—

M	A	K	I	N	G
—	1m	5f	4k	6n	8l

S	T	O	P	P	I	N	G
4n	4l	4e	6g	1d	5l	6n	3g

D	I	F	F	I	C	U	L	T
5k	4m	7b	3j	6m	9b	5b	10m	7j

(6)

11.4.3 Sinking and floating

1 Hot-air ballooning is very popular nowadays. Study the following information and then complete the sentences which follow.

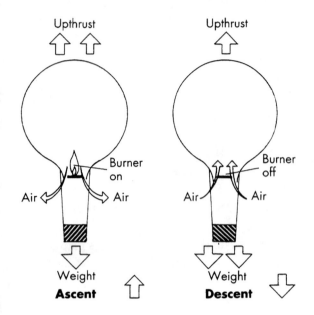

FACTS
- The burner uses propane gas as a fuel.
- The burner heats the air in the envelope to about 100°C.
- Hot air takes up more room than cold air.
- When the burner cuts out, the air in the envelope starts to cool.
- Cold air takes up less room than hot air.
- Cold air is more dense than hot air.

Use the words from the list below to complete the following sentences about hot-air balloons. The words may be used once, or more than once.

decreases   falls   greater
increases   into   less   out

As the air in the balloon is heated its volume INCREASES. This DECREASES the total weight of the balloon, because some of the air is forced OUT of the balloon. Upthrust is now GREATER than weight and so the balloon rises. When the burner is switched off the air in the envelope cools and takes up LESS volume. Air is sucked INTO the balloon which INCREASES the total weight and the balloon FALLS. (8)

2 Density = $\dfrac{\text{mass}}{\text{volume}}$

a) Calculate the density of the following elements. (8)

ELEMENT	MASS (g)	VOLUME (cm³)	DENSITY (g/cm³)
diamond	35	10	3.5
copper	890	100	8.9
magnesium	34	20	1.7
mercury	68	5	13.6
potassium	86	100	0.86
sulphur	63	30	2.1
titanium	90	20	4.5
iron	63.2	8	7.9

b) Which of the elements in the above table is
  i) less dense than water. POTASSIUM (1)
  ii) less dense than sulphur MAGNESIUM and POTASSIUM (2)
  iii) the most dense? MERCURY (1)

3 Below is the Plimsoll Line which is drawn on the side of a ship. The lines show the loading limits for different seas and seasons.

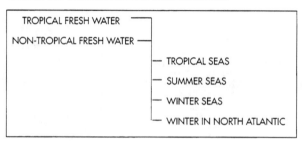

a) What happens to the ship as it is loaded with more goods at the dock-side?

IT SINKS LOWER IN THE WATER (1)

b) Why are there different lines for the different types of water?

THE DIFFERENT WATERS HAVE DIFFERENT DENSITIES (1)

c) In which water will the ship be furthest out of the water for a fixed loading?

WINTER IN NORTH ATLANTIC (1)

11.5.3 Some forces in action

1 Below are 8 jumbled words.

ettrsch   aaeeccointrl   auqssh
ebdn   urnt   aainnblgc
alerotndceei   eiidtcron

First, unjumble the letters and then place the word in the sentence best suited to it. All the sentences concern the effects forces have on the objects which they push or pull.

a) Forces cause objects to move and give them ACCELERATION.

b) Forces stop objects which are moving and cause DECELERATION.

c) Forces distort or BEND objects.

d) Forces twist or TURN objects.

e) Forces SQUASH or compress objects.

f) Forces change the DIRECTION in which an object moves.

g) Forces keep objects still by BALANCING other forces acting on the same object.

h) Forces STRETCH objects. (8)

2

a) Where does the force come from to turn the spanner?

THE PULL OF THE HAND (1)

b) What is the object on which the force acts?

THE PULL ACTS ON THE SPANNER (1)

c) What is the name given to this type of force?

TURNING FORCE or MOMENT (1)

d) Does a spanner with a longer handle make the job easier or harder?

EASIER (1)

e) Explain your answer to part (e)

THE EFFECT OF THE FORCE IS BIGGER THE FURTHER FROM THE NUT BECAUSE ITS MOMENT IS THE FORCE TIMES THE PERPENDICULAR DISTANCE (4)

3 Below is a table of information about some flying insects.

Insect	Complete wingbeats per second	Flying speed (km/h)
Bumblebee	130	11
Cockchafer beetle	50	11
Dragonfly	40	55
Hawk moth	70	35
Honeybee	225	9
Housefly	200	7
Mosquito	600	1.5
Midge	1000	1.5

a) On the graph paper below, draw a bar chart showing the flying speed of each insect. (4)

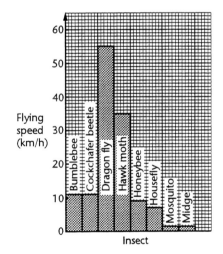

b) If you study the figures for the dragonfly and compare them with those for the midge, what conclusion can you draw?

SLOWER WINGBEATS MEAN FASTER FLYING SPEED (2)

11.6.3 Turning forces

1 Moments can be taken about any point. Look at the girl standing on a plank supported by two trestles. The girl's weight is 240N.

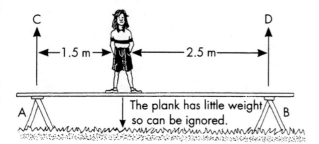

a) If you take moments about point **B**, what is the value of the upward force of the trestle **B**?

ZERO (1)

b) Explain your answer to part (a).

MOMENT IS FORCE TIMES SHORTEST (PERPENDICULAR) DISTANCE OF FORCE FROM THE FULCRUM. B IS THE FULCRUM, SHORTEST DISTANCE OF FORCE D IS ZERO AND SO THE MOMENT IS ZERO. (3)

c) Calculate force **C** taking clockwise and anti-clockwise moments about **B**.

$C \times 4 \text{ m} = 2.5 \text{ m} \times 240 \text{ N}$

$C = 150 \text{ N}$ (2)

d) Now find the upward force **D** at trestle **B**.

TOTAL DOWNWARD FORCE IS 240 N, SO,

$C + D = 240$ N, $D = (240-150)$ N $= 90$ N (2)

2 There are three types of levers. For each type there are three components (parts), load (**L**), effort (**E**) and fulcrum (**F**). For each type of lever place the three components in their correct order.

| First class lever | L F E |
| Second class lever | F L E |
| Third class lever | F E L | (3)

3 On the diagrams below, draw labelled arrows to mark the load, effort and fulcrum on each example, and then name the class of lever to which it belongs.

a) removing a tin lid with a screwdriver

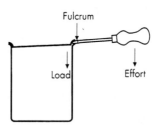

(3)

FIRST class (1)

b) removing a nail

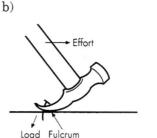

(3)

FIRST class (1)

c) lifting a leaf on a table

(3)

SECOND class (1)

d) using tweezers to lift a small piece of paper

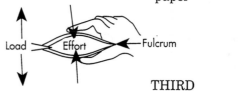

(3)

THIRD class (1)

11.7.3 Work and power

1 Alan is pulling a sledge through the snow. He uses a force of 60 N to pull the sledge the first 8 m on the flat.

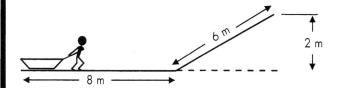

a) How much work has he done?

60 N × 8 m = 480 N m (1)

b) He reaches the slope, and pulls the sledge up it for 6 m in length and only 2 m in height. He uses 90 N of force to get up the slope. What work does he do against gravity to get up the slope?

90 N × 2 m = 180 N m (1)

c) Alan's friend Asif asks Alan to pull the sledge again and now he will time Alan. Alan takes 4 seconds on the flat and 6 seconds up the slope, using the same forces as before. What is his total power?

ON THE FLAT 480/4 = 120 watts (1)

ON THE SLOPE 180/6 = 30 watts (1)

TOTAL POWER 120 + 30 = 150 watts (1)

2 Lynne runs up 3 flights of stairs at school. It takes her 12 seconds. There are 60 steps, and each step is 21 cm high. Lynne's mass is 50 kg (1 kg weighs 10 N).

a) What was the vertical distance Lynne climbed?

60 × 21 = 1260 cm = 12.60 m (1)

b) Find Lynne's weight in newtons.

50 × 10 = 500 N (1)

c) Find Lynne's work done in joules.

500 N × 12.60 m = 6300 J (1)

d) Find Lynne's power.

6300/12 = 525 watts (1)

3 Given the work done in each of the following examples, find the distance moved by the force in the direction of the force.

a) 150 J of work were done in moving a force of 50 N.

150/50 = 3 m (1)

b) 1 kJ of work was done in moving a force of 100 N.

1000/100 = 10 m (1)

c) A removal man did 30 kJ of work in lifting a chair of mass 20 kg to his van

WEIGHT OF CHAIR 20 × 10 N = 200 N

DISTANCE IS 30000/200 = 150 m (2)

d) A climber does 60 kJ of work lifting a rucksack of 15 kg up a rock face.

WEIGHT OF RUCKSACK 15 × 10 = 150 N

DISTANCE IS 60000/150 = 400 m (2)

e) A toy train does 45 J of work in hauling 500 g of train around the track.

WEIGHT OF TRAIN 0.5 × 10 = 5 N

DISTANCE IS 45/5 = 9 m (2)

4 Use the information in each of the following questions to find the time taken for each activity.

a) A model boat has an engine of power 100 watts. Its total weight is 1000 N and it moves a distance of 1.5 metre.

(1000 × 1.5)/100 = 15 seconds (1)

b) A crane lifts a weight of 8000 N over a height of 10 m using a motor of power 800 watts.

(8000 × 10)/800 = 100 seconds (1)

c) A shopkeeper picks up a box of apples weighing 500 N and carries it 20 m using 1000 watts.

(500 × 20)/1000 = 10 seconds (1)

Science Companion © A Porter, M Wood, T Wood and Stanley Thornes (Publishers) Ltd 1995

11.8.3 Under pressure

A pupil was investigating the effect pressure has on some plasticine. She set up two experiments using a heavy iron mass and small wooden cubes. Each of the cubes had a face of 1 cm².

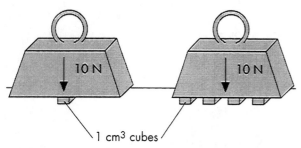

1 cm³ cubes

a) What force does the iron mass exert?

10 newtons. (1)

b) What is the pressure on the plasticine from the cube in the first experiment?

10 newtons per square centimetre. (1)

c) What is the pressure on the plasticine from the cubes in the second experiment?

2.5 newtons per square centimetre. (1)

d) How does the first pressure compare with the second experiment?

LARGER – FOUR TIMES LARGER (2)

e) What results would you expect to see in the plasticine?

IN THE FIRST CASE THE IMPRESSION IN THE PLASTICINE IS DEEPER (1)

2 Use the following words to complete the paragraph below. Words can be used once, more than once or not at all.

> rises   falls   expands
> contracts   more   less
> increases   decreases

When the air inside a hot air balloon is heated, the volume of the air INCREASES. We say that the air EXPANDS. This makes the air LESS dense than before and the balloon RISES. If the air cools down inside the balloon, the speed of the air particles DECREASES. The air inside the balloon takes up MORE volume and the balloon FALLS. (7)

3 Work out the missing words in the passage below and enter them on the grid. 6 across is two words – 3 and 5 letters. (9)

¹P	A	²S	C	³A	L		
	R		Q		L		⁴F
⁵E	Q	U	A	L			O
	S		A				R
⁶S	I	R	I	S	⁷A	A	C
	U		E		R		E
	R		⁸O	N	E		
	E				A		

The French scientist, Blaise (1 across) carried out research into (1 down). Weight is the mass of an object being pulled by the (4 down) of gravity. The units are named after (6 across) Newton.
100 g = (8 across) newton.
The pressure in a liquid is the same, or (5 across) in (3 down) directions. Pressure is measured as a force applied over an (7 down) – usually measured in (2 down) metres.

4 Look at the diagram of the three sealed syringes. The first syringe has a volume of V cm³.

a) What will be the volume of the gas in the second syringe?
$\frac{1}{2}V$ (1)

b) What will be the volume of the gas in the third syringe?
$2V$ (1)

c) Whose law does this illustrate? BOYLE'S LAW (1)

12.1.3 All done with mirrors

1 Say whether the statements below are true or false. Tick the correct box beside each one. (10)

	true?	false?
The reflection in a mirror is a virtual image.	√	
The angle of incidence on flat mirrors is greater than the angle of reflection.		√
The image on the back of a spoon is upside down.		√
Mirrors can be used in electricity power stations.	√	
Optical fibres can curve the rays of light inside them.		√
The speed of light is 300 000 metres per second.		√
Light can travel to the Moon and back in 2½ seconds.	√	
Laser beams are used to measure the distance between the Earth and Moon.	√	
A periscope produces colourful patterns with mirrors.		√
The speed of light is slower in air than in a vacuum.	√	

2 A mirror was used to look at letters of the alphabet.

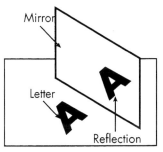

Say whether the reflection shown in the table below is a true or false picture of the letter. (4)

LETTER	REFLECTION	TRUE OR FALSE
B	ᙠ	TRUE
N	И	FALSE
Q	Ọ	FALSE
S	ƨ	TRUE

3 The picture below shows a boy looking into a mirror. On the right hand side draw his reflection. (2)

View from top

mirror   Virtual image

4 The Hubble telescope was launched into orbit around the Earth in 1990. It was designed to beam clear pictures of the stars and planets back to Earth. Part of the telescope used mirrors to magnify and focus light from stars and planets. The telescope is pointed towards an area of the sky. The light from the sky hits the large mirror. The light is reflected on to the second mirror which in turn reflects the light to the receptor. The receptor sends the signals down to Earth.

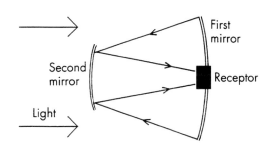

a) What is the advantage in putting a telescope into space rather than having it on the Earth?

AVOIDS THE DISTORTION FROM THE EARTH'S ATMOSPHERE (1)

b) What does 'magnify' mean?

INCREASE IN SIZE (1)

c) What does 'focus' mean?

TO MAKE A SHARP IMAGE (1)

d) Why are the mirrors curved?

TO FOCUS THE LIGHT (1)

12.2.3 Light and shadow

1 The diagram below shows the rays of light which pass through a pinhole camera when it is pointed towards a glowing bulb. An image can be seen on the screen at the back of the camera.

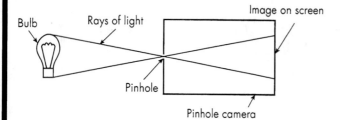

a) Draw a picture of the image you would see on the screen. (1)

b) The camera was moved further away from the bulb. Draw a picture showing how the image would be different from that in part a). (1)

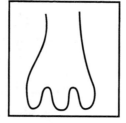

c) Two holes were put in the front of the camera. Draw a picture of what would now be seen on the screen. (1)

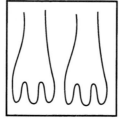

d) What happens to the image if the pinhole is made larger? (*Delete the incorrect word*)

The brightness of the image **increases**

The image becomes more **blurred** (2)

e) How will a lens in front of the pinhole help if the pinhole is too large?

IT MAKES THE IMAGE MORE FOCUSSED (1)

f) How will a lens in front of many pinholes in the camera improve the image?

IT MERGES THE IMAGES TOGETHER (1)

2 Enter the answers to the clues into the crossword grid. (15)

¹S	E	E	²T	H	³R	O	U	⁴G	H
			R		O			L	
		⁵O	P	A	Q	U	E	A	
			N		U			S	
⁶R	A	Y	S		⁷D	⁸E	N	S	⁹E
E					L	Y			Y
F		¹⁰P	U	L	¹¹S	E	¹²S		E
L			C		H		H		L
E			E		A		A		I
¹³C	O	R	N	E	R		D		D
T			T		¹⁴P	R	E	S	S

ACROSS
1. Transparent (two words 3 and 7)
5. This means light will not pass through (6)
6. Another word for beams of light (4)
7. Glass is more _____ than water (5)
10. A lighthouse sends out _____ of light (6)
13. Light will not turn a _____ without a mirror (6)
14. A room will light up when you _____ the lamp switch (5)

DOWN
2. This means light will pass through but it will be diffuse (11)
3. The surface of a lens is (5)
4. A transparent substance (5)
6. What light will do off a mirror (7)
8. The part of our body we use to see (3)
9. When our _____ close we cannot see things (7)
11. A word which means in focus (5)
12. This means not in direct light (5)

3 Delete the incorrect words in the passage below.

Light does not pass through our skin because our skin is **opaque**. One form of light which will pass through skin is **X-rays**. This will not pass through bones which form a **shadow**. (3)

12.3.3 Light

1. Colour in or name the seven colours in the spectrum made when white light passes through a prism. (7)

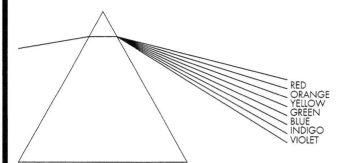

RED
ORANGE
YELLOW
GREEN
BLUE
INDIGO
VIOLET

2. Many radio stations are listed in newspapers with their wavelength (in metres) or their frequency (in kilohertz). You can convert between the two as long as you know what the speed of the radio waves are.

speed = wavelength × frequency
(km/s)   (m)   (kHz)

The speed of the radio waves is 300 000 kilo metres per second. Use this to work out the missing numbers in the following table. (6)

Wavelength (m)	Frequency (kHz)
250	1200
330	909
433	693
247	1215
1515	198
463	648

3. The answers to the clues (a) to (h) will fit into the grid. Use the numbers above the grid and the letters by the side of the grid to complete the phrase under the grid. (9)

a) Laser beams read bar codes at supermarket _____-outs
b) The _____ is the unit of frequency
c) Sun beds use _____ violet light
d) _____ light lies between yellow and blue
e) All colours combine to make _____ light
f) Food can be heated in _____ wave ovens
g) TV signals travel on _____ waves
h) A glass _____ will split light into colours

	1	2	3	4	5
a	C	H	E	C	K
b	H	E	R	T	Z
c	U	L	T	R	A
d	G	R	E	E	N
e	W	H	I	T	E
f	M	I	C	R	O
g	R	A	D	I	O
h	P	R	I	S	M

E	L	E	C	T	R	O	M	A	G	N	E	T	I	C
d3	c2	d4	f3	c3	h2	g5	f1	g2	d1	d5	e5	b4	g4	a1

S	P	E	C	T	R	U	M
h4	h1	b2	a4	e4	f4	c1	h5

4. The table shows the speed of light through different materials.

Material	Speed of light
vacuum	300 000 km per second
glass	200 000 km per second
water	225 000 km per second

The refractive index of a material is the speed of light in a vacuum divided by the speed of light through the material.

Refractive index = $\dfrac{\text{speed of light in a vacuum}}{\text{speed of light in the material}}$

a) Work out the refractive index for glass.

300 000 ÷ 200 000 = 1.5 (1)

b) Work out the refractive index for water.

300 000 ÷ 225 000 = 1.33 (1)

c) The refractive index for ice is 1.31. What is the speed of light through ice?

300 000 ÷ 1.31 = 229 008 (1)

12.4.3 Recording pictures

1 Use clues (a) to (l) below to complete the horizontal words in the spaces which follow. Then find the word in column 4 which reads from top to bottom. (12)

CLUES
a) We can buy or hire our favourite films on this tape.
b) Dr. Edwin Land invented this first instant picture camera.
c) When this gets on a film in a camera, it makes a picture.
d) A camera needs to be in _____ to get a sharp picture.
e) A fast film needs less _____ to make a picture.
f) Film can be developed in a dark _____.
g) This lifts up in an SLR camera when a picture is taken.
h) You need a short _____ time to take a picture of a person.
i) This recording material is found in a cassette.
j) Once a recording is made, we want to use the _____ button on the machine to see what was recorded.
k) Short for photograph.
l) If you buy blank video tape you first _____ something and then watch it.

	1	2	3	4	5	6	7	8
a)		V	I	D	E	O		
b)	P	O	L	A	R	O	I	D
c)		L	I	G	H	T		
d)	F	O	C	U	S			
e)	T	I	M	E				
f)				R	O	O	M	
g)	M	I	R	R	O	R		
h)	E	X	P	O	S	U	R	E
i)				T	A	P	E	
j)	P	L	A	Y	B	A	C	K
k)				P	H	O	T	O
l)				R	E	C	O	R	D

Column 4 is the word DAGUERROTYPE an early type of photograph. (1)

2 Use the following words to complete the passage about recording on video (VHS). Each word may be used once, or more than once.

**capstans   cassette   forwards
heads   magnetic   picture
playback   remove   roller
synchronises**

On loading, the tape is pulled out of the CASSETTE into contact with the recording and PLAYBACK mechanism. As the tape moves round it passes the erase head which uses a MAGNETIC field to REMOVE any previous recording. Two recording HEADS put the picture on the tape by MAGNETIC signal. On playback these recording HEADS turn the signal on the tape back into a PICTURE. The sound recording is put on by a separate head which SYNCHRONISES the sound to the picture. The pinch ROLLER pushes the tape against the CAPSTANS which move the tape FORWARDS.

(12)

3 From the list of words below match **one** to each of the sentences.

**developed   exposure   lens
shutter   viewfinder**

a) The length of time the light shines on a film to make a picture.	EXPOSURE
b) The part of the camera which moves to let a known amount of light to get to the film.	SHUTTER
c) This part of the camera is glass or plastic and focuses the light onto the film.	LENS
d) You look through this when taking a picture.	VIEWFINDER
e) Films have to be _____ using chemicals to produce a picture.	DEVELOPED

(5)

12.5.3 Seeing and hearing

1 Below is a diagram of the eye with eight important parts of the eye labelled. Following the diagram are eight functions of parts of the eye. In the space provided, match the part of the eye to its function.

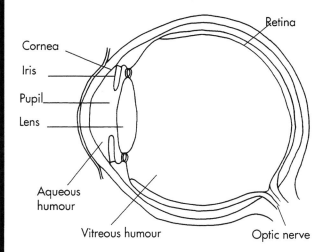

Function of the part	Name of the part
allows light into the eye	**PUPIL**
focuses the light to form a sharp image on the retina	**LENS**
supports the shape of the eye	**VITREOUS HUMOUR**
bends light before it enters the eye	**CORNEA**
contains light-sensitive cells	**RETINA**
feeds the cornea and the lens	**AQUEOUS HUMOUR**
adjusts the size of the pupil to allow different amounts of light to enter	**IRIS**
carries messages from the retina to the brain	**OPTIC NERVE**

(8)

2 Complete the following passage about how we see selecting words from the list below. Each word may be used once only.

brain   cells   cornea   focus   image
lens   rays   reflected   refracted
retina   Sun

We see something when light from the SUN or a lamp is REFLECTED off the surface of the object and reaches our eyes. The light enters the eyes through the CORNEA and LENS. They bend the light inwards to FOCUS it on the RETINA. When light RAYS are bent it is called REFRACTION. There is now an IMAGE of the object on the retina and light-sensitive CELLS send a message to the BRAIN. (11)

3 Name the three main parts into which the ear may be divided. (3)

OUTER, MIDDLE, INNER

4 Describe what each of the following parts of the ear does.

a) *PINNA* COLLECTS SOUND WAVES (1)

b) *EARDRUM* VIBRATES WHEN HIT BY SOUND WAVES (1)

c) *COCHLEA* CONVERTS VIBRATIONS TO NERVE MESSAGES (1)

d) *SEMICIRCULAR CANALS* ALLOW US TO BALANCE (1)

5 What device do some people use to enable them to hear more clearly when they suffer from some types of deafness?

HEARING AID (1)

Science Companion © A Porter, M Wood, T Wood and Stanley Thornes (Publishers) Ltd 1995

12.6.3 How sound is made

1 Below is a word search. Find the NINE words connected with 'How sound is made'. There is one word which is in twice. Write out all NINE words on the lines below and say which ONE is used twice by writing (×2) after it.

	C						V	
L	O	N	G	I	T	U	D	I	N	A	L
	M	E					C	
	P	A	R	T	I	C	L	E	S	U
	R	R		W				U
	E	X	P	A	N	S	I	O	N	M
	S		V					
	S		H	E	A	R		E
V	I	B	R	A	T	I	O	N	A
	O						R	
	N							

LONGITUDINAL PARTICLES HEAR
EXPANSION VIBRATION
COMPRESSION VACUUM WAVE
EAR (× 2)   (10)

2 Solve the clues to find the missing phrase. The first letter of each word fits into the phrase. Write this first letter above the number of the clue to find the phrase.

<u>S O U N D   W A V E</u>
PHRASE 1 2 3 4 5   6 7 8 9   (1)

CLUES

1 There are three states of matter. They are S O L I D, liquid and gas.

2 When a stone is dropped into a pond, waves spread O U T in all directions.

3 When a noise is not heard, we say it is U N H E A R D.

4 A sound is louder when you are N E A R E R than further away.

5 A megaphone can be used to point sound in one D I R E C T I O N.

6 A W H I S P E R is a quiet sound usually made between two people.

7 Sound, rain, and wind all pass through the A I R.

8 Sound cannot travel through a V A C U U M.

9 A longitudinal wave contains alternate places of E X P A N S I O N and compression.   (9)

3 Unscramble the words in each line below and place the correct words in the matching blank. You can use the word clues to help you. Then find the word in the lightly shaded column.

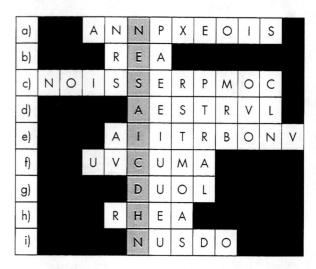

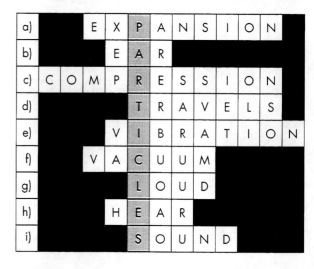

CLUES
a) To make larger.
b) To hear with.
c) Squeezed together.
d) Sound does this through the air.
e) To move backwards and forwards very quickly.
f) No particles, usually no air.
g) The opposite of quiet.
h) Listen.
i) Noise.

The word in the lightly shaded column is
PARTICLES   (1)

12.7.3 Frequency, pitch and volume

1. For a string on a guitar, the musical note it can make gets higher in pitch as the string gets shorter.

   You are given eight straws each of which has been cut to a different length. In the space below draw eight straws of different lengths arranged so that on blowing over them you can produce notes which get higher in pitch as you go from left to right.

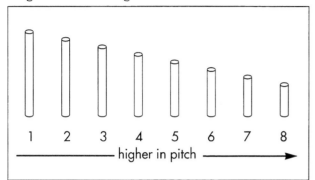

(3)

2. Select words from the list below to complete the following sentences. Use each word once, more than once, or not at all.

   **decreasing   higher   increasing
   lower   lowers   raises**

   a) On stringed musical instruments, the longer the string, the LOWER the musical note.
   b) On a violin, the thinner the string, the HIGHER the musical note.
   c) On a stringed musical instrument raising the tension in a string RAISES the note.
   d) On a trombone you can lower the musical note by INCREASING the length of the tube.
   e) On a church pipe organ, the shorter pipes produce notes of HIGHER pitch. (5)

3. When big speakers are producing loud low notes at pop concerts, why can the speaker cones be seen moving backwards and forwards?

   SOUND IS A LONGITUDINAL WAVE. IT CONTAINS PLACES OF COMPRESSION AND EXPANSION. THE CONES MOVE FORWARD TO COMPRESS AND BACK TO EXPAND THE AIR PARTICLES AS THEY PRODUCE SOUND. (6)

4. Find the ten horizontal words in the grid below by using the clues which follow. Then identify the word in column 3. (10)

   CLUES

   a) Sound travels as longitudinal _____.
   b) In a transverse wave, if the wave has more energy it has a bigger _____.
   c) The _____ of a tuning fork produces a single musical note.

	1	2	3	4	5	6	7	8	9	10	11
a)			W	A	V	E	S				
b)			A	M	P	L	I	T	U	D	E
c)			V	I	B	R	A	T	I	O	N
d)	F	R	E	Q	U	E	N	C	Y		
e)			L	O	U	D	N	E	S	S	
f)			E	N	E	R	G	Y			
g)			N	O	I	S	E				
h)			G	U	I	T	A	R			
i)	P	I	T	C	H						
j)			H	I	G	H					

   d) The number of wavelengths that go by per second.
   e) Opposite of quietness.
   f) The intensity of sound depends on the amount of _____ in the sound.
   g) A lot of sound all mixed up is often described as _____.
   h) A musical instrument usually with six strings used by many 'pop' musicians.
   i) The higher the frequency, the higher the _____ is of a musical note.
   j) Short wavelengths have _____ frequencies.

   The word in column 3 is WAVELENGH (1)

12.8.3 The speed of sound

1 Read the following newspaper report very carefully and then answer the questions which follow.

PENCIL SHAPE FOR NEW 'CONCORDE' TO SILENCE SONIC BOOMS

DESIGNERS at America's National Aeronautics and Space Administration (NASA), claim to have beaten the major environmental objections to a successor for Concorde, by reducing the sonic boom to a rumble.

The boom is caused when a plane breaks the sound barrier. The designers have given their plane a flatter nose, more like that of the high speed train, rather than the famous droop of Concorde. This nose will be built from a new generation of plastics.

Their first design was put forward two years ago, but is now being modified to a pencil shape. This will produce a plane which is one third longer than the present giant passenger plane, the Boeing 747-400.

They claim to have reduced the sonic boom to a rumble. If true, this will enable 'the pencil' to fly over land at supersonic speeds. Concorde is restricted to supersonic speed over the water for the sonic boom is sudden and has been known to cause damage to buildings.

'The pencil' could be flying by 2005 at a cost of £10 billion if built by an international consortium, pooling talent from the United States, Britain, Germany, Japan and the former Soviet Union. If this international consortium holds together then they hope to build around 2000 planes.

World News 19/1/92

a) Who has designed this new supersonic plane?

AMERICA'S NATIONAL AERONAUTICS AND SPACE ADMINISTRATION (1)

b) In what way have the designers changed the shape, from that of Concorde, to lower the noise from the sonic boom?

THE PLANE HAS A FLATTER NOSE, MORE LIKE THAT OF A HIGH SPEED TRAIN, THAN THE FAMOUS DROOP OF CONCORDE. (2)

c) How much larger will this plane be than the Boeing 747-400?

ONE THIRD LONGER (1)

d) From what will the nose of the plane be made?

A NEW GENERATION OF PLASTICS (1)

e) What will this plane be able to do which Concorde cannot, and why?

IT CAN FLY OVER LAND AT SUPERSONIC SPEEDS BECAUSE ITS QUIETER SONIC BOOM WILL NOT CAUSE DAMAGE AS CONCORDE'S DID IN TRIALS. (2)

f) Which countries are to build it?

BRITAIN, GERMANY, JAPAN, THE FORMER SOVIET UNION, AND UNITED STATES (2)

2 Use a calculator, and the equation
speed = frequency × wavelength
to find the values to fit in the lightly shaded boxes and write them in.

Speed (m/s)	Frequency (Hz)	Wavelength (m)
126	9	14
115	5	23
344	10	34.4
344	264.6	1.3
200	50	4
5000	650	7.69
44	440	0.1
132	440	0.3
800	6.4	125
1500	440	3.4

(10)

12.9.3 Noise in the environment

1. Read the following report and then answer the questions which follow using the information in the report wherever possible.

    Noise reduction brings peace and quiet

    Reducing noise at source is often easier than trying to reduce it on its way to the human ear.

    Moving or rotating parts cause vibrations and these in turn make sound. Floors, buildings or vehicles can all magnify this sound because they resonate in time with the vibrations.

    A rotating shaft on an engine causes a lot of vibration in the engine, and so a lot of noise. This in turn produces excess wear on the bearings, and causes a loss of efficiency. If the shaft is removed, balanced, and refitted, it will result in a quieter, smoother and more efficient engine.

    If the engine cannot be made quieter at source, action is taken on the surroundings. An engine can be supported on rubber mountings which absorb the vibrations instead of transmitting them. The engine can be placed in an airtight enclosure if practical. As a last resort, the sound can be allowed 'out' and the operator and nearby people forced to wear ear defenders to prevent permanent damage to their ears.

    At home we can do many simple things to cut down on outside noise spoiling our quietness. Double glazing with a wide air gap of several centimetres helps, and the inside becomes even quieter when heavy curtains are closed. Thick carpets help too. Wall coverings of cork or polystyrene absorb noise and prevent it reflecting around a room.

    Next time you clear a room totally for redecoration, notice how noisy it becomes. Speedy redecoration will soon return the peace and quiet we all like.

    A. Noy (World Sound News 6/1/95)

    a) How can noise be made in some machines?

    MOVING OR ROTATING PARTS (1)

    b) Name **three** things which can magnify unwanted sound.

    FLOORS, BUILDINGS, OR VEHICLES (3)

    c) Name **three** things produced as a result of a badly balanced rotating shaft.

    VIBRATION (NOISE), EXCESS WEAR ON BEARINGS, LOSS OF EFFICIENCY. (3)

    d) How do rubber engine mountings help reduce noise in a motorcar?

    THEY ABSORB VIBRATIONS INSTEAD OF TRANSMITTING THEM. (2)

    e) What must people wear if working in a noisy environment?

    EAR DEFENDERS (1)

2. Why do you think sound recording studios reduce their glass areas to a minimum and have cork or polystyrene ceiling and wall coverings and good carpets?

    GLASS IS A HIGHLY REFLECTIVE SURFACE, IT IS SMOOTH AND SOUND BOUNCES OFF IT WHICH IS BAD NEWS FOR RECORDINGS, CORK, POLYSTYRENE AND CARPETS ALL ABSORB SOUND, STOPPING IT REFLECTING AND MAKE FOR GOOD RECORDINGS. (4)

3. Firing a gun can produce a noise of 160 dB in the ears of the person using the gun.

    a) What do you think would be the result if this noise was allowed into someone's ears?

    PERMANENT DAMAGE (1)

    b) What precautions must people take against noise on shooting ranges?

    WEAR EAR PROTECTORS (DEFENDERS) (1)

12.10.3 A brief history of sound recording

1 Study the information below and then answer the questions which follow.

> Until the 1950s sound recordings were made with just one microphone. Records were described as mono, which stood for monophonic. If the sound was sent out through more than one speaker, all speakers sent out the same sound.
>
> The introduction of a worldwide stereophonic recording system in 1958 meant that two microphones could be used. The sound entering each is different. It is stored on tape or disc and then reproduced through one of two speakers. This is known as stereophonic sound. It is possible to have a singer or one particular instrument just from one of the two speakers, matching their original positions in the recording session.
>
> Today the Dolby system uses many different channels. In cinemas speakers almost surround the audience and can make an actor's voice move across the screen exactly as the actor moves.

a) What is the full name given to recordings made with just one microphone?
MONOPHONIC (1)

b) What shortened version of your answer to part (a) is commonly used to describe this type of recording?
MONO (1)

c) When did a worldwide standard for stereo records come into use?
1958 (1)

d) How many speakers are needed to demonstrate a stereo sound system?
2 (1)

e) Which system is used today in many cinemas to place a sound at a particular point using many sound channels?
DOLBY SYSTEM (1)

2 Read the following facts and use them as well as your own knowledge to answer the questions which follow.

Facts:

CDs — virtually lifelong if used normally.
— produce near perfect sound recordings.

Tapes — need careful handling to last a long time
— wear out as the magnetic particles rub off the plastic tape.

LPs — can be scratched easily.
— wear out because a needle runs over the surface for each playing.

a) Suggest why CDs have taken over from LPs.

LPs WEAR OUT, CAN BE SCRATCHED EASILY, BUT CDs ARE VIRTUALLY FOR LIFE AND PRODUCE BETTER QUALITY RECORDINGS. (4)

b) Why are cassette tapes the cheapest form of music recordings available today?

THEY WEAR OUT EASILY AND ARE EASILY DAMAGED AND SO ARE LIKELY TO NEED FREQUENT REPLACEMENT. (2)

c) Why do CDs last longer than LPs?

WITH AN LP A NEEDLE PHYSICALLY TOUCHES AND WEARS THE SURFACE OF THE RECORD. WITH A CD IT IS 'READ' BY A LIGHT BEAM WHICH DOES NOT WEAR THE SURFACE. ON A CD INFORMATION APPEARS MORE THAN ONCE SO A SCRATCH DOES NOT DAMAGE THE RECORDING. (4)

d) Why does it take longer to find a particular song on a tape than on a CD?

A TAPE HAS TO BE WOUND ALONG TO FIND THE SONG. WITH A CD IT IS POSSIBLE FOR THE SYSTEM TO JUMP ACROSS THE DISC TO THE EXACT SONG REQUIRED. (2)

13.1.3 — Voyage to the Moon

FAMOUS ASTRONAUTS

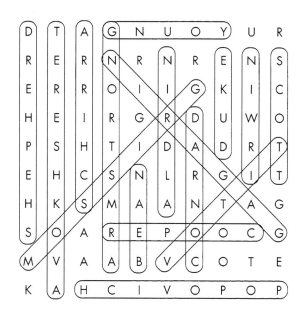

The words in bold can be found on the word-search grid. Circle each word as you find it. (16)

Yuri **Gagarin** was the first human in space. His mission lasted only 1 hour 48 minutes. About four months later Gherman **Titov** spent more than a day in space. The American, Gus **Grissom** entered space in 1961 but did not orbit the Earth. In August 1962 Pavel **Popovich** made 48 orbits of the Earth. The Americans were still far behind the Soviet astronauts. Two months after Popovich, Walter **Schirra** made only 6 orbits but in 1963 Leroy **Cooper** made 22 orbits in America's longest mission so far. The first woman in space was Valentina **Tereshkova** in June 1963.

The first two men on the Moon were part of the Apollo 11 mission. They were Neil **Armstrong** and Edwin **Aldrin** and spent about two and a half hours exploring the Moon. Apollo 12 followed with Charles **Conrad** and Alan **Bean**. Alan **Shepherd** was aboard Apollo 14. The two astronauts in Apollo 15 were David **Scott** and James **Irwin**. Apollo 16 astronauts spent over 20 hours walking and driving about the Moon on a rover vehicle. They were John **Young** and Charles **Duke**.

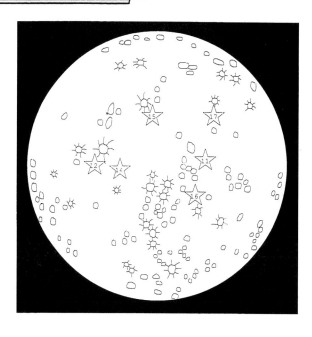

The map above shows the landing sites for all the Apollo missions which made it to the Moon. Complete the missing words in the passage below.

The first Apollo mission to land on the Moon was Apollo <u>11</u> which landed in July 1969. There were <u>6</u> landings on the Moon. The one mission which did not make it to the Moon was Apollo <u>13</u>. Each mission carried two astronauts to the Moon. In all <u>12</u> people have walked on the Moon. The last mission was Apollo <u>17</u> which landed in December 1972. On the Moon the astronauts needed to wear <u>SPACE SUITS</u> because there is no <u>AIR</u> there. The landing sites had to be chosen carefully. They had to avoid any large <u>ROCKS/CRATERS</u> on the surface which could have damaged the lander. Since 1972 there have been <u>NO</u> more Moon landings.

Science Companion © A Porter, M Wood, T Wood and Stanley Thornes (Publishers) Ltd 1995

13.2.3 The planets

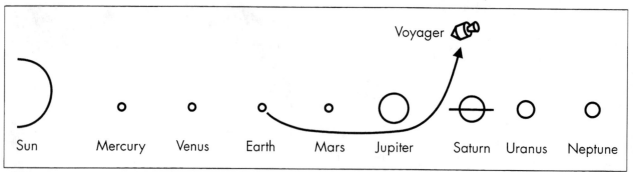

1. The drawing above was fixed on to the side of the Voyager 1 spacecraft. It is meant to show the path that the spacecraft took past some of the planets in the solar system. Put the following labels on the drawing.
   **Earth   Jupiter   Mars   Mercury
   Neptune   Saturn   Sun   Uranus
   Venus   Voyager** (10)

2. The picture below shows the nine planets in order from the Sun and roughly to scale. Use your knowledge or a text book to colour in the planets as accurately as you can. The best pictures to use are those taken by space probes which have visited the planets. (9)

1.S	2.A	T	U	3.R	4.N		5.F
	E		R		E		O	6.C
	R		A		7.P	L	U	T	O
8.M	O	O	N		T		R		M
A			U		U				E
R		9.S	10.U	N				T
S			11.V	E	N	U	S	

3. CROSSWORD – Use the clues below to fill in the crossword above. (11)

   1. The planet with the largest rings (6)
   2. Comes before dynamic, plane and space (4)
   3. The seventh planet from the Sun (6)
   4. Planet named after the god of the sea (7)
   5. The number of inner planets (4)
   6. An icy lump orbiting the Sun (5)
   7. The coldest planet (5)
   8. *across* The Earth's partner (4)
   8. *down* The red planet (4)
   9. The star at the centre of the solar system (3)
   10. Harmful light from this star (2)
   11. The planet roughly the same size as the Earth (5)

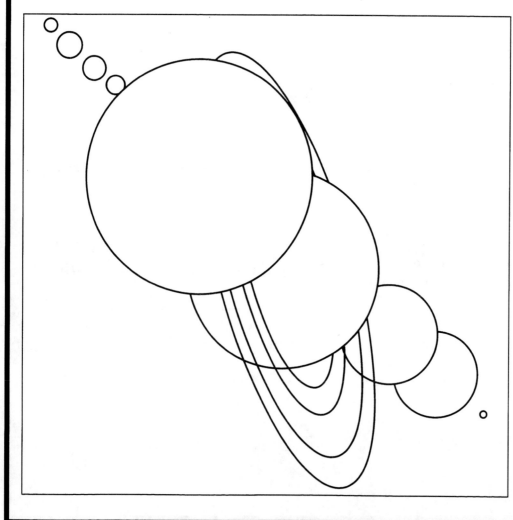

Science Companion © A Porter, M Wood, T Wood and Stanley Thornes (Publishers) Ltd 1995

13.3.3 The Sun as a star

Read the following account of the life of stars and then use the words in the box to complete the flow diagram opposite. (11)

In the space between stars and galaxies there are huge clouds of dust and gas which are mostly remains of previous stars which have exploded. These gas clouds begin to contract as, over a long period of time, gravity pulls the pieces together. As this force makes the particles move faster, the cloud begins to increase in temperature. The centre of what will become the star can be seen and as it spins a disc of material spreads out from its equator. The next stage depends on how much dust and gas were present in the original cloud.

A small amount of material (less than 8% of the Sun's mass) means that the temperature will never rise high enough to start the nuclear reactions which make the star shine. It will remain a brown dwarf.

More than 8% of the Sun's mass means the star will begin to shine. The material close to the star will be blown off, leaving only small rocky planets. Away from the star planets may keep their large gassy atmospheres. The Sun has enough material to shine for 10 000 million years and it is about half way through its life cycle. When it runs out of material to burn it will expand to a size beyond the orbit of the Earth and then collapse to a planet sized white dwarf which will eventually cool down to a black dwarf.

If the star is very large (over 60 times the size of the Sun) then its life is much shorter. Higher temperatures mean that it gets through its fuel supply much faster and when it runs out of fuel (after only perhaps 10 million years) it explodes in a violent way (called a supernova). The material that is left collapses so dramatically that the atoms it is made of collapse in on themselves and all that remains are neutrons. This neutron star is so dense that one cubic centimetre would be about 300 million tonnes.

WORDS TO USE	dust cloud
neutron star	supernova
below 8% Sun mass	contracting cloud
steady use of fuel	brown dwarf
white dwarf	black dwarf
over 60 times Sun mass	rapid use of fuel

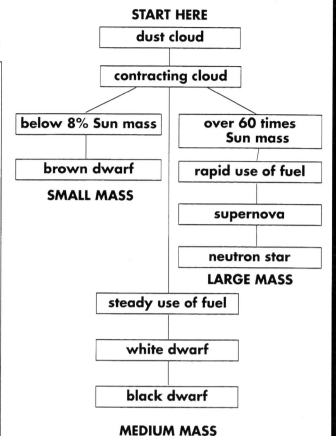

1 How are planets formed from the star?

PLANETS FORM FROM THE DISC OF DUST A STAR GIVES OFF AS IT SPINS (2)

2 What is a brown dwarf?

A STAR WHICH IS NOT MASSIVE ENOUGH TO START SHINING (1)

3 What is a white dwarf?

THE REMAINS OF A STAR WHEN ITS FUEL IS USED UP (1)

4 What is a black dwarf?

A WHITE DWARF WHICH HAS STOPPED SHINING (1)

5 How large does a star have to be (compared with the Sun) in order to explode as a supernova?

OVER SIXTY TIMES LARGER (1)

13.4.3 Our views of the universe

1. The diagram alongside shows the shadow cast by a flagpole at noon on 30th June. Read the three different times below and draw your own shadow on the three diagrams.
   a) 8 am 30th June (2)
   b) 4 pm 30th June (2)
   c) noon 30th December (2)

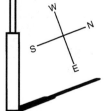

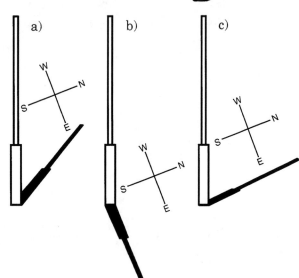

2. Use the words below to complete the passage.

   longer   shorter   higher   lower

   In winter the Sun is LOWER in the sky at noon than in summer. This means that shadows appear LONGER in winter. The Sun takes a SHORTER time to cross the sky which is why the nights are LONGER in winter than in summer. At noon the Sun is HIGHER in the sky than it is in the morning. (5)

3. Put the answers to the clues into the crossword grid. (19)

Across
1. Nicolaus _____ was the first man to say the Earth goes round the Sun.
6. Sir Isaac _____ explained how gravity affected the planets.
7. In 24 hours the Earth turns on its axis _____.
8. The pull of the Moon's gravity has an _____ on the tides on Earth.
10. Shadows are _____ in the evening.
11. The place to see stars is in the _____ .
13. The sun dial, candle and egg _____ are early types of clock.
16. A space craft needs less energy to _____ from the Moon's gravity compared to the Earth's.
17. The time to observe the stars is a _____.
18. The time for the Earth to go round the Sun once.

Down
1. The patterns the stars make in the sky.
2. Modern telescopes are more _____ than the one Galileo used.
3. There are _____ major planets in the solar system.
4. We keep track of the months and years by using _____.
5. The Sun takes longer to cross the sky in _____ than in winter.
9. The Sun is at the _____ of the solar system.
12. Johannes _____ worked out the orbits of the planets round the Sun.
14. When galaxies collide they _____ into one large galaxy.
15. Astronomers use a tele _____ .

¹C	O	²P	E	R	³N	I	⁴C	U	⁵S
O		O			I		A		U
⁶N	E	W	T	O	N		L		M
S		E			E		E		M
T		R			⁷O	N	C		E
⁸E	F	F	E	⁹C	T		D		R
L		U		E			A		
L		¹⁰L	O	N	G	E	R		
A				T			¹¹S	¹²K	Y
¹³T	I	¹⁴M	E	R		¹⁵S		E	
I		E		¹⁶E	S	C	A	P	E
O		R				O		L	
¹⁷N	I	G	H	T		P		E	
S		E			¹⁸Y	E	A	R	

Pupil continuation page

Pupil continuation page